Function Theory on Planar Domains

Function Theory on Planar Domains

A Second Course in Complex Analysis

Stephen D. Fisher
Department of Mathematics
Northwestern University

Dover Publications, Inc.
Mineola, New York

Bibliographical Note

This Dover edition, first published in 2007, is an unabridged republication of the work published by John Wiley & Sons, Inc., New York, in 1983.

Library of Congress Cataloging-in-Publication Data

Fisher, Stephen D., 1941–
Function theory on planar domains : a second course in complex analysis / Stephen D. Fisher.
p. cm.
Originally published: New York : Wiley, c1983, in series: Pure and applied mathematics.
Includes bibliographical references and index.
ISBN 0-486-45768-0 (pbk.)
1. Functions of complex variables. 2. Hardy spaces. I. Title.

QA331.F59 2007
515'.9—dc22

2006052123

Manufactured in the United States of America
Dover Publications, Inc., 31 East 2nd Street, Mineola, N.Y. 11501

To my family

NAOMI, KEFIRA, HANAH, and EFREM

PREFACE

Complex analysis is a member of the first generation of modern mathematics, and substantial portions of today's mathematics can trace their lineage directly back to it. Despite its advanced age, however, the subject remains active and fruitful, death sentences from other quarters notwithstanding. It is my hope that this book will bear witness to this assertion and that each reader will find stimulation and excitement in seeing some of the significant results that complex analysis continues to produce.

The plan of this book is simple. The first chapter and some sections of Chapters 2 and 3 are background material, all of it classical and important in its own right. The remainder of the book presents results in complex analysis from the far, middle, and recent past and even the near present, results that I have selected because of their interest and their merit as substantive mathematics. The selection was necessarily made with my own prejudices and I have been forced to omit many other topics because of the natural limitations of size forced on any such effort. Anyone with a basic graduate course in complex analysis and in real variables–functional analysis has an adequate background to read, understand, and enjoy the topics in this book.

I have frequently used the exercises at the end of each chapter to elaborate on and extend results given in the text; the serious reader would be well advised to work the exercises since mathematics is best understood by doing, rather than only reading. Occasionally in the text there is an opportunity to prove something in two ways; in such cases I have usually opted for the less sophisticated proof, feeling that this was more in keeping with the spirit of the book.

The influence of my teachers at the University of Wisconsin—Frank Forelli, Walter Rudin, Simon Hellerstein, Michael Voichick, and Anatole Beck—is present throughout this book for they not only taught me complex analysis, they also proved many of the theorems to be found in this book. My colleagues at M.I.T. during the fruitful year 1967–1968—Ted Gamelin, John Garnett, Ken Hoffman, Don Wilken, and Larry Zalcman—also influenced me; our analysis seminar was always productive and informative, even if unappreciatedly dry. Many results of theirs are also to be found herein.

I have made some effort at the end of each chapter to give references to related work and to recount the history of the main theorems of that chapter; any omissions of references are inadvertent and regretted by me. In no case should a theorem be attributed to me due solely to an absence of other credits.

In a book of this kind there is inevitably some overlap with other books, especially on the "standard" material, and I have made note of this in the comments at the end of the chapters. A major theme of this work is to elaborate on function theory on a finitely connected planar domain, to show in what ways it resembles and in what ways it differs from function theory on the disc, and to do this without the artifice of saying "one can easily see that. . . ." Of course, there is a great deal more, as well, including a number of theorems on quite arbitrary domains.

This book evolved from courses I taught from time to time. Thanks are especially due to The Technion, Haifa, Israel, for providing me with the opportunity to teach a seminar in the spring of 1980 on much of the material in Chapters 1, 3, and 5. I also wish to thank T. W. Gamelin who read a first draft of the book and provided me with many helpful comments.

Finally, I write usually with the editorial "we" not only because it seems awkward to always write in the first person singular but also because reading a mathematics book is a joint effort between author and reader and the latter is being escorted, as it were, by me through the thickets and meadows of an interesting landscape and we are "seeing" things together.

STEPHEN D. FISHER

Evanston, Illinois
January 1982

CONTENTS

NOTATION AND NUMBERING

In this book, we make use of several notations consistently and thus it is worthwhile to list them here.

Δ, the open unit disc = $\{z : |z| < 1\}$
$\mathbf{T}$, the unit circle = $\{z : |z| = 1\}$
σ, normalized Lebesgue measure on $\mathbf{T}$: $d\sigma = (1/2\pi)\, d\theta$
$\mathbf{S}^2$, the Riemann sphere = $\{(x_1, x_2, x_3) \in \mathbb{R}^3 : x_1^2 + x_2^2 + x_3^2 = 1\}$ $\mathbf{S}^2$ is also the 1-point compactification of the complex plane $\mathbb{C}$; $\mathbf{S}^2$ is $\mathbb{C}$ with the "point" ∞ attached.
Ω always denotes a domain in $\mathbb{C}$ or in $\mathbf{S}^2$; that is, an open connected set.

For a set E, $\mathrm{CL}(E)$ is the closure of E and E^c is the complement of E, relative to $\mathbf{S}^2$. Further, ∂E is the boundary of E and $\mathrm{INT}\, E$ is the interior of E. Finally, $\mathbf{C}(E)$ is the space of continuous complex-valued functions on E, $\mathbf{C}_r(E)$ is the subspace of real-valued continuous functions, and $\|u\|_E$ is the supremum of $|u(x)|$ as x varies over E.

A measure is always a finite regular Borel measure.

A word on the numbering system in this book is also in order. Within each section, say Section 2 of Chapter 3, the propositions and theorems are numbered consecutively using a system of double arabic numbers: 2.1, 2.2, and so on, and reference to them from within that chapter is made accordingly. However, if in another chapter reference is made to, let's say, Theorem 2.2 of Chapter 3, then we write "see Theorem 3.2.2." In this system of triple arabic numbers, the first number refers to the chapter in which the proposition or theorem is found. This should cause no difficulties. A similar convention applies to numbered formulas.

Function Theory on Planar Domains

1

THE DIRICHLET PROBLEM AND HARMONIC MEASURE

1.1. INTRODUCTION

Let Ω be a domain on the Riemann sphere and let u be a continuous real-valued function on $\Gamma = \partial\Omega$. The Dirichlet problem is to find, if possible, a function f which is continuous on $\Omega \cup \Gamma = \text{CL}(\Omega)$ and which satisfies the following conditions:

1. f is harmonic on Ω; that is, $\Delta f = 0$ on Ω.
2. $f = u$ on Γ.

There certainly are cases in which this problem is not solvable; for example, if $\Omega = \{z : 0 < |z| < 1\}$ and $u(0) = 1$, $u(e^{it}) = 0$, $0 \leqslant t \leqslant 2\pi$. However, there are a wide variety of domains for which it is solvable since there are quite reasonable conditions that are sufficient for solvability. The standard approach is by the method of Otto Perron and makes use of subharmonic functions. We begin in Section 2 with some material on the Poisson integral and proceed to the definition of subharmonic functions and some of their basic properties in Section 3; in Section 4, we use subharmonic functions to "solve" the Dirichlet problem. Related matters, including harmonic measure, the Green's function, and logarithmic capacity are covered in Sections 5, 6, and 7.

1.2. THE POISSON FORMULA AND SOME PRELIMINARIES

It is worthwhile to begin by recalling the Poisson integral formula and some of its basic properties. Set

$$P(r, \theta) = \frac{1 - r^2}{1 - 2r\cos\theta + r^2}, \qquad 0 < r < 1, \qquad 0 \leqslant \theta \leqslant 2\pi \qquad (2.1)$$

$P(r, \theta)$ is the Poisson kernel and it is a simple matter to verify that

$$P(r,\theta) = \mathrm{Re}\left(\frac{1+z}{1-z}\right), \qquad z = re^{i\theta} \tag{2.2a}$$

$$P(r,\theta) > 0 \tag{2.2b}$$

$$\frac{1}{2\pi}\int_{-\pi}^{\pi} P(r,\theta)\,d\theta = 1, \qquad 0 < r < 1 \tag{2.2c}$$

$$\text{for } \delta > 0,\ \lim_{r\to 1} \max\{P(r,\theta) : \delta \leqslant |\theta| \leqslant \pi\} = 0 \tag{2.2d}$$

Now let u be a continuous function on the unit circle $\mathbf{T}$, $\mathbf{T} = \{e^{i\theta} : -\pi \leqslant \theta \leqslant \pi\}$, and set

$$P_u(re^{it}) = \frac{1}{2\pi}\int_{-\pi}^{\pi} P(r, t-\theta)u(e^{i\theta})\,d\theta$$

The function P_u is a harmonic function of $z = re^{it}$ as (2.2a) shows. The significant thing about P_u is what $P_u(z)$ does as z tends to a point of $\mathbf{T}$.

Theorem 2.1. *$P_u(z) \to u(\lambda)$ as $z \to \lambda$, $\lambda \in \mathbf{T}$; that is, P_u is continuous on $\Delta \cup \mathbf{T}$ and coincides on $\mathbf{T}$ with u.*

Proof. The proof is the standard application of an approximate identity argument and makes use of (2.2b), (2.2c), and (2.2d). Let $\lambda = e^{i\psi}$. Then

$$P_u(re^{it}) - u(e^{i\psi}) = \frac{1}{2\pi}\int_{-\pi}^{\pi} P(r, t-\theta)\left[u(e^{i\theta}) - u(e^{i\psi})\right]d\theta$$

$$= \int_{|\theta-\psi|<\delta} + \int_{|\theta-\psi|\geqslant\delta}$$

Given $\varepsilon > 0$, choose $\delta > 0$ so that whenever $|\theta - \psi| < \delta$ it follows that $|u(e^{i\theta}) - u(e^{i\psi})| < \varepsilon/2$; from now on we view t as being restricted by $|t - \psi| < \delta/2$. Next, for this δ, let r_0 be chosen so that whenever $r_0 \leqslant r < 1$ it follows that

$$\max\left\{P(r,s) : \frac{\delta}{2} \leqslant |s| \leqslant \pi\right\} < \frac{\varepsilon}{4M} \tag{2.3}$$

where $M = \max\{|u(\theta)| : |\theta| \leqslant \pi\}$. In the preceding identity for $P_u(re^{it}) - u(e^{i\psi})$ we estimate the first integral by $\varepsilon/2$ since (2.2b) and (2.2c) hold. We estimate the second integral by $2M\max\{P(r,s) : \delta/2 \leqslant s \leqslant \pi\} \leqslant \varepsilon/2$ because of (2.2b), (2.2d) and (2.3). Thus, the distance from $P_u(re^{it})$ to $u(e^{i\psi})$ is at most ε when $|\psi - t| < \delta/2$ and $r \geqslant r_0$, which is precisely what was to be shown.

Definition. Let μ be a measure on $\mathbf{T}$ and set

$$P_\mu(re^{it}) = \int_{\mathbf{T}} P(r, t-\theta)\, d\mu(\theta) \tag{2.4}$$

The function P_μ, which is harmonic in Δ, behaves relatively well at the unit circle $\mathbf{T}$.

Theorem 2.2. *Let $d\mu = v\, d\theta + d\alpha$ be the Lebesgue decomposition of μ where $v \in L^1(\mathbf{T}, d\theta)$ and $d\alpha$ is singular with respect to $d\theta$. Then*

$$\lim_{r\to 1} P_\mu(re^{it}) = 2\pi v(t) \quad \text{a.e. } dt \tag{2.5}$$

Proof. This proof is like that of Theorem 2.1 but, since we are only looking at radial limits, it is somewhat simpler. The measure $d\mu$ is given by the function μ of bounded variation on $[-\pi, \pi]$ and (2.5) will follow if we show that

$$\lim_{r\to 1} P_\mu(re^{i\theta_0}) = 2\pi\mu'(\theta_0) \tag{2.6}$$

at each point θ_0 at which μ is differentiable. Note that if $d\mu = 1/(2\pi)\, d\theta$, then $P_\mu \equiv 1$ and so (2.6) holds. Hence, we may subtract a constant multiple of $1/(2\pi)\, d\theta$ from $d\mu$ and so assume that $\int d\mu = 0$; that is, $\mu(-\pi) = \mu(\pi)$.

One integration by parts leads to

$$P_\mu(re^{i\theta}) = \int_{-\pi}^{\pi} P'(r, \theta - t)\mu(t)\, dt$$

Now $P'(r, \psi) = [2r(1-r^2)\sin\psi]/[(1 - 2r\cos\psi + r^2)^2]$ and is an odd function of ψ. Thus, we obtain

$$\begin{aligned} P_\mu(re^{i\theta}) &= \int_{-\pi}^{\pi} P'(r, \theta - t)\mu(t)\, dt \\ &= \int_{-\pi}^{0} + \int_{0}^{\pi} \\ &= \int_0^{\pi} P_r'(t)[\mu(\theta - t) - \mu(\theta + t)]\, dt \\ &= \int_{-\pi}^{\pi} [-\sin t] P_r'(t)\left[\frac{\mu(\theta + t) - \mu(\theta - t)}{2\sin t}\right] dt \end{aligned}$$

The functions

$$K(r, t) = \frac{-1}{2\pi r}(\sin t) P_r'(t), \qquad 0 < r < 1$$

form an approximate identity; that is, (2.2b)–(2.2d) hold with $K(r, t)$ in place of $P(r, t)$. The function

$$F(t) = \frac{\mu(\theta_0 + t) - \mu(\theta_0 - t)}{2 \sin t}$$

is continuous at $t = 0$ with value $F(0) = \mu'(\theta_0)$. Hence

$$\lim_{r \to 1} 2\pi r \int_{-\pi}^{\pi} K(r, t) F(t)\, dt = 2\pi F(0)$$

$$= 2\pi\mu'(\theta_0)$$

which is the desired conclusion.

Proposition 2.3. A harmonic function u in Δ has the representation

$$u(re^{i\theta}) = \int_{\mathbf{T}} P(r, \theta - t)\, d\mu(t) \tag{2.7}$$

for some measure μ on $\mathbf{T}$ if and only if

$$\sup\left\{ \int_{-\pi}^{\pi} |u(re^{i\theta})|\, d\theta : r < 1 \right\} \text{ is finite} \tag{2.8}$$

If (2.7) holds, then μ is uniquely determined.

Proof. Suppose (2.7) holds. Then

$$\frac{1}{2\pi} \int_{-\pi}^{\pi} |u(re^{i\theta})|\, d\theta \leqslant \frac{1}{2\pi} \int_{\mathbf{T}} \int_{-\pi}^{\pi} P(r, \theta - t)\, d\theta\, d|\mu|(t)$$

$$= \int_{\mathbf{T}} d|\mu(t)| = \|\mu\|, \text{ the total variation of } \mu$$

Here we have made use of (2.2b) and (2.2c). Conversely, suppose (2.8) holds. Let μ_ρ be the measure on $\mathbf{T}$ given by

$$d\mu_\rho(t) = \frac{1}{2\pi} u(\rho e^{it})\, dt, \qquad 0 < \rho < 1$$

so that (2.8) says exactly that the total variation of μ_ρ is uniformly bounded for $0 < \rho < 1$. There thus is a measure μ on $\mathbf{T}$ which is a weak-* cluster point of

$\{\mu_\rho\}$. Hence,

$$\begin{aligned} u(re^{i\theta}) &= \lim_{\rho\to 1} u(\rho re^{i\theta}) \\ &= \lim_{\rho\to 1} \frac{1}{2\pi}\int_{-\pi}^{\pi} P(r,\theta-t)u(\rho e^{it})\,dt \\ &= \lim_{\rho\to 1}\int_{\mathbf{T}} P(r,\theta-t)\,d\mu_\rho(t) \\ &= \int_{\mathbf{T}} P(r,\theta-t)\,d\mu(t) \end{aligned}$$

As for uniqueness, suppose μ_1 also satisfies (2.7). The difference $\mu-\mu_1$ then annihilates all $P(r,\theta-t)$ and we must show that this implies $\mu-\mu_1$ is zero. By taking real and imaginary parts we need only show that if ν is a real measure with

$$0=\int_{\mathbf{T}} P(r,\theta-t)\,d\nu(t), \qquad 0<r<1, \qquad \theta\in[0,2\pi] \tag{2.9}$$

then $\nu=0$. From (2.9) and (2.2a) we have

$$0=\operatorname{Re}\int_{\mathbf{T}}\frac{e^{it}+z}{e^{it}-z}\,d\nu(t), \qquad |z|<1$$

so that the analytic function

$$\int_{\mathbf{T}}\frac{e^{it}+z}{e^{it}-z}\,d\nu(t)=h(z)$$

is identically constant and hence 0 since $h(0)=0$. However,

$$h(z)=\int_{\mathbf{T}} d\nu+2\sum_{1}^{\infty} z^n\left\{\int_{\mathbf{T}} e^{-int}\,d\nu(t)\right\}$$

Hence,

$$0=\int_{\mathbf{T}} e^{-int}\,d\nu(t), \qquad n=0,1,2,\ldots$$

Since ν is real, this implies ν is the zero measure.

Corollary 2.4. *If u is a positive harmonic function on Δ then there is a unique non-negative measure μ on $\mathbf{T}$ with*

$$u(re^{i\theta}) = \int_{\mathbf{T}} P(r, \theta - t)\, d\mu(t) \tag{2.10}$$

Proof. Since u is positive it satisfies (2.8):

$$u(0) = \frac{1}{2\pi} \int_{-\pi}^{\pi} u(re^{i\theta})\, d\theta$$

Thus, (2.10) holds for some measure μ; note that Proposition 2.3 actually yielded the information that

$$\|\mu\| \leqslant \sup\left\{ \frac{1}{2\pi} \int_{-\pi}^{\pi} |u(re^{i\theta})|\, d\theta : 0 < r < 1 \right\}$$

In this case, this supremum is $u(0)$; hence

$$\|\mu\| \leqslant u(0) = \int_{\mathbf{T}} d\mu$$

so that μ is non-negative.

1.3. SUBHARMONIC FUNCTIONS

Definition. A function $u(z)$ defined for z in a domain Ω on the sphere is *subharmonic* on Ω if it satisfies the following conditions:

$$-\infty \leqslant u(z) < \infty, \qquad z \in \Omega \tag{3.1a}$$

u is upper semicontinuous on Ω; that is,

$$u(a) \geqslant \limsup\{u(z) : z \to a\} \text{ for all } a \in \Omega \tag{3.1b}$$

if the closed disc $\{z : |z - p| \leqslant r\}$ lies in Ω, then

$$u(p) \leqslant \frac{1}{2\pi} \int_{-\pi}^{\pi} u(p + re^{it})\, dt \tag{3.1c}$$

Evidently every real-valued harmonic function on Ω is subharmonic and if both u and $-u$ are subharmonic then u is harmonic, since in this case u will be continuous and equality will hold in (3.1c). It is also evident that the sum and the maximum of two subharmonic functions are also subharmonic, as is a positive multiple of a subharmonic function. We gather some simple facts about subharmonic functions in the next several propositions.

Proposition 3.1. Let u be subharmonic on Ω and let ϕ be a monotonically increasing convex function on $\mathbb{R}$. Then $\phi(u(z))$ is subharmonic on Ω.

Proof. If we set $v(z) = \phi(u(z))$ then it is apparent that v satisfies (3.1a) and (3.1b). Further, we have

$$\begin{aligned} v(p) = \phi(u(p)) &\leqslant \phi\left(\frac{1}{2\pi}\int_{-\pi}^{\pi} u(p + re^{it})\,dt\right) \\ &\leqslant \frac{1}{2\pi}\int_{-\pi}^{\pi} \phi(u(p + re^{it}))\,dt \\ &= \frac{1}{2\pi}\int_{-\pi}^{\pi} v(p + re^{it})\,dt \end{aligned}$$

since ϕ is both increasing and convex.

EXAMPLE.

Let f be holomorphic on Ω. Then both $\log|f|$ and $|f|^q$, $0 < q < \infty$, are subharmonic on Ω.

This is reasonably straightforward. Clearly (3.1a) and (3.1b) are satisfied. Further, (3.1c) is direct for $|f|^q$ if $1 \leqslant q < \infty$ since

$$\begin{aligned} |f(p)| &= \left|\frac{1}{2\pi}\int_{-\pi}^{\pi} f(p + re^{it})\,dt\right| \\ &\leqslant \left(\frac{1}{2\pi}\int_{-\pi}^{\pi} |f(p + re^{it})|^q\,dt\right)^{1/q} \end{aligned}$$

by Hölder's inequality. However, this argument won't work for $0 < q < 1$, so we first show $\log|f|$ is subharmonic and then apply Proposition 3.1 with $\phi(t) = e^{qt}$.

Suppose that $\{z : |z - p| \leqslant r\}$ is in Ω; we assume first that $f \neq 0$ on $|z - p| = r$. Let $z_1, \dots, z_N$ be the zeros of f in $|z - p| < r$, and put

$$g(z) = f(z)\prod_{1}^{N} \frac{r^2 - \overline{(z_j - p)}\,(z - p)}{r(z - z_j)}, \qquad |z - p| \leqslant r$$

Then g is holomorphic on $\{|z - p| \leqslant r\}$, $|g(z)| = |f(z)|$ if $|z - p| = r$, and $|g(z)| > |f(z)|$ if $|z - p| < r$. Further, g is zero-free in $|z - p| \leqslant r$ so that $\log|g|$ is harmonic there. Thus,

$$\begin{aligned} \log|f(p)| < \log|g(p)| &= \frac{1}{2\pi}\int_{-\pi}^{\pi} \log|g(p + re^{it})|\,dt \\ &= \frac{1}{2\pi}\int_{-\pi}^{\pi} \log|f(p + re^{it})|\,dt \end{aligned}$$

Finally, if f does vanish on $|z-p|=r$ choose $r_n \uparrow r$ so that $f \neq 0$ on $|z-p|=r_n$. Thus,

$$\log|f(p)| \leqslant \limsup_{n\to\infty} \frac{1}{2\pi}\int_{-\pi}^{\pi} \log|f(p+r_n e^{it})|\,dt$$

$$\leqslant \frac{1}{2\pi}\int_{-\pi}^{\pi} \limsup_{n\to\infty} \log|f(p+r_n e^{it})|$$

$$= \frac{1}{2\pi}\int_{-\pi}^{\pi} \log|f(p+re^{it})|\,dt$$

which establishes the desired inequality.

Proposition 3.2. Let $\{u_n\}$ be a sequence of subharmonic functions on Ω and suppose that $u_1(z) \geqslant u_2(z) \geqslant \cdots$, $z \in \Omega$ and $\lim_{n\to\infty} u_n(z_0) = L > -\infty$ for some $z_0 \in \Omega$. Then $u(z) = \lim_{n\to\infty} u_n(z)$ is subharmonic on Ω.

Proof. Suppose that $p \in \Omega$ is a point at which $-\infty < u(p)$. If the disc $\{z : |z-p| \leqslant r\}$ lies in Ω, then

$$u(p) \leqslant u_n(p) \leqslant \frac{1}{2\pi}\int_{-\pi}^{\pi} u_n(p+re^{it})\,dt$$

But the latter integrals converge to the integral of u by the monotone convergence theorem. Hence, u satisfies (3.1c) at p. Moreover, $u(p+re^{it}) > -\infty$ for almost all points (dt) on the circle $|z-p|=r$.

Next let a be any point of Ω at which $u(a) > -\infty$. Then, given $\varepsilon > 0$, there is a large n for which $u(a) > u_n(a) - \varepsilon$. Thus,

$$u(a) > u_n(a) - \varepsilon \geqslant \limsup_{z\to a} u_n(z) - \varepsilon$$

$$\geqslant \limsup_{z\to a} u(z) - \varepsilon$$

and so u is upper semicontinuous at a. If $u(a) = -\infty$, then given a large positive number M, we know that $u_n(a) < -M$ for all large n. Hence,

$$\limsup_{z\to a} u(z) \leqslant \limsup_{z\to a} u_n(z) \leqslant -M$$

so that $\limsup_{z\to a} u(z) = -\infty$, in the case $u(a) = -\infty$.

Proposition 3.3. Suppose $u \in C^2(\Omega)$. If $\Delta u \geqslant 0$ in Ω, then u is subharmonic on Ω.

Proof. Suppose the disc $D = \{z : |z - p| \leqslant r\}$ lies in Ω. Green's theorem gives

$$\iint_D \Delta u \, dx \, dy = \int_{-\pi}^{\pi} r \frac{\partial}{\partial t} u(p + te^{i\theta}) \, d\theta$$

$$= r \frac{d}{dt} \left(\int_{-\pi}^{\pi} u(p + te^{i\theta}) \, d\theta \right) \Bigg|_{t=r}$$

However, the first integral is non-negative so that

$$\frac{1}{2\pi} \int_{-\pi}^{\pi} u(p + te^{i\theta}) \, d\theta$$

is an increasing function of t. As $t \to 0$, this integral converges to $u(p)$, by continuity. Hence, u satisfies (3.1c).

We need a simple fact about upper semicontinuous functions which we isolate here; its proof is left as an exercise.

Lemma 3.4. Let $\mathbf{K}$ be a compact set and let u be a function on $\mathbf{K}$ with values in $[-\infty, \infty)$. Then u is upper semicontinuous if and only if there is a sequence $\{f_n\}$ of continuous functions on $\mathbf{K}$ with $f_1 \geqslant f_2 \geqslant \cdots$ and $\lim f_n(z) = u(z)$, $z \in \mathbf{K}$.

The most important fact about subharmonic functions and the origin of their name is contained in the next result.

Theorem 3.5. *An upper semicontinuous function u on Ω with values in $[-\infty, \infty)$ is subharmonic if and only if, whenever $\mathbf{K}$ is a compact subset of Ω and h is a function continuous on $\mathbf{K}$ and harmonic on* INT $\mathbf{K}$ *with $h \geqslant u$ on $\partial\mathbf{K}$, then $h \geqslant u$ on* INT $\mathbf{K}$ *as well.*

Proof. One direction of the theorem is straightforward. Suppose u has the property described in the second half of the theorem. If $D = \{z : |z - p| \leqslant r\}$ lies in Ω, then let $\{f_n\}$ be a sequence of continuous functions on the compact set $\gamma = \{z : |z - p| = r\}$ which decrease to u. Again denote by f_n the function continuous on $\{z : |z - p| \leqslant r\}$, harmonic on $\{z : |z - p| < r\}$ which equals f_n on γ. Then, we have $f_n \geqslant u$ on γ so that $f_n \geqslant u$ inside γ as well, by the hypothesis. Thus,

$$u(p) \leqslant f_n(p) = \frac{1}{2\pi} \int_{-\pi}^{\pi} f_n(p + re^{it}) \, dt$$

But the integrals converge to $\frac{1}{2\pi} \int_{-\pi}^{\pi} u(p + re^{it}) \, dt$ by the monotone convergence theorem. Thus, u satisfies (3.1c).

Conversely, let u be subharmonic in Ω, let $\mathbf{K}$ be a compact set in Ω and let h be continuous on $\mathbf{K}$, harmonic on $\text{INT}\,\mathbf{K}$, and $h \geqslant u$ on $\partial\mathbf{K}$. Set $v = u - h$ so that we wish to show $v \leqslant 0$ on $\text{INT}\,\mathbf{K}$. If not, let

$$m = \text{lub}\{v(z) : z \in \mathbf{K}\}$$

and

$$\mathbf{E} = \{z \in \mathbf{K} : v(z) = m\}$$

Since v is upper semicontinuous, $\mathbf{E}$ is nonempty and closed. Since $v \leqslant 0$ on $\partial\mathbf{K}$, we also know that $\mathbf{E}$ is a subset of $\text{INT}\,\mathbf{K}$. Let p be a point in $\partial\mathbf{E}$ and choose $r > 0$ so small that $\{z : |z - p| \leqslant r\}$ lies in $\text{INT}\,\mathbf{K}$. Then on some arc of the circle $\gamma = \{z : |z - p| = r\}$ we have $v \leqslant m - \delta$, $\delta > 0$, and everywhere on γ we have $v \leqslant m$. But then we obtain the contradiction

$$m = v(p) \leqslant \frac{1}{2\pi}\int_{-\pi}^{\pi} v(p + re^{it})\,dt < m$$

This proves that $v \leqslant 0$ in $\text{INT}\,\mathbf{K}$, as desired.

Proposition 3.6. If $u \in C^2(\Omega)$, then u is subharmonic if and only if $\Delta u \geqslant 0$ on Ω.

Proof. Suppose u is subharmonic and $\{z : |z - p| \leqslant r\}$ lies in Ω. If $0 < r_1 < r_2 \leqslant r$, let v be the harmonic function on $|z - p| < r_2$ which agrees with u on $|z - p| = r_2$. Then $u \leqslant v$ in $|z - p| < r_2$ so that

$$\begin{aligned}\int_{-\pi}^{\pi} u(p + r_1e^{i\theta})\,d\theta \leqslant \int_{-\pi}^{\pi} v(p + r_1e^{i\theta})\,d\theta &= 2\pi v(p)\\ &= \int_{-\pi}^{\pi} v(p + r_2e^{i\theta})\,d\theta\\ &= \int_{-\pi}^{\pi} u(p + r_2e^{i\theta})\,d\theta\end{aligned}$$

Thus, $\int_{-\pi}^{\pi} u(p + te^{i\theta})\,d\theta$ is an increasing function of t for $0 < t \leqslant r$ and so

$$\begin{aligned}0 \leqslant r\frac{d}{dt}\int_{-\pi}^{\pi} u(p + te^{i\theta})\,d\theta &= \int_{-\pi}^{\pi} r\left(\frac{\partial}{\partial t}u(p + te^{i\theta})\right)d\theta\\ &= \iint_{|z-p|\leqslant r} \Delta u\,dx\,dy\end{aligned}$$

by Green's theorem. This clearly implies $\Delta u \geqslant 0$ in Ω.

There is a "maximum modulus" theorem for subharmonic functions.

Proposition 3.7. Suppose there is a number $M < \infty$ such that

$$\limsup\{u(z): z \to \zeta\} \leqslant M \quad \text{for all } \zeta \in \partial\Omega \tag{3.2}$$

Then $u(z) \leqslant M$ for all $z \in \Omega$. If $u(z_0) = M$ for some $z_0 \in \Omega$, then $u \equiv M$ in Ω.

The proof is virtually the same as that of Theorem 3.5 and we leave it as an exercise.

As a finale to this section we note that subharmonicity is preserved by conformal maps, a result we will have need of later. Specifically, let ϕ be a one-to-one holomorphic mapping of a domain Ω onto a domain Ω_1 and suppose u_1 is subharmonic in Ω_1. Then $u(z) = u_1(\phi(z))$ is subharmonic in Ω, as is easily verified by use of Theorem 3.5.

1.4. SOLUTION OF THE DIRICHLET PROBLEM

The fundamental result needed to attack the Dirichlet problem is this.

Proposition 4.1. Let $\mathfrak{F}$ be a family of subharmonic functions satisfying these two conditions:

$$\text{whenever } u, v \in \mathfrak{F}, \text{ then } \max(u, v) \text{ also lies in } \mathfrak{F} \tag{4.1}$$

$$\text{if}\{z : |z - p| \leqslant r\} \subset \Omega \text{ and if } u \in \mathfrak{F}, \text{ then the function} \tag{4.2}$$

$$s(u, z) = \begin{cases} u(z) & \text{if } |z - p| \geqslant r, \qquad z \in \Omega \\ P_u(z) & \text{if } |z - p| < r \end{cases}$$

also lies in $\mathfrak{F}$.

Set

$$v(z) = \sup\{u(z) : u \in \mathfrak{F}\} \tag{4.3}$$

Then either $v \equiv +\infty$ in Ω or v is harmonic in Ω.

Proof. Recall that $P_u(z)$ is the Poisson extension of u to the disc which, in this case is the disc $|z - p| < r$.

Suppose first that there is some point $z_0 \in \Omega$ at which $v(z_0) = \infty$. Choose elements $u_1, u_2, \ldots$ of $\mathfrak{F}$ so that $\{u_n(z_0)\}$ increases to ∞. By (4.1) we may replace u_n by $v_n = \sup\{u_1, \ldots, u_n\}$ and still have $v_n \in \mathfrak{F}$. However, $v_1 \leqslant v_2 \leqslant \cdots$ on all of Ω and $v_n(z_0) \to \infty$. Further, in the disc $D = \{|z - z_0| \leqslant r\}$ we can replace v_n by P_{v_n} and the resulting function $s(v_n, z)$ is again in $\mathfrak{F}$ by (4.2).

But

$$s(v_n, z_0) = \frac{1}{2\pi}\int_{-\pi}^{\pi} v_n(z_0 + re^{it})\, dt$$

$$\geqslant v_n(z_0)$$

so that $s(v_n, z_0) \to \infty$ as $n \to \infty$ and so $s(v_n, z) \to \infty$ for all z, $|z - z_0| < r$. Hence, $v(z) = \infty$ if $|z - z_0| < r$. This implies that the set $\Omega' = \{z \in \Omega : v(z) = \infty\}$ is open. A moment's thought shows that the same argument implies that Ω' is also closed. Hence, $\Omega' = \Omega$ and $v = \infty$ throughout Ω.

Suppose now that v is finite at all points of Ω; we shall show that v is harmonic in Ω. Let a be a point of Ω and let D be a disc centered at a whose closure lies in Ω. Let $\{u_n\}$ be a sequence of elements of $\mathfrak{F}$ with $u_n(a) \to v(a)$. By (4.1) we may replace u_n by $\max\{u_1, \ldots, u_n\}$ and still remain in $\mathfrak{F}$, so there is no loss in assuming $u_1 \leqslant u_2 \leqslant \cdots$. Using the disc D, we may employ (4.2) and assume as well that each u_n is harmonic in D. Hence, the sequence $\{u_n\}$ increases on Ω to a function U which is harmonic in D and equals $v(a)$ at a. Now take b to be any point of D, $b \neq a$. We may apply the same sort of reasoning to find a sequence $\{w_n\}$ of elements of $\mathfrak{F}$ with $w_1 \leqslant w_2 \leqslant \cdots$ on Ω and $w_n(b) \to v(b)$. Let r_n be the function in $\mathfrak{F}$ which is harmonic in D and which equals $\max\{u_n, w_n\}$ on ∂D. Then $w_n(b) \leqslant r_n(b)$ since w_n is subharmonic and likewise $u_n(a) \leqslant r_n(a)$. Further, $\{r_n\}$ increases to a function R which is necessarily harmonic in D with $U(z) \leqslant R(z)$, $z \in D$. But we also know that $U(a) = v(a) \leqslant R(a)$ and $U(b) \leqslant v(b) \leqslant R(b)$. Hence $R - U$ is a non-negative harmonic function which vanishes at a and so is identically zero. Consequently, $R(b) = U(b)$ as well so that $v(b) = U(b)$. Thus, $v \equiv U$ in D and so v is harmonic in Ω.

Although Theorem 4.1 produces a harmonic function in Ω it does not make any reference to behavior at $\partial\Omega$, which is the essential issue in the Dirichlet problem. This aspect is attacked by means of a barrier, which we now define.

Definition. Let $x \in \partial\Omega$. There is a barrier at x if for each small $\delta > 0$ it is possible to find a function $b(z)$, which may depend on δ, such that

$$-b \text{ is subharmonic in } \Omega \tag{4.4a}$$

$$b \geqslant 0 \text{ in } \Omega \tag{4.4b}$$

$$b(z) \geqslant 1 \text{ if } z \in \Omega \quad \text{and} \quad |z - x| \geqslant \delta \tag{4.4c}$$

$$b(z) \to 0 \text{ if } z \in \Omega \quad \text{and} \quad z \to x \tag{4.4d}$$

The notion of a barrier and Theorem 4.1 can now be combined to give a point-by-point solution to the Dirichlet problem.

Let h be a bounded function on $\partial\Omega$ and let $\mathfrak{F}(h)$ consist of all subharmonic functions u on Ω which satisfy

$$\limsup\{u(z) : z \in \Omega, z \to \zeta\} \leqslant h(\zeta), \qquad \text{all } \zeta \in \partial\Omega \tag{4.5}$$

Then set

$$v(z) = v_h(z) = \sup\{u(z) : u \in \mathfrak{F}(h)\} \tag{4.6}$$

Theorem 4.2. *The function v given in* (4.6) *is harmonic on Ω. Further, if h is continuous at $x \in \partial\Omega$ and if there is a barrier at x, then*

$$\lim_{z \to x} v(z) = h(x) \tag{4.7}$$

Corollary 4.3. *If there is a barrier at each point of $\partial\Omega$, then the Dirichlet problem is solvable for Ω.*

Proof. We now proceed to the proof of Theorem 4.2. We know from Proposition 3.7 that each function in $\mathfrak{F}(h)$ is bounded above by $M = \sup\{h(\zeta) : \zeta \in \partial\Omega\}$ and so the function v given by (4.6) is harmonic on Ω.

Let $\varepsilon > 0$ be given; choose $\delta > 0$ so small that $|h(x) - h(y)| < \varepsilon/2$ if $y \in \partial\Omega$ and $|x - y| < \delta$. Let b be the barrier for this δ. Consider

$$s(z) = h(x) - \varepsilon - 2Mb(z), \qquad z \in \Omega$$

Suppose $y \in \partial\Omega$ and $|y - x| \leqslant \delta$; then

$$\limsup\{s(z) : z \to y\} \leqslant h(x) - \varepsilon < h(y)$$

If $y \in \partial\Omega$ and $|y - x| > \delta$, then

$$\limsup\{s(z) : z \to y\} \leqslant h(x) - 2M < h(y)$$

Thus, $s \in \mathfrak{F}(h)$ so that $v(z) \geqslant s(z)$ for all $z \in \Omega$. As a consequence, we have

$$\begin{aligned}\liminf\{v(z) : z \to x\} &\geqslant \liminf\{s(z) : z \to x\} \\ &\geqslant h(x) - \varepsilon\end{aligned}$$

Since ε is arbitrary, we have

$$\liminf\{v(z) : z \to x\} \geqslant h(x) \tag{4.8}$$

Consider next the family $\mathfrak{F}(-h)$ and put

$$w(z) = -\sup\{u(z) : u \in \mathfrak{F}(-h)\}$$

Then w is harmonic in Ω and

$$\liminf\{-w(z): z \to x\} \geqslant -h(x)$$

or, equivalently,

$$\limsup\{w(z): z \to x\} \leqslant h(x) \tag{4.9}$$

Finally, if $u_1 \in \mathfrak{F}(h)$ and $u_2 \in \mathfrak{F}(-h)$, then $u_1 + u_2$ is subharmonic in Ω and

$$\begin{aligned}\limsup\{u_1(z) + u_2(z): z \to \zeta\} &\leqslant \limsup u_1 + \limsup u_2 \\ &\leqslant h(\zeta) + (-h(\zeta)) = 0\end{aligned}$$

so that $u_1 + u_2 \leqslant 0$ in Ω. Hence, $v - w \leqslant 0$ in Ω. This immediately gives the desired conclusion at x, for

$$\begin{aligned}h(x) &\geqslant \limsup\{w(z): z \to x\} \\ &\geqslant \limsup\{v(z): z \to x\} \\ &\geqslant \liminf\{v(z): z \to x\} \\ &\geqslant h(x)\end{aligned}$$

Thus, $\lim\{v(z): z \to x\} = h(x)$, as asserted.

Quite obviously now all that remains to be done to solve the Dirichlet problem is to give some sort of a condition that guarantees that there is a barrier; luckily there is a relatively good one that is strictly geometric. Recall that a continuum is a closed connected set consisting of more than one point.

Theorem 4.4. *Let Ω be a domain and let $x \in \partial\Omega$. If there is a continuum in the complement of Ω which contains x, then there is a barrier at x.*

Proof. Let x' be another point in the continuum. There is a linear fractional transformation which sends x to ∞ and x' to 0. We may thus restrict ourselves to the case when $x = \infty$ and the continuum $\mathbf{C}$ in the complement of Ω contains both 0 and ∞.

There is a single-valued branch of $\log z$ in the domain $\mathfrak{D} = \mathbf{C} \setminus \mathbf{C}$; note that $\Omega \subset \mathfrak{D}$. Let $\mathfrak{R}$ be the image of $\mathfrak{D}$ under $\log z$ so that $\mathfrak{R}$ is also a domain. There is no loss of generality in assuming that $\mathfrak{R}$ meets the imaginary axis (otherwise, just replace $\log z$ by $\log z - \alpha$ for some $\alpha \in \mathbb{R}$.) Let us write

$$\mathfrak{R} \cap \{it : t \in \mathbb{R}\} = \bigcup_{j=1}^{\infty} (i\alpha_j, i\beta_j)$$

where

$$\alpha_1 < \beta_1 < \alpha_2 < \beta_2 < \cdots \quad \text{and} \quad \sum_{j=1}^{\infty} (\beta_j - \alpha_j) \leqslant 2\pi \tag{4.10}$$

(there may, of course, be only a finite number of the intervals (α_j, β_j); this is not relevant.)

Define

$$h_j(z) = \arg\left(\frac{z - i\alpha_j}{z - i\beta_j}\right), \qquad \operatorname{Re} z > 0, \qquad j = 1, 2, \ldots$$

and then

$$h(z) = -\frac{1}{\pi} \sum_{j=1}^{\infty} h_j(z), \qquad \operatorname{Re} z > 0 \tag{4.11}$$

Then each h_j is harmonic on $\operatorname{Re} z > 0$ and h is the increasing limit of the partial sums of its series, so h is also harmonic on $\operatorname{Re} z > 0$. Further, $-1 < h(z) < 0$.

If $x \in (\alpha_j, \beta_j)$ for some j and if $\{z_m\}$ is a sequence in $\operatorname{Re} z > 0$ with $z_m \to ix$, then $h_j(z_m) \to \pi$ while $h_k(z_m) \to 0$ as $m \to \infty$, $k \neq j$. Hence, h is continuous at ix with $h(ix) = -1$. Finally, if $\operatorname{Re} z_m > 0$ and $|z_m| \to \infty$ then $h(z_m) \to 0$. Set

$$g(z) = \begin{cases} -1 & \text{if } \operatorname{Re} z \leqslant 0, \quad z \in \mathfrak{R} \\ h(z) & \text{if } \operatorname{Re} z > 0, \quad z \in \mathfrak{R} \end{cases} \tag{4.12}$$

Then g is continuous in $\mathfrak{R}$, subharmonic on $\mathfrak{R}$, $-1 \leqslant g(z) \leqslant 0$, and $g(z) \to 0$ if $\operatorname{Re} z > 0$ and $|z| \to \infty$.

Now set

$$G(z) = g(\log z), \qquad z \in \mathfrak{D}$$

Then G is subharmonic in $\mathfrak{D}$, $-1 \leqslant G \leqslant 0$, and $G(z) \to 0$ as $|z| \to \infty$. However, it may happen that $G \to 0$ at some finite boundary point. To compensate for this, let $\{t_n\}$ be real numbers increasing to ∞ so that the lines $\operatorname{Re} z = t_n$ all meet $\mathfrak{R}$. Let g_n be the function corresponding to the line $\operatorname{Re} z = t_n$ and put

$$H(z) = \sum_{n=1}^{\infty} \frac{1}{2^n} g_n(\log z), \qquad z \in \mathfrak{D}$$

The series is uniformly convergent so H is actually continuous on $\mathfrak{D}$, subharmonic there, and $-1 \leqslant H \leqslant 0$. Moreover,

$$\lim\{H(z): |z| \to \infty; z \in \mathcal{D}\} = 0$$

If y is any finite point in $\partial\mathcal{D}$, then $\log y$ is a finite point in $\partial\mathcal{R}$ and so $g_n(\log y) = -1$ for all $n \geqslant n_0$. Hence

$$\limsup\{H(z) : z \in \mathcal{D}, z \to y\} < 0$$

Thus, if we are given a (large) M and we put

$$\rho = \sup_{|y| \leqslant M} \{\limsup\{H(z) : z \in D, z \to y\}\}$$

then $\rho < 0$ so that the function $\rho^{-1}H(z)$ is the desired function for the barrier.

Corollary 4.5. *If each component of $\partial\Omega$ is nontrivial, then the Dirichlet problem is solvable in Ω.*

We should note that the existence of a barrier at each point of $\partial\Omega$ is not only sufficient to solve the Dirichlet problem in Ω, it is also necessary. To see this let x be a point in $\partial\Omega$ and let u be a continuous function on $\partial\Omega$ with $u(x) = 0$ and $0 < u(y) \leqslant 1$ for all $y \in \partial\Omega$, $y \neq x$. If h is the harmonic extension to Ω of u, then $\rho^{-1}h(z)$ is a barrier at x for an appropriately small ρ.

1.5. THE GREEN'S FUNCTION OF A DOMAIN

Suppose that Ω is a domain on the sphere and that $p \in \Omega$. A function $g(z; p)$ is a Green's function for Ω with pole (or singularity) at p, $p \neq \infty$, if

$$g(z; p) \text{ is harmonic on } \Omega - \{p\} \tag{5.1a}$$

$$g(z; p) + \log|z - p| \text{ is harmonic near } p \tag{5.1b}$$

$$\lim\{g(z; p): z \to \zeta\} = 0 \quad \text{for all } \zeta \in \partial\Omega \tag{5.1c}$$

If $p = \infty$, then (5.1b) is modified to

$$g(z; \infty) - \log|z| \text{ is harmonic near } \infty \tag{5.1b$'$}$$

We begin by compiling a (short) list of some features of the Green's function.

Proposition 5.1. (a) The Green's function is unique, if it exists.

(b) If ϕ is a one-to-one holomorphic mapping of a domain Ω onto a domain Ω_1 and if Ω_1 has a Green's function $g_1(z; p_1)$ with singularity at $p_1 = \phi(p)$, then $g(z; p) = g_1(\phi(z); p_1)$ is the Green's function for Ω with singularity at p.

Proof. (a) If $h(z; p)$ is another function on Ω with properties (5.1a), (5.1b), (5.1c), then the difference $g(z; p) - h(z; p)$ is harmonic on all of Ω and tends to zero at every point of $\partial\Omega$. Hence, it is identically zero.

(b) Clearly $g(z; p)$ has properties (5.1a) and (5.1c). To check (5.1b) let us assume that neither p nor p_1 is the point at ∞. Then, near p,

$$g(z; p) + \log|z - p| = g_1(\phi(z); p_1) + \log|z - p|$$

$$= u(z) - \log|\phi(z) - \phi(p)| + \log|z - p|$$

where u is harmonic near p. However, because $\phi'(p) \neq 0$ we have

$$\phi(z) - \phi(p) = (z - p)\psi(z)$$

where $\psi(z)$ does not vanish near $z = p$, and hence $\log|\psi|$ is harmonic near p. Thus,

$$g(z; p) + \log|z - p| = u(z) - \log|\psi(z)|$$

near $z = p$ and so g satisfies (5.1b). If p or p_1 is ∞, the proof is similar.

Theorem 5.2. *Let Ω be a domain for which the Dirichlet problem is solvable and let $p \in \Omega$. Then Ω has a Green's function with pole at p.*

Proof. By an initial linear fractional transformation we may assume $p = \infty$ and that $\partial\Omega$ is compact. Let h be the harmonic function on Ω whose values on $\partial\Omega$ are $-\log|z|$, and set

$$g(z; \infty) = h(z) + \log|z|, \qquad z \in \Omega$$

Then g is harmonic on $\Omega \setminus \{\infty\}$, g vanishes on $\partial\Omega$, and near ∞, $g(z; \infty) - \log|z| = h(z)$ so that g satisfies (5.1b′).

It turns out that there is a "Green's function" for most every domain, whether or not the Dirichlet problem is solvable but it is necessary to relax condition (5.1c). The construction of the Green's function is by means of a "regular exhaustion" of Ω, which we now define.

Definition. Let Ω be a domain. A regular exhaustion of Ω is a sequence $\{\Omega_n\}$ of subdomains of Ω satisfying

$$\text{CL}(\Omega_n) \subset \Omega_{n+1}, \qquad n = 1, 2, \ldots \tag{5.2a}$$

$$\bigcup_{n=1}^{\infty} \Omega_n = \Omega \tag{5.2b}$$

each component of $\partial\Omega_n$ is nontrivial (5.2c)

Proposition 5.3. Each domain has a regular exhaustion.

Proof. There is no loss in assuming $\infty \in \Omega$ so that $\partial\Omega$ is compact. Cover $\partial\Omega$ by a finite number of open discs of radius $1/n$ and let D_n be the complement of the union of the closures of these discs. Let Ω_n be the component of D_n which contains ∞.

Let Ω be any proper subdomain of the Riemann sphere and let $p \in \Omega$. Let $\{\Omega_n\}$ be any regular exhaustion of Ω; there is no loss in assuming that $p \neq \infty$ and that p lies in all the Ω_n. Let $g_n(z; p)$ be the Green's function for Ω_n with pole at p; this exists because of Theorem 5.2 and assumption (5.2c). If $m > n$, then $g_m - g_n$ is harmonic on Ω_n and is positive on $\partial\Omega_n$ and hence positive on all of Ω_n. Hence, the sequence $\{g_n(z; p)\}$ is increasing on $\Omega_j \setminus \{p\}$ for each j and so converges, uniformly on compact subsets of $\Omega \setminus \{p\}$ to either $+\infty$ identically or to a harmonic function $g(z; p)$. If the former occurs we say that Ω has no Green's function; more on this is given in Section 7 of this chapter. If the latter occurs, then it is a simple matter to check that $g(z; p)$ is independent of the particular choice of the regular exhaustion $\{\Omega_m\}$ and that $g(z; p)$ satisfies (5.1a) and (5.1b). Finally, if Ω already is known to have a Green's function satisfying (5.1a), (5.1b), (5.1c), then again it is immediate that the $g(z; p)$ arrived at by this process must coincide with it.

Proposition 5.4. Let g be the Green's function for Ω. Then for all points $p \neq q$ in Ω we have

$$g(p, q) = g(q, p) \tag{5.3}$$

Proof. Assume first that Ω is a bounded domain each of whose boundary components is nontrivial. The function

$$g(z; q) + \log|z - q|$$

is harmonic on Ω and continuous at $\partial\Omega$ with boundary values equal to $\log|z - q|$, $z \in \partial\Omega$. Let $I(z; q) = \log|z - q|$ and let ℓ be the linear functional: $\ell(u) = \tilde{u}(p)$, where $\tilde{u}$ is the harmonic extension to Ω of u. Since I is a harmonic function of q, so is $\ell(I(\cdot; q))$. But $\ell(I(\cdot; q)) = g(p; q) + \log|p - q|$. Consequently, $g(p; q)$ is a harmonic function of q for $q \in \Omega$, $q \neq p$.

Next, set $d(p, q) = g(p; q) - g(q; p)$. d is a harmonic function of both p and q except possibly where $p = q$. From the foregoing we see that $g(p; q)$ has a logarithmic pole at p (as a function of q) and, of course, $g(q; p)$ also has a logarithmic pole at p as a function of q. Hence, $d(p, z)$ is harmonic in a neighborhood of $z = p$. Further, if $x \in \partial\Omega$, we have

$$\liminf\{d(p, q) : q \to x\} = \liminf\{g(p, q) : q \to x\} \geqslant 0$$

and hence $d(p, q) \geqslant 0$ in Ω for each p and all q. On the other hand,

$$\limsup\{d(p, q): p \to x\} = \limsup\{-g(q, p): p \to x\} \leqslant 0$$

so that $d(p, q) \leqslant 0$ in Ω for each q and all p. Thus, $d \equiv 0$ and we see that $g(p, q) = g(q, p)$.

For the general case let $\{\Omega_m\}$ be a regular exhaustion of Ω; there's no loss in assuming that $p, q \in \Omega_1$. If g_m is the Green's function for Ω_m then we know $\{g_m(z; w)\}$ increases to $g(z; w)$, $z, w \in \Omega$ and $z \neq w$. Thus,

$$g(p, q) = \lim g_m(p, q) = \lim g_m(q, p) = g(q, p)$$

1.6. HARMONIC MEASURE

Let Ω be a domain on the sphere for which the Dirichlet problem is solvable and let p be a point of Ω. For each real-valued continuous function u on $\partial\Omega = \Gamma$ we can associate the real number $\tilde{u}(p)$, where $\tilde{u}$ is the harmonic extension to Ω of u. The rule $u \mapsto \tilde{u}(p)$ is linear and, because of the maximum principle, we know $|\tilde{u}(p)| \leqslant \|u\|_\Gamma = \sup\{|u(z)|: z \in \Gamma\}$. Thus, the Riesz representation theorem implies that there is a unique real measure ω_p on Γ with

$$\tilde{u}(p) = \int_\Gamma u \, d\omega_p, \qquad u \in C_r(\Gamma) \tag{6.1}$$

This measure is the *harmonic measure* on Γ for p. Note as well that if $u \geqslant 0$ then $\tilde{u}(p) \geqslant 0$ and so ω_p is a non-negative measure; further, the total mass of ω_p is 1 since

$$\|\omega_p\| = \int_\Gamma 1 \, d\omega_p = \tilde{1}(p) = 1$$

Clearly ω_p depends on Ω but we shall suppress this in the notation except when needed.

The next theorem relates the measures ω_p and ω_q for $p, q \in \Omega$.

Theorem 6.1. *If $p, q \in \Omega$, then ω_p and ω_q are boundedly mutually absolutely continuous. Indeed, if $\mathbf{K}$ is a compact set in Ω, then there is a constant M with*

$$\omega_q(E) \leqslant M\omega_p(E)$$

for all $q \in \mathbf{K}$ and for all measurable sets E in Γ.

Proof. We first show that ω_p and ω_q are mutually absolutely continuous. Suppose E is a closed set in Γ of ω_p-measure 0. Let u be a continuous function on Γ with $u = 1$ on E and $0 \leqslant u < 1$ off E. Let v_n be the harmonic extension to Ω of u^n. Then $\{v_n\}$ is a decreasing sequence of positive harmonic functions on Ω and

$$v_n(p) = \int_\Gamma u^n \, d\omega_p \to \omega_p(E) = 0$$

Hence, $v_n \to 0$ on all of Ω so that

$$0 = \lim v_n(q) = \lim \int_\Gamma u^n \, d\omega_q = \omega_q(E)$$

For $p \in \Omega$ fixed, let us write $d\omega_q = A_q \, d\omega_p$, where A_q is a non-negative function in $L^1(\omega_p)$. We wish to show that A_q is actually in $L^\infty(\omega_p)$ and, indeed, its sup norm is uniformly bounded as q varies over K.

The key ingredient in the proof is *Harnack's inequality*: there are positive numbers c_1 and c_2 depending only on $\mathbf{K}$ and p with the property that if v is a positive harmonic function on Ω satisfying

$$v(p) = 1, \quad \text{then} \quad c_1 \leqslant v(q) \leqslant c_2, \qquad q \in \mathbf{K}$$

It suffices to prove this when $\mathbf{K}$ is a closed disc containing p. Let $\mathfrak{D}$ be an open disc containing both p and $\mathbf{K}$. Suppose the assertion about the upper bound c_2 is false. Then there is a sequence $\{v_n\}$ of positive harmonic functions on Ω and a sequence $\{q_n\}$ of points in $\mathbf{K}$ with $v_n(p) = 1$ but $v_n(q_n) > n$. There is no loss in assuming that $q_n \to q \in \mathbf{K}$. Consider the holomorphic functions

$$f_n = \exp[-v_n - i{}^*v_n]$$

where ${}^*v_n$ is the harmonic conjugate of v_n in $\mathfrak{D}$ vanishing at p. These functions f_n are bounded by 1 and so at least a subsequence converges uniformly on compact sets in $\mathfrak{D}$ to a holomorphic function f. Since $f_n(p) = e^{-1}$ for all n, we have $f(p) = e^{-1}$ and so f is not zero anywhere on $\mathfrak{D}$. But $|f_n(q_n)| \leqslant e^{-n}$ so that $f(q) = 0$, a contradiction. In a similar fashion the lower bound c_1 can be shown to exist.

With this observation, the proof of the theorem is quite easy. Let E be a closed set in Γ and suppose $\omega_p(E) = \sigma > 0$. Again let u be a continuous function on Γ with $u = 1$ on E and $0 \leqslant u < 1$ off E. As before, let v_n be the harmonic extension to Ω of u^n and let v be the limit of the sequence $\{v_n\}$. Then v is a positive harmonic function in Ω and

$$v(p) = \lim \int_\Gamma u^n \, d\omega_p = \int_E d\omega_p = \omega_p(E) = \sigma$$

Thus,

$$\sigma c_1 \leqslant v(q) \leqslant \sigma c_2 \quad \text{for all } q \in \mathbf{K}.$$

However,

$$v(q) = \lim v_n(q) = \lim \int u^n \, d\omega_q$$

$$= \int_E d\omega_q = \omega_q(E)$$

Consequently,

$$\sigma c_1 \leqslant \omega_q(E) \leqslant \sigma c_2, \qquad q \in \mathbf{K}$$

Equivalently,

$$c_1 \leqslant \frac{\omega_q(E)}{\omega_p(E)} \leqslant c_2 \tag{6.2}$$

Since (6.2) holds for all closed sets E in Γ and since the measures are regular, (6.2) also holds for all measurable sets and this is exactly the assertion of the theorem.

Proposition 6.2. Suppose Ω_1 and Ω_2 are two domains and h is a holomorphic function on Ω_1 which maps $\mathrm{CL}(\Omega_1)$ homeomorphically onto $\mathrm{CL}(\Omega_2)$. Suppose the Dirichlet problem is solvable in Ω_1 (and hence in Ω_2); let $p_1 \in \Omega_1$ and put $p_2 = h(p_1) \in \Omega_2$. Let ω_1 be harmonic measure on $\partial\Omega_1$ for p_1, and define a measure μ on $\partial\Omega_2$ by the rule

$$\mu(E) = \omega_1(h^{-1}(E)), \qquad E \subset \partial\Omega_2 \tag{6.3}$$

Then μ is harmonic measure on $\partial\Omega_2$ for the point p_2.

Proof. The proof is shorter than the statement. The rule (6.3) is equivalent to the statement

$$\int_{\partial\Omega_2} u \, d\mu = \int_{\partial\Omega_1} u \circ h \, d\omega_1, \qquad u \in C(\partial\Omega_2) \tag{6.4}$$

Hence,

$$\int_{\partial\Omega_2} u \, d\mu = \widetilde{u \circ h}(p_1) = \tilde{u}(h(p_1)) = \tilde{u}(p_2)$$

which is just what was to be proved.

Theorem 6.3. *Harmonic measure has no atoms.*

Proof. Fix some $p \in \Omega$ and let ω be harmonic measure for p. There is no loss in assuming that $\infty \in \Omega$ and that $0 \in \Gamma = \partial\Omega$. We shall show that ω has no mass at 0; indeed, that $\log|z|$ is in $L^1(\Gamma, \omega)$.

Let $g(z; \infty)$ be the Green's function with pole at ∞ and put $u(z) = g(z; \infty) - \log|z|$. Then u is harmonic in Ω and u is continuous at all points of $\Gamma \setminus \{0\}$ with value equal to $-\log|x|$. Let $\{f_n\}$ be a sequence of continuous functions on Γ with $f_1 \leqslant f_2 \leqslant \cdots$, $\lim f_n(x) = -\log|x|$ if $x \in \Gamma$, $x \neq 0$, and $\{f_n(0)\}$ increasing to ∞. Let u_n be the harmonic extension of f_n to Ω. Then $u_1 \leqslant u_2 \leqslant \cdots$ in Ω and $u_n(z) \leqslant u(z)$ if $z \in \Omega$ since $u_n \leqslant u$ on $\Gamma \setminus \{0\}$ and surely $u_n(z) < u(z)$ if $|z|$ is small enough. Thus,

$$u(p) \geqslant \lim u_n(p) = \lim \int_\Gamma f_n \, d\omega$$

$$= \lim[f_n(0)\,\omega(\{0\})] + \int_\Gamma -\log|x| \, d\omega(x)$$

which proves that $\omega(\{0\}) = 0$ and $\log|x|$ is in $L^1(\Gamma, \omega)$.

When the domain Ω has a nice boundary, harmonic measure is particularly easy to understand. We shall have great use for the next few results in Chapters 4 and 5.

Theorem 6.4. *Suppose Ω is bounded by a finite number of disjoint analytic simple closed curves. For each $p \in \Omega$ we have*

$$d\omega_p = \frac{-1}{2\pi}\frac{\partial}{\partial n} g(\cdot\,; p)\, ds \tag{6.5}$$

where $g(\cdot\,; p)$ is the Green's function for Ω with pole at p, $\partial/\partial n$ is the derivative in the direction of the outward normal at Γ, and ds is arc length.

Proof. The proof is an application of Green's theorem. Let h be a smooth function on Γ and for small $\delta > 0$ let

$$\Omega_\delta = \Omega \setminus \{z : |z - p| \leqslant \delta\}$$

Let u be the harmonic extension to Ω of h and let $v(z) = g(z; p)$. Then by Green's theorem on Ω_δ,

$$\iint_{\Omega_\delta} [u\Delta v - v\Delta u]\, dx\, dy = \int_{\partial\Omega_\delta} \left[u\frac{\partial v}{\partial n} - v\frac{\partial u}{\partial n}\right] ds \tag{6.6}$$

Now the left-hand side of (6.6) is zero since both u and v are harmonic on Ω_δ.

Further, $v = 0$ on Γ, so (6.6) simplifies to

$$\int_\Gamma h \frac{\partial v}{\partial n}\, ds = \int_{|z-p|=\delta} \left[u \frac{\partial v}{\partial n} - v \frac{\partial u}{\partial n} \right] ds$$

Let us set $z = p + re^{it}$, $0 \leqslant r \leqslant \delta$, and $0 \leqslant t \leqslant 2\pi$. Then the normal derivative is just the radial derivative and $g(z;\, p) = -\log r + G(z)$ where G is harmonic near p. Thus,

$$\frac{\partial g}{\partial n} = -\frac{1}{r} + \text{continuous term}$$

Since u is also continuous near p, we have

$$\int_{|z-p|=\delta} u \frac{\partial g}{\partial n}\, ds = -\int_0^{2\pi} u(p + \delta e^{it})\, dt + 0(\delta)$$

$$\to -2\pi u(p) \quad \text{as } \delta \to 0$$

Finally, the other term

$$-\int_{|z-p|=\delta} g \frac{\partial u}{\partial n}\, ds$$

behaves like $-\delta \log \delta$ as $\delta \to 0$ and so goes to zero as $\delta \to 0$.

It is convenient to put here two more facts related to harmonic measure. Let us assume that $\Gamma = \partial\Omega$ consists of $m + 1$ disjoint analytic simple closed curves. Let $p \in \Omega$ and let $h(z) = h(z;\, p)$ be the harmonic conjugate of $g(z;\, p)$, the Green's function for Ω and p. Of course, h is not single-valued since it has periods about the holes in Ω (we will explore this type of difficulty later in some detail) but locally $g + ih$ is analytic and its derivative is single-valued on Ω. We write $Q = g + ih$ and Q' for its complex derivative.

Proposition 6.5. $d\omega_p(\zeta) = \dfrac{i}{2\pi} Q'(\zeta)\, d\zeta$

Proof. Let n be a unit outward normal at $\zeta \in \Gamma$, and let τ be a unit tangent at ζ. By the Cauchy–Riemann equations

$$\frac{\partial h}{\partial n}(\zeta) = \frac{\partial g}{\delta \tau}(\zeta) = 0$$

since g vanishes on Γ. Next, note that $n = -i\tau = -i\, d\zeta/|d\zeta|$ and

$$Q'(\zeta) = \lim_{t \downarrow 0} \frac{Q(\zeta + tn) - Q(\zeta)}{tn}$$

Further,

$$\frac{\partial g}{\partial n} = \lim_{t \downarrow 0} \frac{g(\zeta + tn) - g(\zeta)}{t}$$

Thus,

$$iQ'(\zeta)\, d\zeta = i \lim_{t \downarrow 0} \frac{Q(\zeta + tn) - Q(\zeta)}{tn} d\zeta$$

$$= -\lim_{t \downarrow 0} \frac{g(\zeta + tn) - g(\zeta)}{t} |d\zeta|$$

$$= -\frac{\partial g}{\partial n} ds$$

which establishes the result in light of Theorem 6.4.

Proposition 6.6. Let $\Gamma = \partial\Omega$ consist of $m + 1$ disjoint analytic simple closed curves, let $p \in \Omega$, and let $Q(z) = g(z; p) + ih(z; p)$ where h is the (multiple-valued) harmonic conjugate of g. Then

(a) Q' does not vanish on Γ
(b) Q' has precisely m zeros in Ω, counting multiplicities

Proof. We know that Q' has a single pole of order one at p. Further, from Proposition 6.5 we also know that $iQ'(z)\, dz$ is a non-negative measure on Γ. Thus, the total change in $1/2\pi \arg Q'(z)$ as z traverses Γ once is precisely $m - 1$. But this must be exactly the number of zeros of Q' minus the number of poles, and hence (b) is proven. To see (a) note that the Cauchy–Riemann equations imply that h is increasing locally on Γ and thus Q must be one-to-one in a neighborhood of each $\zeta \in \Gamma$.

Definition. The m zeros of Q' in Ω are the *critical points* of the Green's function $g(\cdot\,; p)$. They obviously depend on p but we suppress this in the notation.

The critical points of the Green's function will reappear in several different contexts in the succeeding material.

1.7. LOGARITHMIC CAPACITY

This section contains a quick trip through some of the most basic theorems in potential theory with the view of gaining some knowledge which will be helpful later on, as well as to become more familiar with the functions from the first part of the chapter.

Let $\mathbf{E}$ be a compact set in the plane and let $\mathcal{P}(\mathbf{E})$ denote the set of probability measures on $\mathbf{E}$: $\nu \in \mathcal{P}(\mathbf{E})$ if ν is a non-negative measure of total mass 1. For $\nu \in \mathcal{P}(\mathbf{E})$, set

$$p_\nu(z) = -\int_{\mathbf{E}} \log|z-\zeta|\,d\nu(\zeta), \qquad z \in \mathbb{C} \tag{7.1}$$

p_ν is the *potential generated* by ν. Some properties of the potential follow.

$$p_\nu \text{ is harmonic off } \mathbf{E} \text{ except at } \infty \text{ where } p_\nu(z) + \log|z| \text{ is harmonic} \tag{7.2}$$

$$-p_\nu \text{ is subharmonic on } \mathbb{C} \tag{7.3}$$

$$p_\nu(z) \leqslant \sup\{p_\nu(\zeta) : \zeta \in E\} \quad \text{if } z \notin \mathbf{E} \tag{7.4}$$

It is easy to verify (7.2). As for (7.3), let

$$L_n(w) = \max\{\log|w|, -n\}$$

Then L_n is subharmonic and continuous on $\mathbb{C}$ and $L_1 \geqslant L_2 \geqslant \cdots$. Set

$$q_n(z) = \int L_n(z-\zeta)\,d\nu(\zeta)$$

Then q_n is continuous, subharmonic, and $\{q_n(z)\}$ decreases to $-p_\nu(z)$ for each z. Thus, (7.3) holds by Proposition 3.2.

To see that (7.4) holds, there is nothing lost in assuming that $\sup\{p_\nu(\zeta): \zeta \in \mathbf{E}\}$ is finite. Let M denote this supremum and note that since M is finite, ν can have no point masses. Given $\varepsilon > 0$ and $\zeta_0 \in \mathbf{E}$ choose $\delta > 0$ so that the mass of ν on the disc $\mathbf{V} = \{\zeta : |\zeta - \zeta_0| \leqslant \delta\}$ is no more than ε. Put

$$p_1(z) = -\int_{\mathbf{V}} \log|z-\zeta|\,d\nu(\zeta)$$

$$p_2(z) = -\int_{\mathbf{E}\setminus\mathbf{V}} \log|z-\zeta|\,d\nu(\zeta).$$

p_2 is continuous on $\mathbf{V}$. Let $z \notin \mathbf{E}$ and let ζ' be a point of $\mathbf{E}$ at minimum distance to z. Then $\zeta' \to \zeta_0$ as $z \to \zeta_0$ and further, if $\zeta \in \mathbf{E}$, we have

$$|\zeta - \zeta'| \leqslant |\zeta' - z| + |z - \zeta| \leqslant 2|z-\zeta|$$

This gives

$$\begin{aligned} p_1(z) &\leqslant \nu(\mathbf{V})\log 2 + p_1(\zeta') \\ &< \varepsilon \log 2 + p_1(\zeta') \end{aligned}$$

Thus, for z near ζ_0, we find

$$\begin{aligned} p_\nu(z) = p_1(z) + p_2(z) &< \varepsilon \log 2 + p_1(\zeta') + p_2(\zeta') + \varepsilon \\ &= \varepsilon(1 + \log 2) + p_\nu(\zeta') \\ &\leqslant \varepsilon(1 + \log 2) + M \end{aligned}$$

This proves that $\lim\sup\{p_\nu(z):\ z \to \zeta_0\} \leqslant M$ for all $\zeta_0 \in \mathbf{E}$ and hence (7.4) holds.

Proposition 7.1. Let Ω be a domain containing ∞ on which the Dirichlet problem is solvable and let ω_∞ be harmonic measure on $\partial\Omega$ for ∞. Let $g(z, \infty)$ be the Green's function for Ω with pole at ∞ and put

$$\gamma = \lim\{g(z, \infty) - \log|z| : |z| \to \infty\} \tag{7.5}$$

Then the potential generated by ω_∞ satisfies

$$-\int \log|z - \zeta|\, d\omega_\infty(\zeta) = \begin{cases} \gamma - g(z; \infty), & z \in \Omega \\ \gamma, & z \notin \Omega \cup \partial\Omega \end{cases} \tag{7.6}$$

and

$$-\int \log|z - \zeta|\, d\omega_\infty(\zeta) \leqslant \gamma \quad \text{for all } z \tag{7.7}$$

Proof. Fix any point $a \in \Omega$, $a \neq \infty$. Set

$$w(z) = \int \log|a - \zeta|\, d\omega_z(\zeta) - \log|a - z| + g(z; \infty)$$

The function w has a logarithmic pole at a, is harmonic on $\Omega \setminus \{a\}$, including at ∞, and w vanishes identically on $\partial\Omega$. Thus, $w(z) = g(z; a)$. However, $g(z; a) = g(a; z)$ by Proposition 5.4 so we have

$$-\int \log|a - \zeta|\, d\omega_z(\zeta) = \{g(z; \infty) - \log|z - a|\} - g(a; z) \tag{7.8}$$

In (7.8) let $|z| \to \infty$; the term in the curly brackets approaches γ so that

$$-\int \log|a - \zeta|\, d\omega_\infty(\zeta) = \gamma - g(a; \infty)$$

which is the desired equality in (7.6) for $a \in \Omega$. Next suppose $a \notin \Omega \cup \partial\Omega$; put

$$w(z) = -\int \log|a - \zeta|\, d\omega_z(\zeta) + \log|z - a| - g(z; \infty)$$

Then w is harmonic in Ω and identically zero on $\partial\Omega$. Thus w vanishes identically and, in particular

$$0 = w(\infty) = -\int \log|a - \zeta|\, d\omega_\infty(\zeta) - \gamma$$

which is the desired equality in (7.6) for $a \notin \Omega \cup \partial\Omega$. Lastly, (7.7) follows from (7.6) and the fact that a potential is lower semicontinuous.

Definition. The number γ defined by (7.5) is the *Robin's constant* of Ω, or the *Robin's constant* of $\mathbf{E} = \Omega^c$.

Now let $\mathbf{E}$ be any compact set in $\mathbb{C}$ and let Ω be the unbounded component of the complement of $\mathbf{E}$. Let $\{\Omega_n\}$ be any regular exhaustion of Ω and let $\{g_n(z; p)\}$ be the corresponding Green's function with pole at p. We noted in Section 5 that the sequence $\{g_n(z; p)\}$ is increasing with n for each z and converges uniformly on compact subsets of Ω to either $+\infty$ identically or to a function $g(z; p)$ which possesses properties (5.1a) and (5.1b). If the latter occurs for some $p \in \Omega$ then it occurs for all $p \in \Omega$ and the function $g(z; p)$ is independent of the sequence $\{\Omega_n\}$. In this case we know that

$$\gamma_n = \lim\{g_n(z; \infty) - \log|z| : |z| \to \infty\}$$

increases to

$$\gamma = \lim\{g(z; \infty) - \log|z| : |z| \to \infty\} \tag{7.9}$$

and this number is surely finite. Furthermore, if $\omega_\infty^{(n)}$ is harmonic measure on $\partial\Omega_n$ for ∞, then at least some subsequence of $\{\omega_\infty^{(n)}\}$ converges weak-* in the space of measures to a probability measure ν supported on $\mathbf{E}$ (indeed, on $\partial\Omega$) which satisfies

$$p_\nu(z) = \begin{cases} \gamma - g(z; \infty), & z \in \Omega \\ \gamma, & z \notin \Omega \cup \partial\Omega \end{cases} \tag{7.10}$$

In this case we define the *logarithmic capacity of* **E** by

$$\text{cap}(\mathbf{E}) = e^{-\gamma} \tag{7.11}$$

What, however, happens in the former case, when $g_n(z;\, p)$ increases to $+\infty$ for each z, $p \in \Omega$? In this case we must have $\{\gamma_n\}$ increasing to $+\infty$ [or else $g_n(z, \infty)$ would stay bounded] and we set

$$\text{cap}(\mathbf{E}) = 0$$

Note well that the capacity of **E** depends only on the unbounded component of the complement of **E**. We can phrase this in the following way:

$$\text{cap}(\mathbf{E}) = \text{cap}(\hat{\mathbf{E}}) \tag{7.12}$$

where $\hat{\mathbf{E}}$ is the *polynomial hull* of **E** defined by the rule

$$z \in \hat{\mathbf{E}} \text{ if and only if } |p(z)| \leqslant \max\{|p(\zeta)| : \zeta \in \mathbf{E}\}$$

for all polynomials p. We now show that $\hat{\mathbf{E}}$ is obtained from **E** by the device of adding to **E** all the bounded components of the complement of **E**. First, if $\mathcal{O}$ is a bounded component of $\mathbf{E}^c$, then the maximum principle immediately implies that each point of $\mathcal{O}$ lies in $\hat{\mathbf{E}}$. Second, if $|z_0| > \max\{|\zeta| : \zeta \in \mathbf{E}\}$, then $z_0 \notin \hat{\mathbf{E}}$ since the polynomial $p(z) = z$ is larger at z_0 than it is on **E**. Let Ω be the unbounded component of the complement of **E**. Clearly the set of those points of Ω which are not in $\hat{\mathbf{E}}$ is open. We shall also show that $\Omega \cap \hat{\mathbf{E}}$ is open and thus $\Omega \cap \hat{\mathbf{E}}$ is both open and closed and consequently empty. Let $a \in \Omega \cap \hat{\mathbf{E}}$; then the functional $p \mapsto p(a)$ defined on all polynomials is bounded, with norm 1, since $|p(a)| \leqslant \|p\|_{\mathbf{E}}$. Hence, it can be extended to a linear functional on $C(\mathbf{E})$ of norm 1; this implies that there is a measure λ on **E** of total variation 1 with

$$\int_{\mathbf{E}} p\, d\lambda = p(a), \qquad p \text{ a polynomial}$$

The function

$$u(b) = \int_{\mathbf{E}} \frac{z - a}{z - b}\, d\lambda(z)$$

is continuous for $b \in \Omega$ and equals 1 at a; let b be any point near a at which $u(b) \neq 0$. If p is a polynomial then so is

$$\frac{p(z) - p(b)}{z - b}(z - a)$$

and we find

$$0 = \int_{\mathbf{E}} \frac{p(z) - p(b)}{z - b}(z - a)\, d\lambda(z)$$

or

$$p(b) = c\int_{\mathbf{E}} p(z)\frac{z - a}{z - b}\, d\lambda(z), \qquad c = (u(b))^{-1}$$

Consequently,

$$|p(b)| \leqslant C\|p\|_{\mathbf{E}}$$

where C is a constant depending only on a and b, not p. Replace p by p^n, take nth roots, and then let $n \to \infty$. We find

$$|p(b)| \leqslant \|p\|_{\mathbf{E}}$$

That is, $b \in \Omega \cap \hat{\mathbf{E}}$ if b is near a. This shows $\Omega \cap \hat{\mathbf{E}}$ is open and, as above, $\Omega \cap \hat{\mathbf{E}}$ must be empty. In summary, we have this result.

Proposition 7.2. Let $\mathbf{E}$ be compact, and set

$$\hat{\mathbf{E}} = \{z : |p(z)| \leqslant \|p\|_{\mathbf{E}} \text{ for all polynomials } p\}$$

Then $\hat{\mathbf{E}}$ is obtained by adding to $\mathbf{E}$ all the bounded components of $\mathbf{E}^c$. Furthermore, $\operatorname{cap}(\mathbf{E}) = \operatorname{cap}(\hat{\mathbf{E}})$.

We now derive another approach to the logarithmic capacity of a compact set.

Let $\mathbf{E}$ be a compact set and for each positive integer n set

$$M_n = \inf\left\{ \sup_{z \in \mathbf{E}} \prod_1^n |z - z_j| : z_1, \ldots, z_n \in \mathbf{C} \right\} \tag{7.13}$$

That is, M_n is the smallest that the sup norm on $\mathbf{E}$ of a monic polynomial of degree n can be made. We leave it as an exercise to show that the infimum in (7.13) is actually a minimum and that if $z_1^*, \ldots, z_n^*$ are points for which

$$M_n = \sup_{z \in \mathbf{E}} \prod_1^n |z - z_j^*|$$

then $z_1^*, \ldots, z_n^*$ lie in the closed convex hull of $\mathbf{E}$. Next we observe that

$$M_{n+m} \leqslant M_n M_m \quad \text{for all } n, m \tag{7.14}$$

since the union best points for n and the best points for m form one possible set of points for $n + m$. We now need a simple fact:

$$\text{suppose } c_n + c_m \geqslant c_{n+m} \text{ for all } n, m; \text{ then } \lim c_m/m \text{ exists although it may be } -\infty \tag{7.15}$$

To see (7.15), let $\alpha > \inf\{c_m/m\}$. Then choose an integer s with $\alpha > c_s/s$. For $m > s$ we write $m = us + v$, where $0 \leqslant v < s$. Hence

$$c_m = c_{us+v} \leqslant c_{us} + c_v \leqslant uc_s + c_v$$

Thus,

$$\frac{c_m}{m} \leqslant \frac{u}{us + v} c_s + \frac{1}{m} c_v$$

As $m \to \infty$, we know $u \to \infty$; we then have for large m

$$\frac{c_m}{m} \leqslant \frac{c_s}{s} + \varepsilon < \alpha + \varepsilon$$

Consequently,

$$\inf_m \frac{c_m}{m} \leqslant \limsup\left\{\frac{c_m}{m} : m \to \infty\right\} \leqslant \frac{c_s}{s} \leqslant \alpha$$

Since this holds for all $\alpha > \inf\{c_m/m\}$ we find that

$$\lim\left\{\frac{c_m}{m} : m \to \infty\right\} = \inf\left\{\frac{c_m}{m}\right\}$$

With this in mind and in view of (7.14) we see that

$$\rho(\mathbf{E}) = \lim\{M_m^{1/m} : m \to \infty\} \tag{7.16}$$

exists, although it may be 0.

The function ρ is a set function, defined at least on compact sets and ρ has these two properties

$$\text{if } \mathbf{E} \subset \mathbf{F}, \quad \text{then } \rho(\mathbf{E}) \leqslant \rho(\mathbf{F}) \tag{7.17}$$

$$\text{if } \{\mathbf{E}_n\} \text{ is a decreasing sequence of compact sets with } \mathbf{E} = \cap \mathbf{E}_n, \text{ then } \rho(\mathbf{E}) = \lim \rho(\mathbf{E}_n) \tag{7.18}$$

The proof of (7.17) is trivial so we examine (7.18). Given $\varepsilon > 0$ there is a large integer s such that

$$\varepsilon + \rho(\mathbf{E}) \geqslant M_s^{1/s}(\mathbf{E})$$

However, with this s fixed, we know that

$$M_s^{1/s}(\mathbf{E}) \geqslant M_s^{1/s}(\mathbf{E}_n) - \varepsilon$$

for all large enough n and also that

$$M_s^{1/s}(\mathbf{E}_n) \geqslant \rho(\mathbf{E}_n), \qquad \text{all } s$$

since $\rho(\mathbf{G}) = \inf_r\{M_r^{1/r}(\mathbf{G})\}$ for any compact set $\mathbf{G}$. Putting these inequalities together we find

$$2\varepsilon + \rho(\mathbf{E}) \geqslant \rho(\mathbf{E}_n), \qquad n \text{ large}$$

so that (7.18) holds.

Theorem 7.3. $\rho = e^{-\gamma} = \mathrm{cap}(\mathbf{E})$

Proof. Let $z_1^*, \ldots, z_n^*$ be an optimal choice in (7.13). Then

$$\log M_n \geqslant \sum_1^n \log|z - z_j^*| \qquad \text{if} \qquad z \in \mathbf{E}$$

and so if we define

$$v(z) = -\frac{1}{n}\sum_1^n \log|z - z_j^*|$$

wc find

$$v(z) \geqslant -\frac{1}{n}\log M_n, \qquad z \in \mathbf{E}$$

Put $\mathbf{F} = \{z \in \mathbb{C} : v(z) \geqslant -(1/n)\log M_n\}$. Then $\mathbf{F}$ is compact and $\mathbf{E} \subset \mathbf{F}$. Consequently,

$$\gamma(\mathbf{E}) \geqslant \gamma(\mathbf{F}) \tag{7.19}$$

However, we notice that $-v(z) - (1/n)\log M_n$ has all the properties of the Green's function of the complement of $\mathbf{F}$ with pole at ∞ so that

$$-v(z) - \frac{1}{n}\log M_n = g(z, \infty; \mathbf{F}^c)$$

Thus,

$$\gamma(\mathbf{F}) = \lim\{g(z, \infty; \mathbf{F}^c) - \log|z| : |z| \to \infty\}$$

$$= -\frac{1}{n}\log M_n$$

Consequently,

$$\gamma(\mathbf{E}) \geqslant \gamma(\mathbf{F}) = -\frac{1}{n}\log M_n$$

which implies

$$\operatorname{cap}(\mathbf{E}) = e^{-\gamma(\mathbf{E})} \leqslant \rho(\mathbf{E})$$

We prove the reverse inequality first for a compact set **E** with the property that Ω, the unbounded component of the complement of **E**, is regular for the Dirichlet problem; that is, the Dirichlet problem is solvable in Ω. Suppose, then that Ω is such a domain, $\infty \in \Omega$, and **E** is the complement of Ω. Let $g(z, \infty)$ be the Green's function for Ω with pole at ∞ and recall (7.6)

$$-\int \log|z - \zeta|\, d\omega_\infty(\zeta) = \gamma - g(z, \infty), \qquad z \in \Omega \tag{7.20}$$

Choose a compact set **K** in Ω so close to $\partial\Omega$ that

$$g(z, \infty) < \varepsilon, \qquad z \in \mathbf{K} \tag{7.21}$$

Since ω_∞ is a probability measure on $\partial\Omega \subset \mathbf{E}$ there is a sequence $\{\nu_n\}$ of measures of the form

$$\nu_n = \frac{1}{n}\sum_{j=1}^{n} \delta_{j,n}$$

where $\delta_{j,n}$ is the unit point mass at a point $\zeta_{j,n} \in \partial\Omega$, such that $\{\nu_n\}$ converges weak-* to ω_∞ in the space of measures on $\partial\Omega$; the points ζ_{jn} need not be distinct. Thus,

$$\int \log|z - \zeta|\, d\nu_n(\zeta) \to \int \log|z - \zeta|\, d\omega_\infty(\zeta)$$

uniformly on compact subsets of Ω and, in particular, on **K**. If **K** is chosen so that **E** lies in $\hat{\mathbf{K}}$ then

$$\max\{|q(z)| : z \in \mathbf{K}\} \geqslant \max\{|q(z)| : z \in \mathbf{E}\} \tag{7.22}$$

for any polynomial q. Set

$$Q_n(z) = \prod_{j=1}^{n} (z - \zeta_{jn})$$

Then

$$\left|\frac{1}{n}\log|Q_n(z)| - \int \log|z-\zeta|\,d\nu_n(\zeta)\right| < \varepsilon, \qquad z \in \mathbf{K}$$

for n large enough and so, combining this inequality with (7.20), (7.21) and (7.22) we find

$$\rho(\mathbf{E}) \leqslant \max_{z\in\mathbf{E}}|Q_n(z)|^{1/n} \leqslant e^{-\gamma+2\varepsilon}$$

Hence, Theorem 7.3 holds in the special case when the Dirichlet problem is solvable in the unbounded component of the complement of **E**. For the general case let $\{\Omega_n\}$ be a regular exhaustion of Ω and let $\mathbf{E}_n$ be the complement of Ω_n. Then, by definition

$$e^{-\gamma_n} \to e^{-\gamma}$$

and $\rho(\mathbf{E}_n) \to \rho(\mathbf{E})$ by (7.18). This finishes the proof.

Theorem 7.4. *Let* **E** *be a compact set of capacity zero. Then there is an element* $\sigma \in \mathcal{P}(\mathbf{E})$ *such that*

$$\lim\{p_\sigma(z) : z \to \mathbf{E}\} = \infty \tag{7.23}$$

Proof. We begin by noting that the proof of Theorem 7.3 showed that when computing M_n and ρ attention could be restricted to monic polynomials all of whose zeros lie in **E** (indeed, in $\partial\Omega$, where Ω is the unbounded component of the complement of **E**.) Let $z_{1n},\ldots,z_{nn}$ be points of **E** such that

$$\max_{z\in\mathbf{E}}\left(\prod_{j=1}^{n}|z-z_{jn}|\right)^{1/n} < \varepsilon_n$$

where $\varepsilon_n \to 0$ as $n \to \infty$, since $\operatorname{cap}(\mathbf{E}) = 0$. Thus,

$$-\frac{1}{n}\sum_{j=1}^{n}\log|z-z_{jn}| > -\log\varepsilon_n, \qquad z \in \mathbf{E}$$

Choose $n = n_k \to \infty$ such that $\varepsilon_{n_k} < \exp[-e^k]$ and put

$$\sigma_k = \frac{1}{n_k}\sum_{j=1}^{n_k}\delta_{j,n_k}, \qquad k = 1,2,3,\ldots$$

where δ_{j,n_k} is the unit point mass at z_{j,n_k}. Finally, put

$$\sigma = \sum_{1}^{\infty}\frac{1}{2^k}\sigma_k$$

so that $\sigma \in \mathcal{P}(\mathbf{E})$ and for $z \subset \mathbf{E}$ we have

$$p_\sigma(z) = -\int \log|z - \zeta|\, d\sigma(\zeta) \geqslant \sum_1^\infty \frac{1}{2^k} e^k = \infty$$

By lower semicontinuity we find

$$\liminf\{p_\sigma(z) : z \to \zeta, \zeta \in \mathbf{E}\} \geqslant p_\sigma(\zeta) = \infty$$

The final theorems of this section show that a compact set of capacity zero is "removable" for bounded harmonic functions and that this property characterizes such sets.

Theorem 7.5. *Let* $\mathbf{E}$ *be a compact set and* Ω *any domain containing* $\mathbf{E}$. *If* $\mathbf{E}$ *has logarithmic capacity zero then each function* u *which is bounded and harmonic in* $\Omega \setminus \mathbf{E}$ *extends to be harmonic in* Ω.

Proof. Let Ω' be a domain for which the Dirichlet problem is solvable with $\mathbf{E} \subset \Omega'$ and $\mathrm{CL}(\Omega') \subset \Omega$. Let h be the function harmonic on Ω', continuous on $\mathrm{CL}(\Omega')$, with $h = u$ on $\partial\Omega'$ and put $v = u - h$. Then v is bounded and harmonic on $\Omega' \setminus \mathbf{E}$ and $v = 0$ on $\partial\Omega'$. Let σ be the measure constructed in Theorem 7.4 and let $p(z)$ be the potential generated by σ. The function $v - \varepsilon p$ is harmonic on $\Omega' \setminus \mathbf{E}$ and is nonpositive on $\partial(\Omega' \setminus \mathbf{E}) \subseteq \partial\Omega' \cup \mathbf{E}$ (we may assume that the diameter of Ω is at most one with no loss of generality). Hence, $v \leqslant \varepsilon p$ on $\Omega' \setminus \mathbf{E}$ and, letting $\varepsilon \to 0$ we find that $v \leqslant 0$ on $\Omega' \setminus \mathbf{E}$. In a like fashion by considering $v + \varepsilon p$ we see that $v \geqslant 0$ on $\Omega' \setminus \mathbf{E}$. Hence, $v \equiv 0$ on $\Omega' \setminus \mathbf{E}$ and so $u = h$ on $\Omega' \setminus \mathbf{E}$. But h is harmonic across $\mathbf{E}$ and thus u is as well.

Theorem 7.6. *Let* $\mathbf{E}$ *be a compact set. Suppose* $\mathbf{E}$ *has the property that whenever* Ω *is a domain containing* $\mathbf{E}$ *and* u *is bounded and harmonic on* $\Omega \setminus \mathbf{E}$, *then* u *is actually harmonic on* Ω. *Then* $\mathbf{E}$ *has logarithmic capacity zero.*

Proof. The hypothesis certainly implies $\mathbf{E}$ contains no continuum (otherwise there is a nonconstant bounded analytic function in the complement of that continuum). Hence, $\Omega = S^2 \setminus \mathbf{E}$ is connected. Suppose cap($\mathbf{E}$) is positive; then Ω has a (generalized) Green's function $g(z, \infty)$ with pole at ∞. By hypothesis $g(z, \infty)$ extends to be harmonic across $\mathbf{E}$ and hence $g(z, \infty)$ is harmonic on $\mathbb{C}$ with a logarithmic pole at ∞. Let h be the harmonic conjugate of $g(z, \infty)$ on $\mathbb{C}$ and consider

$$F = \exp[-g - ih]$$

F is a bounded entire function and hence is constant; this contradicts the fact that $g(z, \infty)$ is not identically ∞. Thus, cap($\mathbf{E}$) = 0.

ADDITIONAL READINGS AND NOTES

The material of Chapter 1 is classical and may be found in many texts. The paperback book of Fuchs (1967) has a nice development of the topics of this chapter and others as well; the presentation here is very much like his. The book of Tsuji (1959) is a compendium of results on function and potential theory in the plane. Helms (1969) gives an exposition of potential theory in Euclidean n-space. In the exercises which follow two more notions of capacity are developed and shown to coincide with the logarithmic capacity presented in Section 1.7; the concept of the transfinite diameter of a compact set was introduced and studied by Fekete. Among other things, he showed that the logarithmic capacity (equivalently, the transfinite diameter, τ) is the only nonzero set function defined on the compact sets in the plane which satisfies these four properties:

1. $\tau(\mathbf{E}) \leqslant \tau(\mathbf{F})$ if $\mathbf{E} \subseteq \mathbf{F}$
2. If $a \in \mathbb{C}$ and $a\mathbf{E} = \{az : z \in \mathbf{E}\}$, then $\tau(a\mathbf{E}) = |a|\tau(\mathbf{E})$
3. If $\varepsilon > 0$ and $\mathbf{E}_\varepsilon = \{z : \operatorname{dist}(z, \mathbf{E}) \leqslant \varepsilon\}$, then $\tau(\mathbf{E}_\varepsilon) \to \tau(\mathbf{E})$ as $\varepsilon \to 0$
4. If Q is a monic polynomial of degree k and if $\mathbf{E}^*$ consists of all roots of $Q(z) = w$ as w ranges over $\mathbf{E}$, then $\tau(\mathbf{E}^*) = (\tau(\mathbf{E}))^{1/k}$

See Fekete (1923).

EXERCISES

1. Prove the assertion made in the introduction to this chapter, that the Dirichlet problem is not solvable on the punctured disc, $0 < |z| < 1$.
2. If u is subharmonic on $a < |z| < b$, then

$$I(r) = \int_{-\pi}^{\pi} u(re^{it})\, dt, \qquad a < r < b$$

is a convex function of $\log r$.

3. Let Ω be a domain for which the Dirichlet problem is solvable and let E be a ω-measurable set in $\Gamma = \partial\Omega$. If u is subharmonic on Ω and

$$\limsup\{u(z) : z \to \zeta\} \leqslant M, \qquad \text{all } \zeta \in E$$

$$\limsup\{u(z) : z \to \lambda\} \leqslant L, \qquad \text{all } \lambda \in \Gamma \setminus E$$

then

$$u(z) \leqslant M\omega_z(E) + L\omega_z(\Gamma \setminus E), \qquad z \in \Omega$$

4. If, in problem 3, $E = \Gamma$ and if $u(z_0) = M$ for some $z_0 \in \Omega$, show $u \equiv M$ in Ω.
5. Prove a function u on a compact set $\mathbf{K}$ with values in $[-\infty, \infty)$ is upper semicontinuous if and only if there are continuous functions $\{f_j\}$ on $\mathbf{K}$ with $f_1 \geqslant f_2 \geqslant \cdots$ and $f_j(x) \to u(x)$, $x \in \mathbf{K}$.
6. Suppose Ω is simply connected and $\Gamma = \partial\Omega$ has two or more points. Show that the Dirichlet problem is solvable in Ω.
7. Let Ω be bounded and simply connected and let $g(z; z_0)$ be the Green's function for Ω with pole at z_0. If H is the harmonic conjugate of $G(z) = g(z; z_0) + \log|z - z_0|$ on Ω, show that $F(z) = (z - z_0)\exp[-G(z) - iH(z)]$ is a one-to-one mapping of Ω onto the open unit disc Δ, with $F(z_0) = 0$.
8. Find the Green's function for the open unit disc Δ with pole at $a \in \Delta$. Do the same for the annulus $\{z : 1 < |z| < \rho\}$ with pole at $r \in (1, \rho)$.
9. Let $\mathbf{E}$ be compact. Show that there are points $z_1^*, \ldots, z_n^*$ such that

$$M_n = \sup\left\{\prod_1^n |z - z_j^*| : z \in \mathbf{E}\right\}$$

Show further that $z_1^*, \ldots, z_n^*$ lie in the convex hull of $\mathbf{E}$.
10. Let $\mathbf{E}$ be a compact, connected set with connected complement Ω (relative to the sphere) and let $f(z) = Az + b_0 + b_1/z + \cdots$ be the Riemann mapping of Ω onto $\{w : |w| > 1\}$ with $f(\infty) = \infty$, $A > 0$. Show that $\operatorname{cap}(\mathbf{E}) = 1/|A|$. If $\mathbf{E} = [a, b]$ in $\mathbb{R}$, show $\operatorname{cap}(\mathbf{E}) = \frac{1}{4}(b - a)$.
11. Let Ω be a domain for which the Dirichlet problem is solvable and let $\mathbf{E} \subset \partial\Omega$ be compact. If $\operatorname{cap}(\mathbf{E}) = 0$, then the harmonic measure of $\mathbf{E}$ is also zero.
12. Let Ω_1 and Ω_2 be two domains on which the Dirichlet problem is solvable and suppose $\Omega_1 \subset \Omega_2$. Let $p \in \Omega_1$ and let ω_1 and ω_2 be harmonic measure on $\partial\Omega_1$ and $\partial\Omega_2$, respectively, for p. If E is a compact set in $\partial\Omega_1 \cap \partial\Omega_2$, show that $\omega_1(E) \leqslant \omega_2(E)$.

There are at least two other ways to arrive at the logarithmic capacity of a compact set $\mathbf{E}$. These are developed in the next several exercises.

Let $z_1, \ldots, z_n$ be points of $\mathbf{E}$ and set

a. $P(z_1, \ldots, z_n) = \prod_{j<k} |z_j - z_k|$
b. $P_n = \sup\{P(z_1, \ldots, z_n) : z_1, \ldots, z_n \in \mathbf{E}\}$
c. $d_n = P_n^{\varepsilon_n}$, $\varepsilon_n = 2/n(n-1)$

13. Show that $d_1 \geqslant d_2 \geqslant \cdots$. Let $\tau(\mathbf{E}) = \lim d_n$; τ is the *transfinite diameter* of $\mathbf{E}$.

14. Show that $\rho(\mathbf{E}) \leqslant \tau(\mathbf{E})$ [ρ given by (7.16)]

Now let $\nu \in \mathcal{P}(\mathbf{E})$ and set

a. $I(\nu) = -\iint_{\mathbf{EE}} \log|z - \zeta| \, d\nu(z) \, d\nu(\zeta)$

b. $\gamma_0(\mathbf{E}) = \inf\{I(\nu) : \nu \in \mathcal{P}(\mathbf{E})\}$

15. Show there is a $\nu_0 \in \mathcal{P}(\mathbf{E})$ with $I(\nu_0) = \gamma_0(\mathbf{E})$.

16. Show that $e^{-\gamma_0(\mathbf{E})} = \tau(\mathbf{E})$; use the technique of Theorem 7.3.

17. Show that $\gamma(\mathbf{E}) \geqslant \gamma_0(\mathbf{E})$; use (7.6) and (7.7).

18. Show that $\gamma(\mathbf{E}) \leqslant \gamma_0(\mathbf{E})$ by showing that if ν_0 is the measure from problem 15, then its potential ρ_{ν_0} is equal a.e. ν_0 to $\gamma_0(\mathbf{E})$ on $\mathbf{E}$. HINT: let μ be any probability measure on E for which $I(\mu) < \infty$; expand $I[\nu_0 + \varepsilon(\mu - \nu_0)]$ and note this must exceed $I(\nu_0)$ for all $\varepsilon > 0$, ε small.

2

UNIFORMIZATION AND CONDITIONAL EXPECTATION

2.1. INTRODUCTION

There are two approaches to function theory on a multiply connected planar domain. The first is to work directly on the domain and examine the analytic functions in their own customary setting. This approach is quite natural but difficulties do appear when the functions must be analyzed on or near the boundary of the domain, due to the possible complicated topological nature of the boundary. The second approach is to "lift" the function theory of the domain to the unit disc by means of the uniformizer. With this technique the boundary is well behaved (what could be nicer than the unit circle?) but new difficulties crop up with the requirement that the functions must be invariant under the group of linear fractional transformations that fix the uniformizer. In general, we deal in this book with function theory on the domain itself, that is, we utilize the first approach; but it is definitely worthwhile to consider the second approach and in this chapter we set the stage for doing this. We begin by proving that every planar domain can be "uniformized" by either the unit disc or the plane. This is, of course, the famous uniformization theorem of Koebe. This theorem plays the role for a multiply connected planar domain that the Riemann mapping theorem does for a simply connected planar domain. Namely, it allows the function theory on the multiply connected domain to be "lifted" to the unit disc.

2.2. THE UNIFORMIZATION THEOREM

We must begin by stating two theorems from topology but only giving a reference for their proof since it would take us too far afield to prove them here.

Theorem 2.1. *Let Ω be an arcwise connected, locally arcwise connected, and locally simply connected topological space. Then there is an arcwise connected topological space $\mathfrak{X}$ and a mapping T from $\mathfrak{X}$ onto Ω with these properties:*

(*a*) *$\mathfrak{X}$ is simply connected; that is, $\pi_1(\mathfrak{X}) = 0$*
(*b*) *T is an open map*
(*c*) *Each point $z \in \Omega$ has a neighborhood $\mathfrak{U}$ such that $T^{-1}(\mathfrak{U})$ is the union of disjoint open sets each of which is mapped homeomorphically by T onto $\mathfrak{U}$*
(*d*) *If α is a path in Ω and $Tx = \alpha(0)$, then there is a unique path β in $\mathfrak{X}$ with $\beta(0) = x$ and $T \circ \beta = \alpha$*

Definition. Let Ω, $\mathfrak{X}$, and T be as in Theorem 2.1. A *deck transformation* of $\mathfrak{X}$ is a homeomorphism h of $\mathfrak{X}$ onto $\mathfrak{X}$ with $T \circ h = T$.

Theorem 2.2. *The set $\mathfrak{G}$ of deck transformations is a group and $\mathfrak{G}$ is isomorphic to $\pi_1(\Omega)$. Further, if x_1 and x_2 are points of $\mathfrak{X}$ with $Tx_1 = Tx_2$, then there is a unique $h \in \mathfrak{G}$ with $h(x_1) = x_2$. Finally, if u is a continuous function on $\mathfrak{X}$ such that $u \circ h = u$ for all $h \in \mathfrak{G}$, then there is a unique continuous function v on Ω with $u = v \circ T$.*

For a proof of these and related facts, see Singer and Thorp (1967, chapter 3).

Now let us take Ω to be a domain in the complex plane so that Ω clearly satisfies the hypotheses of Theorem 2.1. In this case we can make $\mathfrak{X}$ into a Riemann surface by using T to transfer the local coordinates to $\mathfrak{X}$; that is, a complex-valued function f on $\mathfrak{X}$ is analytic or harmonic at $x \in \mathfrak{X}$ if the function $f \circ \tau$ is analytic or harmonic on a sufficiently small neighborhood of $T(x)$ where τ is the homeomorphism of a neighborhood of $T(x)$ onto a neighborhood of x assured by (c) of Theorem 2.1.

Theorem 2.3. *$\mathfrak{X}$ is conformally equivalent to either the open unit disc or to the whole complex plane.*

Proof. Since Ω is open, we know that $\mathfrak{X}$ is noncompact. Let $\{\mathfrak{X}_n\}$ be a sequence of open connected subsets of $\mathfrak{X}$ with these properties:

$$\mathrm{CL}(\mathfrak{X}_n) \subset \mathfrak{X}_{n+1}, \qquad n = 1, 2, \ldots \tag{2.1a}$$

$$\bigcup_n \mathfrak{X}_n = \mathfrak{X} \tag{2.1b}$$

$$\text{each component of } \partial\mathfrak{X}_n \text{ is nontrivial} \tag{2.1c}$$

The existence of such subsets is a topological problem, unrelated to function theory. Now fix a point $x_0 \in \mathfrak{X}$ which we may suppose to be in all of the sets $\mathfrak{X}_n$.

We may now carry over to $\mathfrak{X}_n$ all the machinery of Sections 2–6 of Chapter 1, basically without change, and assert that the Dirichlet problem is solvable on $\mathfrak{X}_n$. The only point worthy of mention is that of a barrier, but since this is fundamentally a local concept it, too, can be lifted to $\mathfrak{X}$ by means of T. With these preliminaries out of the way, we let $g_n(z; x_0)$ be the Green's function for $\mathfrak{X}_n$ with pole at x_0. If $m > n$, then $g_m - g_n$ is harmonic on $\mathfrak{X}_n$ and positive on $\partial\mathfrak{X}_n$ so that $g_m - g_n$ is positive on all of $\mathfrak{X}_n$. Hence, the sequence $\{g_n(z; x_0)\}$ is increasing on $\mathfrak{X} \setminus \{x_0\}$ and consequently one of two mutually exclusive eventualities occurs:

$$g_n(z; x_0) \text{ increases to } \infty \text{ uniformly on compact sets in } \mathfrak{X} \setminus \{x_0\} \tag{2.2}$$

or

$$g_n(z; x_0) \text{ increases to } g(z; x_0) \text{ uniformly on compact sets in } \mathfrak{X} \setminus \{x_0\} \tag{2.3}$$

where g is harmonic in $\mathfrak{X} \setminus \{x_0\}$ and in local coordinates $g(z; x_0) + \log|z - x_0|$ is harmonic near x_0. We handle these cases one at a time, beginning with the second, (2.3).

The monodromy theorem allows us to conclude that there is a single-valued analytic function $f(z; x_0)$ on $\mathfrak{X}$ with a simple zero at x_0 and no other zeros on $\mathfrak{X}$ such that

$$\log|f(z; x_0)| = -g(z; x_0) \tag{2.4}$$

Now let x_1 be another point of $\mathfrak{X}$, distinct from x_0 and let $f(z; x_1)$ be the analytic function associated with x_1 as just outlined. We now show that

$$\lambda f(z; x_1) = \frac{f(z; x_0) - a}{1 - \bar{a} f(z; x_0)}, \qquad a = f(x_1; x_0) \tag{2.5}$$

where λ is a constant of modulus 1. To see this, let $h(z)$ denote the right-hand side of (2.5). Then

$$\log|h(z)| + g_n(z; x_1) \leqslant 0, \qquad z \in \mathfrak{X}_n$$

by the maximum principle. Hence

$$\log|h(z)| \leqslant -g(z; x_1) \qquad z \in \mathfrak{X} \tag{2.6}$$

so that $|h(z)| \leqslant |f(z; x_1)|$ and, in particular,

$$|f(x_0; x_1)| \geqslant |h(x_0)| = |f(x_1; x_0)|$$

Reversing the roles of x_0 and x_1 we find $|f(x_0; x_1)| = |f(x_1; x_0)|$. Hence,

$\log|h(x_0)| = -g(x_0; x_1)$ so that (2.6) and the maximum principle imply that

$$h(z) = \lambda f(z; x_1), \qquad z \in \Omega$$

for some unimodular constant λ, which is precisely (2.5). Hence, h vanishes only at x_1 and there has a simple zero; equivalently, $f(z; x_0)$ is univalent on $\mathfrak{X}$. Hence, $f(z; x_0)$ maps $\mathfrak{X}$ in a one-to-one manner onto a domain Λ inside Δ; Λ is necessarily simply connected since $\mathfrak{X}$ is, so that it is in turn mapped onto the open unit disc Δ by a one-to-one holomorphic map ϕ. The composition $\phi(f(z; x_0))$ then maps $\mathfrak{X}$ onto Δ and so finishes the first case, when (2.3) holds.

Now we assume that (2.2) holds. We first show that $\mathfrak{X}$ carries no nonconstant bounded holomorphic function. For if f were such a function, then we could assume that $f(x_0) = 0$ and $|f(x)| \leqslant 1$ for all $x \in \mathfrak{X}$. The maximum principle then implies

$$g_n(z; x_0) + \log|f(z)| \leqslant 0, \qquad z \in \mathfrak{X}_n$$

and hence that the sequence $\{g_n(z; x_0)\}$ is bounded above by $-\log|f(z)|$, a contradiction.

Let $\mathcal{V}$ be a small neighborhood of x_0, chosen so that T maps $\mathcal{V}$ homeomorphically onto $\{\zeta : |\zeta - T(x_0)| < r\}$. Let

$$m_n = \max\{g_n(z; x_0) : z \in \partial\mathcal{V}\}$$

and put $v_n(z) = g_n(z; x_0) - m_n$. Then

$$v_n(z) \leqslant 0, \qquad z \in \mathfrak{X}_n, \qquad z \notin \mathcal{V} \tag{2.7}$$

Further,

$$\max\{v_n(z) : z \in \partial\mathcal{V}\} \geqslant -m_1 \tag{2.8}$$

Some subsequence of $\{v_n\}$ then converges uniformly on compact subsets of $\mathfrak{X} \setminus \{x_0\}$ to a harmonic function v which is nonpositive and not identically $-\infty$ on $\mathfrak{X} \setminus \mathcal{V}$ and which has a logarithmic singularity at x_0. (On $\mathcal{V}$ consider the sequence $\{v_n(z) + \log|Tz - Tx_0|\}$ which is harmonic on $\mathcal{V}$ and less than or equal to $\log r$ on $\partial\mathcal{V}$ and hence on all of $\mathcal{V}$.)

Again the monodromy theorem allows us to conclude that there is a single-valued holomorphic function $f(z) = f(z; x_0)$ with $\log|f(z)| = -v(z)$. Hence, f has a simple zero at x_0, no other zero on $\mathfrak{X}$ and,

$$|f(z)| = \exp[-v(z)] \geqslant 1, \qquad z \notin \mathcal{V} \tag{2.9}$$

Suppose g is another single-valued holomorphic function on $\mathfrak{X}$ with a simple zero at x_0, no other zero on $\mathfrak{X}$, and $|g| \geqslant \delta > 0$ off some neighborhood of x_0.

Then

$$h = \frac{1}{f} - \frac{A}{g}$$

is holomorphic at x_0 for an appropriate complex number A and h is bounded off some neighborhood of x_0. Thus, h is identically constant on $\mathfrak{X}$ and so

$$g = Af(1 - cf)^{-1}$$

for some $c \in \mathbf{C}$; that is, $g = \phi \circ f$ where ϕ is some linear fractional transformation.

According to (2.9) if $|w| < 1$, then $f(z)$ assumes the value w just once on $\mathfrak{X}$ and this at a point of $\mathcal{V}$, say $f(x_1) = w$. The function $f(z) - f(x_1)$ thus has a simple zero at x_1 and is bounded away from zero off $\mathcal{V}$; consequently, $f(z; x_1)$ is related to $f(z; x_0)$ by a linear fractional transformation. Let $\mathfrak{Y}$ consist of all points $y \in \mathfrak{X}$ such that

$$f(z; y) = \phi(f(z; x_0)), \qquad z \in \mathfrak{X} \tag{2.10}$$

for some linear fractional transformation ϕ. We know that an entire neighborhood of x_0 lies in $\mathfrak{Y}$. Also if $y \in \mathfrak{Y}$ and y' is near to y then the argument above shows that

$$f(z; y') = \psi(f(z; y)), \qquad z \in \mathfrak{X}$$

for some linear fractional transformation ψ and hence $y' \in \mathfrak{Y}$ also. Thus, $\mathfrak{Y}$ is open. But this argument also shows that $\mathfrak{Y}$ is closed since if $y_n \in \mathfrak{Y}$ and $y_n \to y$, then eventually y is in a suitable neighborhood of y_n so that

$$f(z; y_n) = \psi_n(f(z; y)), \qquad z \in \mathfrak{X}$$

and hence $y \in \mathfrak{Y}$. Thus, $\mathfrak{Y} = \mathfrak{X}$. This is enough to prove that $f(z) = f(z; x_0)$ is one-to-one. For suppose $f(p; x_0) = f(q; x_0)$. We have

$$f(z; x_0) = \phi(f(z; q)), \qquad z \in \mathfrak{X}$$

and so

$$f(q; x_0) = f(p; x_0) = \phi(f(p; q))$$

But

$$f(q; x_0) = \phi(f(q; q)) = \phi(0)$$

so that $\phi(0) = \phi(f(p; q))$ which implies $0 = f(p; q)$ and, in turn, that $p = q$.

Thus, $f(z)$ effects a one-to-one holomorphic mapping of $\mathfrak{X}$ onto a simply connected domain $\mathfrak{D}$ in the plane. $\mathfrak{D}$ cannot be a proper subset of $\mathbf{C}$ since then $\mathfrak{D}$ would be conformal with Δ and hence support nonconstant bounded analytic functions, a contradiction. Hence, $\mathfrak{D} = \mathbf{C}$ and the proof is finished.

Remark. The reader already familiar with the uniformization theorem will have noticed that the possibility that $\mathfrak{X}$ is the Riemann sphere has been omitted. This is because we assumed that Ω was a planar domain; such an Ω clearly does not have a compact covering surface.

2.3. CONDITIONAL EXPECTATION AND THE SPACE N

Henceforth we assume that Ω is a domain on the sphere whose uniformizing surface is the open unit disc Δ. This is certainly true if Ω supports nonconstant bounded analytic functions or even nontrivial H^p functions; see Chapter 3. In this case, the set $\mathfrak{G}$ of deck transformations consists exactly of those holomorphic functions h which map Δ homeomorphically onto itself (and which preserve T). Of course, such an h is a linear fractional transformation of the form

$$h(z) = \lambda \frac{z + a}{1 + \bar{a}z}, \qquad |\lambda| = 1, \qquad a \in \Delta \tag{3.1}$$

Each such h (independent of preserving T) is a homeomorphism of the unit circle onto itself so that we can use h to define a measure on $\mathbf{T}$ by

$$\mu(E) = \sigma\big(h^{-1}(E)\big)$$

where

$$d\sigma = \frac{1}{2\pi} d\theta$$

Equivalently,

$$\int_{\mathbf{T}} u \, d\mu = \int_{\mathbf{T}} u \circ h \, d\sigma, \qquad u \in C(\mathbf{T}) \tag{3.2}$$

Hence, μ is exactly harmonic measure for $a = h(0)$:

$$\mu = P_a \sigma$$

where we have temporarily adopted the briefer notation

$$P_a(e^{ix}) = P(r_0, t_0 - x), \qquad a = r_0 e^{it_0}$$

Thus, (3.2) becomes

$$\int_{\mathbf{T}} uP_a\,d\sigma = \int_{\mathbf{T}} u\circ h\,d\sigma, \qquad u\in C(\mathbf{T}), \qquad a = h(0) \tag{3.3}$$

Now let $\Sigma/\mathfrak{G}$ be the sigma subfield of the Lebesgue measurable sets Σ which are invariant under $\mathfrak{G}$:

$$X\in\Sigma/\mathfrak{G} \quad \text{iff} \quad X\in\Sigma \quad \text{and} \quad \sigma(X\Delta h(X)) = 0 \tag{3.4}$$

for all $h\in\mathfrak{G}$. (Here Δ stands for the symmetric difference of the two sets to the left and right of it.) Let $L^p/\mathfrak{G}$ consist of those functions in $L^p(\sigma,\mathbf{T},\Sigma)$ which are invariant under $\mathfrak{G}$:

$$f\in L^p/\mathfrak{G} \quad \text{iff} \quad f\in L^p(\sigma,\mathbf{T},\Sigma) \quad \text{and} \quad f\circ h = f \quad \text{a.e}\ \sigma \tag{3.5}$$

for all $h\in\mathfrak{G}$. The following proposition is needed; its proof is left as an exercise.

Proposition 3.1. The set of $\Sigma/\mathfrak{G}$-measurable functions consists precisely of those Σ-measurable functions f which satisfy $f\circ h = f$ a.e. σ for all $h\in\mathfrak{G}$.

We may now define the *conditional expectation*. Let $f\in L^1$ and consider the functional on $\Sigma/\mathfrak{G}$ given by

$$X\mapsto\int_X f\,d\sigma, \qquad X\in\Sigma/\mathfrak{G}$$

This functional defines a (complex-valued) measure on $\Sigma/\mathfrak{G}$ which is absolutely continuous with respect to σ, hence given by an element of $L^1/\mathfrak{G}$ which is written as $\mathfrak{E}f$. Thus, $\mathfrak{E}f$ is the unique element of $L^1/\mathfrak{G}$ defined by the rule

$$\int_X \mathfrak{E}f\,d\sigma = \int_X f\,d\sigma, \qquad X\in\Sigma/\mathfrak{G} \tag{3.6}$$

Proposition 3.2. The conditional expectation is a linear projection of L^p onto $L^p/\mathfrak{G}$ for $1\leqslant p\leqslant\infty$; for $p=2$ it is precisely the orthogonal projection of L^2 onto $L^2/\mathfrak{G}$. Further,

$$\mathfrak{E}(fg) = f\mathfrak{E}(g), \qquad f\in L^\infty/\mathfrak{G}, \qquad g\in L^1 \tag{3.7}$$

Proof. We begin by showing that $\mathfrak{E}$ is the orthogonal projection of L^2 onto $L^2/\mathfrak{G}$. The relationship (3.6) which defines $\mathfrak{E}$ implies

$$\int_{\mathbf{T}} \chi_X(\mathfrak{E}f - f)\,d\sigma = 0$$

where χ_X is the characteristic function of $X \in \Sigma/\mathfrak{G}$. Hence,

$$\int_{\mathbf{T}} g(\mathcal{E}f - f)\, d\sigma = 0, \qquad f \in L^2, \qquad g \in L^2/\mathfrak{G} \tag{3.8}$$

and this is exactly the statement that $\mathcal{E}$ is the orthogonal projection of L^2 onto $L^2/\mathfrak{G}$.

To show (3.7) let $f \in L^\infty/\mathfrak{G}$, let $u \in L^2/\mathfrak{G}$, and let v be orthogonal to $L^2/\mathfrak{G}$, $v \in L^2$. Then $fu \in L^2/\mathfrak{G}$ so

$$0 = \int_{\mathbf{T}} fuv\, d\sigma \tag{3.9}$$

But (3.9) implies that fv is orthogonal to $L^2/\mathfrak{G}$. Thus, when we write

$$fg = f\mathcal{E}g + f(g - \mathcal{E}g), \qquad g \in L^2$$

we see the first summand is in $L^2/\mathfrak{G}$ and the second is in the orthogonal complement of $L^2/\mathfrak{G}$ and hence is annihilated by $\mathcal{E}$;

$$\begin{aligned} \mathcal{E}(fg) &= \mathcal{E}(f\mathcal{E}g) + \mathcal{E}(f(g - \mathcal{E}g)) \\ &= f\mathcal{E}g \end{aligned}$$

Definition. The space H^2 consists of those functions $g \in L^2$ for which

$$\hat{g}(-n) = \int_{\mathbf{T}} g(e^{it}) e^{int}\, dt = 0, \qquad n = 1, 2, \ldots$$

That is, $g \in H^2$ if and only if $g(e^{it}) = \Sigma_0^\infty a_n e^{int}$ where $\Sigma_0^\infty |a_n|^2 < \infty$. H_0^2 is the subspace of those g with $a_0 = 0$.

For L^2 we clearly have the direct sum decomposition

$$L^2 = H_0^2 \oplus \mathbf{C} \oplus \overline{H_0^2}$$

where the third term consists of all the complex conjugates of H_0^2 functions.

Definition. Let $u \in L^2$ with Fourier series

$$u(e^{i\theta}) = \sum_{-\infty}^{\infty} c_n e^{in\theta}$$

Set

$$*u(e^{i\theta}) = i\left(\sum_1^\infty c_{-n} e^{-in\theta} - \sum_1^\infty c_n e^{in\theta}\right) \tag{3.10}$$

$*u$ is the harmonic conjugate of u in the sense that

$$u(e^{i\theta}) + i\,{*u}(e^{i\theta}) = c_0 + 2\sum_1^\infty c_n e^{in\theta}$$

lies in H^2 and, in the case when u is real-valued, $*u$ is the unique real-valued function with vanishing mean-value with this property.

For $h \in \mathfrak{G}$ set

$$v_h = \mathcal{E}(*P_a), \qquad a = h(0) \tag{3.11}$$

where P_a is the Poisson kernel for a.

Proposition 3.3. The map $h \mapsto v_h$ is a homomorphism from $\mathfrak{G}$ into $L^2/\mathfrak{G}$:

$$v_{hg} = v_h + v_g, \qquad g, h \in \mathfrak{G} \tag{3.12a}$$

$$v_I = 0 \tag{3.12b}$$

Proof. We begin by showing that for $h \in \mathfrak{G}$ and $u \in L^2/\mathfrak{G}$,

$$*u \circ h - *u + \int u v_h \, d\sigma = 0 \tag{3.13}$$

Suppose with no loss of generality that u is real-valued. Then $*(u \circ h)$ is uniquely determined and, of course, is just $*u$ since $(u \circ h) = u$ a.e. σ. However, $*u \circ h$ has the property that

$$u \circ h + i\,{*u} \circ h = (u + i\,{*u}) \circ h$$

lies in H^2 and hence

$$*u = *u \circ h - \int_{\mathbf{T}} *u \circ h \, d\sigma$$

Equivalently,

$$\begin{aligned} *u \circ h - *u &= \int_{\mathbf{T}} *u \circ h \, d\sigma \\ &= \int_{\mathbf{T}} *u P_a \, d\sigma, \qquad a = h(0) \\ &= -\int_{\mathbf{T}} u\, {*P_a} \, d\sigma \\ &= -\int_{\mathbf{T}} u \mathcal{E}(*P_a) \, d\sigma \end{aligned}$$

which is just what (3.13) asserts. Next, if $g, h \in \mathfrak{G}$, then

$$*u \circ (g \circ h) - *u = (*u \circ g - *u) \circ h + *u \circ h - *u$$

$$= -\int_{\mathbf{T}} uv_g \, d\sigma - \int_{\mathbf{T}} uv_h \, d\sigma$$

But also

$$*u \circ (g \circ h) - *u = -\int_{\mathbf{T}} uv_{gh} \, d\sigma$$

Since this holds for all $u \in L^2/\mathfrak{G}$ we see that (3.12a) holds and thus also (3.12b).

Definition. Let $\mathbf{N}$ be the complex linear span of the functions $\{v_h : h \in \mathfrak{G}\}$. Let $H^2/\mathfrak{G} = H^2 \cap L^2/\mathfrak{G}$ and let $H_0^2/\mathfrak{G}$ be those $H^2/\mathfrak{G}$ functions with mean-value zero.

Proposition 3.4. Denote by $\mathbf{N}_2$ the closure of $\mathbf{N}$ in $L^2/\mathfrak{G}$. Then

$$L^2/\mathfrak{G} = \mathbf{N}_2 \oplus H_0^2/\mathfrak{G} \oplus \mathbf{C} \oplus \overline{H_0^2/\mathfrak{G}} \tag{3.14}$$

Proof. Suppose $u \in L^2/\mathfrak{G}$ and u is orthogonal to $\mathbf{N}$. Then

$$0 = \int uv_h \, d\sigma = *u \circ h - *u, \qquad h \in \mathfrak{G}$$

so that $*u \in L^2/\mathfrak{G}$ as well. Hence,

$$2u = \left(u + i*u - \int u \, d\sigma\right) + 2\int u \, d\sigma + \left(u - i*u - \int u \, d\sigma\right)$$

The first summand is in $H_0^2/\mathfrak{G}$, the second is in $\mathbf{C}$, and the third is the complex conjugate of an element of $H_0^2/\mathfrak{G}$. Further, if $f \in H_0^2/\mathfrak{G}$ and $h \in \mathfrak{G}$, then

$$\int_{\mathbf{T}} fv_h \, d\sigma = *f - *f \circ h = 0$$

since both f and $*f$ are invariant under $\mathfrak{G}$. This completes the proof.

Let u be a real-valued harmonic function on Ω and let γ be an element of $\pi_1(\Omega, p)$, $T(0) = p$. In a small neighborhood of p, u has a uniquely determined harmonic conjugate $*u$ if we specify that $*u(p) = 0$. Analytic continuation of the function element $u + i{}^*u$ along γ results in the function element $u + i{}^*u + ic(\gamma)$ where $c(\gamma)$ is a real scalar. Clearly, if γ_1 is another element of $\pi_1(\Omega, p)$, then $c(\gamma + \gamma_1) = c(\gamma) + c(\gamma_1)$; that is, c is a homomorphism of $\pi_1(\Omega, p)$ into $\mathbb{R}$. It is also apparent that $c(\gamma)$ depends only on the homotopy class of γ since analytic continuation of $u + i{}^*u$ along homotopic curves results in the same function element at p.

Let's transfer this to the unit disc Δ. Suppose v is a real-valued harmonic function on Δ which is invariant under the group $\mathfrak{G}$. Let ${}^*v$ be the harmonic conjugate of v with ${}^*v(0) = 0$. Suppose now that a is a point of Δ at which $T(a) = T(0) = p$. Then there is a unique $h \in \mathfrak{G}$ with $h(0) = a$. Let γ be the element of $\pi_1(\Omega, p)$ corresponding to h and let γ_1 be the lift of γ to Δ by T. Then ${}^*v(a)$ is obtained by analytic continuation of $v + i{}^*v$ along the curve γ_1. However, there is a real-valued harmonic function u on Ω with $u \circ T = v$, so that analytic continuation of $v + i{}^*v$ along the curve γ_1 corresponds to analytic continuation of the function element $u + i{}^*u$ along γ. This latter continuation produces $u + i{}^*u + ic(\gamma)$ and so ${}^*v(a) = {}^*v(0) + c(\gamma) = c(h)$ where we see now that c is a homomorphism of $\mathfrak{G}$ into $\mathbb{R}$. We collect all of this in the following proposition.

Proposition 3.5. Each real-valued harmonic function v on Δ which is invariant under $\mathfrak{G}$ gives rise to a homomorphism c_v of $\mathfrak{G}$ into $\mathbb{R}$.

The converse of Proposition 3.5 is true if $\mathfrak{G}$ is finitely generated (Ω is finitely connected) but need not be true in general. It is a difficult question; see also Section 6 of Chapter 4 and Section 6 of Chapter 5.

2.4. HARMONIC MEASURE AND $L^1/\mathfrak{G}$

There is another way to arrive at $L^p/\mathfrak{G}$ and the conditional expectation. For simplicity let us take Ω to be a bounded domain in $\mathbb{C}$ for which the Dirichlet problem is solvable. Let T be the uniformizer of Ω, put $T(0) = z_0$, and let ω be harmonic measure on $\partial\Omega$ for z_0. The mapping T has radial limits T^* a.e. $d\theta$ on $\mathbf{T}$ and $T^* \in L^\infty$; see Section 3 of Chapter 3. Since T is a covering map it must follow that $T^*(e^{i\theta})$ lies in $\partial\Omega$ a.e. If u is continuous on $\partial\Omega$, then $u \circ T^*$ is in $L^\infty(\mathbf{T}, d\theta)$ and so the linear functional

$$u \mapsto \int_{\mathbf{T}} u \circ T^* \, d\sigma, \qquad d\sigma = \frac{1}{2\pi} d\theta$$

is well defined, bounded, and even positive. However,

$$\begin{aligned}\int_{\mathbf{T}} u \circ T^* \, d\sigma &= (\widetilde{u \circ T})(0) = \tilde{u}(T(0)) \\ &= \tilde{u}(z_0)\end{aligned}$$

where $\tilde{u}$ is the harmonic extension to Ω of u. Hence, we have the relationship

$$\int_{\mathbf{T}} u \circ T^* \, d\sigma = \int_{\partial\Omega} u \, d\omega, \qquad u \in C(\partial\Omega) \tag{4.1}$$

It follows that the image of **T** under T^* is a set of full harmonic measure and that (4.1) continues to hold for all $u \in L^1(\omega, \partial\Omega)$. If E is a measurable set in $\partial\Omega$, then $X = (T^*)^{-1}(E)$ is in $\Sigma/\mathfrak{G}$ and (4.1) implies

$$\omega(E) = \sigma\big((T^*)^{-1}(E)\big) = \sigma(X) \tag{4.2}$$

Conversely, each element X of $\Sigma/\mathfrak{G}$ arises in just this way from a measurable set E in $\partial\Omega$.

ADDITIONAL READINGS AND NOTES

The uniformization theorem was proved by Koebe (1907); the proof given here is modeled on a paper by Heins (1949). The subject of Riemann surfaces, which is touched on here only slightly, is rich; see Farkas and Kra (1980) or Springer (1957) for some of the basics. The conditional expectation in its applications to function theory on multiply connected domains is from a paper of Forelli (1966); we will have more on this subject in Chapter 4. The group $\mathfrak{G}$ is an example of a Fuchsian group, another area of importance in complex analysis; see the book of Farkas and Kra (1980) and Lerner (1966).

EXERCISES

1. It's a fact that the uniformizer is unique up to homeomorphisms of X; that is, if T_1 is another function from X onto Ω satisfying conclusions (a)–(d) of Theorem 2.1, then there is a homeomorphism h of X onto X such that $T_1 = T \circ h$. Use this fact to show that if Ω is a domain in the plane with Δ as its covering space and if $z_0 \in \Omega$ is fixed, then there is just one uniformizer T with $T(0) = z_0$ up to rotation; any other uniformizer T_1 of Ω with $T_1(0) = z_0$ has the form $T_1(z) = T(\lambda z)$ for some unimodular constant λ.
2. Show that the uniformizer T is one-to-one if and only if the domain Ω is simply connected.

In the next five problems we suppose that Ω is *not* simply connected, but that it is covered by Δ.

3. Show $\mathfrak{G}$ has infinitely many elements.
4. If $c \in \Omega$, then $T^{-1}(c)$ is infinite.

5. If Ω is an annulus, then show $\mathfrak{G}$ is singly generated and hence abelian. If Ω is not doubly connected, show $\mathfrak{G}$ is not abelian.
6. If $f \in H^2/\mathfrak{G}$ is not constant, show f is not continuous on $\mathbf{T}$.
7. A point $\lambda \in \mathbf{T}$ is in the limit set L of $\mathfrak{G}$ if there are points $\{z_n\}$ in Δ with
   a. $z_n \to \lambda$
   b. $T(z_n) = c$ for all n and some $c \in \Omega$

   Show that L is an F_σ set in $\mathbf{T}$ and that L is invariant under $\mathfrak{G}$. (Actually, L is closed; see Lehrner (1966).)
8. Prove Proposition 3.1.
9. Extend equation (4.1) to other points; namely, if $a = re^{i\theta} \in \Delta$ and $p = T(a) \in \Omega$, then

$$\int_\Gamma u\,d\omega_p = \frac{1}{2\pi}\int_{\mathbf{T}} (u \circ T^*)(e^{it})P(r, \theta - t)\,dt, \qquad u \in L^1(\Gamma, \omega)$$

3

THE HARDY SPACES $H^p(\Omega)$

3.1. INTRODUCTION

In this chapter we introduce and begin our study of the Hardy spaces $H^p(\Omega)$, a family of Banach spaces consisting of functions analytic on the planar domain Ω which satisfy a growth condition there. These spaces extend and generalize the classical Hardy spaces on the open unit disc Δ. We give a number of specific properties of $H^p(\Delta)$ in Section 3 and it will be one goal of this chapter and the next to determine which of these properties are also valid for the H^p spaces of more general domains. The especially interesting space $H^\infty(\Omega)$, the space of bounded holomorphic functions on Ω, is studied in the final three sections of this chapter and in Chapters 5 and 6.

Let's begin at the beginning—with the definition of $H^p(\Omega)$.

Definition. Let $0 < p < \infty$; a holomorphic function f on the domain Ω is in $H^p(\Omega)$ if the subharmonic function $|f(z)|^p$ has a harmonic majorant on Ω; that is, there is a harmonic function $v(z)$ with

$$|f(z)|^p \leqslant v(z), \qquad z \in \Omega$$

The function f is in $H^\infty(\Omega)$ if it is both holomorphic on Ω and bounded there.

Clearly, $H^\infty(\Omega) \subset H^p(\Omega)$ and if $H^\infty(\Omega)$ contains nonconstant functions then so does $H^p(\Omega)$. The converse of this last statement does not hold; see the notes at the end of the chapter.

3.2. BASIC PROPERTIES OF $H^p(\Omega)$

Recall that whenever f is analytic on Ω, then $u = |f|^p$ is subharmonic on Ω and, in this case, continuous. The assumption that f is in $H^p(\Omega)$, $0 < p < \infty$, then is that the continuous subharmonic function $|f|^p$ has a harmonic majorant.

We now show that it has a unique least harmonic majorant; that is, there is a unique harmonic function u_f satisfying

$$|f(z)|^p \leqslant u_f(z), \qquad z \in \Omega \tag{2.1}$$

and

$$u_f(z) \leqslant v(z), \qquad z \in D \tag{2.2}$$

if v is any harmonic majorant of $u = |f|^p$. For this we only use the facts that u is subharmonic on Ω and continuous there, and even the latter assumption is not needed, only convenient. To begin let $\{\Omega_m\}$ be a regular exhaustion of Ω; let u_m be the harmonic extension to Ω_m of the restriction of u to $\partial\Omega_m$, $m = 1, 2, \ldots$. If $n > m$, then on $\partial\Omega_m$ we know that $u_m = u \leqslant u_n$ because $\partial\Omega_m \subset \Omega_n$. Hence, $u_m \leqslant u_n$ on all of Ω_m. Thus $\{u_m\}$ is an increasing sequence of harmonic functions on Ω and so increases either to $+\infty$ identically or to a harmonic function u_f. The former case is impossible, since if v is any harmonic majorant of u we have

$$u_m(a) = \int_{\partial\Omega_m} u_m \, d\omega_{m,a} = \int_{\partial\Omega_m} u \, d\omega_{m,a}$$

$$\leqslant \int_{\partial\Omega_m} v \, d\omega_{m,a} = v(a)$$

for all points $a \in \Omega_m$ where $\omega_{m,a}$ is harmonic measure on $\partial\Omega_m$ for a. Thus, u_m is pointwise bounded above by v on Ω_k, $k \leqslant m$, so that $u_f \leqslant v$ on Ω_m and hence u_f is the unique least harmonic majorant of u on Ω.

Definition. Fix a point $z_0 \in \Omega$ and define

$$\|f\| = \begin{cases} \left(u_f(z_0)\right)^{1/p}, & 0 < p < \infty \\ \sup\{|f(z)| : z \in \Omega\}, & p = \infty \end{cases} \tag{2.3}$$

where u_f is the least harmonic majorant of $|f|^p$.

Theorem 2.1. *The function defined in* (2.3) *is a norm on* $H^p(\Omega)$ *if* $1 \leqslant p \leqslant \infty$. *The resulting topology is not dependent on the choice of* $z_0 \in \Omega$.

Proof. For $p = \infty$ there is nothing new here so we concentrate on the case $1 \leqslant p < \infty$. Let $\{\Omega_m\}$ be a regular exhaustion of Ω and assume with no loss that $z_0 \in \Omega_1$. Let ω_m be harmonic measure on $\partial\Omega_m$ for the point z_0. Suppose $f, g \in H^p(\Omega)$. Then we know that

$$\left(\int |f + g|^p \, d\omega_m\right)^{1/p}$$

increases to $\|f+g\|$ as $m \to \infty$. However,

$$\left(\int |f+g|^p \, d\omega_m\right)^{1/p} \leqslant \left(\int |f|^p \, d\omega_m\right)^{1/p} + \left(\int |g|^p \, d\omega_m\right)^{1/p}$$

$$\leqslant \|f\| + \|g\|$$

which shows that $\|f+g\| \leqslant \|f\| + \|g\|$. Further, it's clear that for $f \in H^p(\Omega)$ and $\lambda \in \mathbf{C}$, we have

$$\|\lambda f\| = |\lambda| \|f\|$$

since $|\lambda|^p u_f$ is obviously the least harmonic majorant of $|\lambda f|^p$. Finally, we must show that although the norm depends on the point z_0 which is chosen, the resulting topology does not. If z_1 is another point of Ω, recall from Section 6 of Chapter 1 that there are positive numbers A and B with

$$A \leqslant v(z_1) \leqslant B$$

if v is a positive harmonic function on Ω with $v(z_0) = 1$. Thus,

$$A \leqslant \frac{u_f(z_1)}{u_f(z_0)} \leqslant B$$

which is exactly what we needed to show.

Proposition 2.2. Let f be analytic on Ω and let $0 < p < \infty$. Then $f \in H^p(\Omega)$ if and only if for each regular exhaustion $\{\Omega_m\}$ of Ω there is a constant C with

$$\int_{\partial\Omega_m} |f|^p \, d\omega_{m,a} \leqslant C, \qquad m = 1, 2, \ldots \tag{2.4}$$

where $\omega_{m,a}$ is harmonic measure on $\partial\Omega_m$ for some point $a \in \Omega_1$.

Proof. We have virtually proved this already. If $f \in H^p(\Omega)$, then we know that the numbers

$$\int_{\partial\Omega_m} |f|^p \, d\omega_{m,a}$$

increase to $u_f(a)$, which is finite. Conversely, suppose (2.4) holds. Set u_m to be harmonic extension to Ω_m of the restriction to $\partial\Omega_m$ of $|f|^p$. Then $\{u_m\}$ is an increasing sequence of harmonic functions, $|f|^p \leqslant u_m$ on Ω_k if $k \leqslant m$, and (2.4) implies that $\{u_m(a)\}$ is bounded. Hence, $\{u_m\}$ increases to a harmonic majorant of $|f|^p$ and so $f \in H^p(\Omega)$.

Corollary 2.3. *If $q < p$, then $H^p(\Omega) \subset H^q(\Omega)$.*

Proposition 2.4. Let $\mathbf{K}$ be a compact subset of Ω. There is a constant C depending only on $\mathbf{K}$ such that

$$|f(z)| \leqslant C\|f\|, \qquad z \in \mathbf{K}, \qquad f \in H^p(\Omega)$$

C does not depend on p or f.

Proof. Let Ω' be subdomain of Ω with z_0 and $\mathbf{K}$ in its interior and chosen so that the Dirichlet problem is solvable on Ω'. Let u be the harmonic extension to Ω' of the restriction to $\partial\Omega'$ of $|f|^p$. Then for $z \in \mathbf{K}$

$$|f(z)|^p \leqslant u(z) = \int_{\partial\Omega'} u \, d\omega'_z$$

where ω'_z is harmonic measure on $\partial\Omega'$ for z. We know from Theorem 1.6.1 that $d\omega'_z = w \, d\omega'_{z_0}$ where w is an L^∞ function whose sup norm is bounded by some constant C independent of z so long as $z \in \mathbf{K}$. Thus,

$$\begin{aligned} |f(z)|^p &\leqslant \int_{\partial\Omega'} |f(\zeta)|^p w(\zeta) \, d\omega'_{z_0}(\zeta) \\ &\leqslant C\int_{\partial\Omega'} |f(\zeta)|^p \, d\omega'_{z_0}(\zeta) \\ &\leqslant Cu_f(z_0) \end{aligned}$$

This is just what was to be proved.

Theorem 2.5. *$H^p(\Omega)$ is a Banach space for $1 \leqslant p \leqslant \infty$.*

Proof. The case $p = \infty$ is routine and is left to the reader. Suppose, then, that $\{f_k\}$ is a Cauchy sequence in $H^p(\Omega)$. We then know from Proposition 2.4 that $\{f_k\}$ is uniformly bounded on compact subsets of Ω and so by the classical normal families theorem some subsequence converges uniformly on compact subsets of Ω to a holomorphic function f. Let $\{\Omega_m\}$ be a regular exhaustion of Ω with $z_0 \in \Omega_1$. Given $\varepsilon > 0$ we know

$$\int_{\partial\Omega_m} |f_n - f_k|^p \, d\omega_m < \varepsilon$$

if $n, k \geqslant N = N(\varepsilon)$ and $m = 1, 2, \ldots$. Let $k \to \infty$ through the subsequence. Then

$$\int_{\partial\Omega_m} |f_n - f|^p \, d\omega_m \leqslant \varepsilon$$

if $n \geqslant N$ and $m = 1, 2, \ldots$. Hence, $f \in H^p(\Omega)$ and $f_n \to f$ in $H^p(\Omega)$.

Proposition 2.6. Let ϕ be a one-to-one holomorphic mapping of a domain Ω onto a domain Λ. Then $H^p(\Omega)$ and $H^p(\Lambda)$ are isometrically isomorphic by the rule

$$g = f \circ \phi, \qquad f \in H^p(\Lambda), \qquad g \in H^p(\Omega)$$

Proof. The essential fact is that ϕ carries the least harmonic majorant of $|f|^p$ to the least harmonic majorant of $|g|^p$.

3.3. H^p ON THE UNIT DISC

This section is intended as either a (very) short course in H^p theory on the unit disc Δ or, if the reader is familiar already with this topic, it can be skipped or read as a refresher. Obviously it is brief and no effort is made either for completeness or generality.

We begin by noting that a function f holomorphic in Δ is in $H^p(\Delta)$, $0 < p < \infty$, if and only if the integral means

$$\left(\frac{1}{2\pi}\int_{-\pi}^{\pi} |f(re^{it})|^p\, dt\right)^{1/p} = M_p(f; r), \qquad 0 < r < 1$$

remain uniformly bounded as r varies in the interval $(0, 1)$; this is just Proposition 2.2. However, $|f(z)|^p$ is subharmonic, so that $M_p(f; r)$ is an increasing function of r (see the proof of Proposition 1.3.6) and so f lies in $H^p(\Delta)$ precisely when limit $\{M_p(f; r): r \to 1\}$ is finite. This is the usual definition of $H^p(\Delta)$.

The next step is to examine the zeros of an $H^p(\Delta)$ function.

Theorem 3.1. *Let $f \in H^p(\Delta)$, $0 < p \leqslant \infty$, $f \not\equiv 0$. Let $z_1, z_2, \ldots$ be the zeros of f in Δ repeated according to their respective multiplicities. If f has infinitely many zeros, then they satisfy*

$$\sum_1^{\infty} (1 - |z_j|) < \infty \tag{3.1}$$

If the points $z_1, z_2, \ldots$ satisfy (3.1) *set*

$$B(z) = \prod_{j=1}^{\infty} \left(\frac{-\bar{z}_j}{|z_j|}\right)\left(\frac{z - z_j}{1 - \bar{z}_j z}\right) \tag{3.2}$$

Then B is a holomorphic function in Δ bounded by 1 *which vanishes precisely at the points $\{z_j\}$. Further,*

$$f = BF \tag{3.3}$$

where $F \in H^p(\Delta)$, $\|F\|_p = \|f\|_p$, and F has no zeros in Δ.

Proof. There is no loss in assuming $f(0) \neq 0$. Let $z_1, \ldots, z_n$ be zeros of f and put

$$B_n(z) = \prod_1^n \left(\frac{-\bar{z}_j}{|z_j|} \right) \left(\frac{z - z_j}{1 - \bar{z}_j z} \right) \tag{3.4}$$

and $F_n = f/B_n$. If $r = |z|$ is near to 1, then

$$\frac{1}{2\pi} \int_{-\pi}^{\pi} |F_n(re^{it})|^p \, dt \leqslant (1 - \varepsilon)^{-p} \frac{1}{2\pi} \int_{-\pi}^{\pi} |f(re^{it})|^p \, dt$$

$$\leqslant (1 - \varepsilon)^{-p} \|f\|_p^p$$

since $|B_n(z)|$ is continuous on $\{|z| \leqslant 1\}$ with value 1 identically on **T**. Hence, $F_n \in H^p(\Delta)$ and $\|F_n\|_p \leqslant \|f\|_p$. However, $|F_n(z)| \geqslant |f(z)|$ for all $z \in \Delta$ so that $\|F_n\|_p \geqslant \|f\|_p$ as well, which shows that $\|F_n\|_p = \|f\|_p$. From this we see that

$$\frac{|f(0)|}{\prod_1^n |z_j|} = |F_n(0)| \leqslant C\|F_n\|_p = C\|f\|_p$$

where C is some constant; see Proposition 2.4. Hence,

$$\prod_1^\infty |z_j| > 0$$

which is equivalent to (3.1). The assertion about $B(z)$ given by (3.2) is also direct. We have

$$1 - \left(\frac{-\bar{z}_j}{|z_j|} \right) \left(\frac{z - z_j}{1 - \bar{z}_j z} \right) = (1 - |z_j|) \left(\frac{|z_j| + z\bar{z}_j}{|z_j|(1 - \bar{z}_j z)} \right)$$

Thus, if $|z| \leqslant r < 1$ we have

$$\sum_{j=1}^\infty \left| 1 - \left(\frac{-\bar{z}_j}{|z_j|} \right) \left(\frac{z - z_j}{1 - \bar{z}_j z} \right) \right| \leqslant \frac{2}{(1 - r)} \sum_1^\infty (1 - |z_j|)$$

which implies, by standard theorems in complex analysis, that $B(z)$ given by (3.2) has all the stated properties. Further, $\{B_n\}$ converges uniformly on compact subsets of Δ to B. Hence, $\{F_n\}$ converges uniformly on compact subsets of Δ to F and so

$$\frac{1}{2\pi} \int_{-\pi}^{\pi} |F(re^{it})|^p \, dt = \lim \frac{1}{2\pi} \int_{-\pi}^{\pi} |F_n(re^{it})|^p \, dt$$

$$\leqslant \lim \|F_n\|_p^p \leqslant \|f\|_p^p$$

However, $|F(z)| \geqslant |f(z)|$ for all z so that $\|F\|_p \geqslant \|f\|_p$ and hence equality holds.

Proposition 3.2. A holomorphic function f on Δ with power series

$$f(z) = \sum_0^\infty a_n z^n$$

is in $H^2(\Delta)$ if and only if

$$\sum_0^\infty |a_n|^2 < \infty \tag{3.5}$$

In this case, $\|f\|_2 = (\Sigma_0^\infty |a_n|^2)^{1/2}$.

Proof. This result is really just a computation. For $r < 1$ we have

$$\frac{1}{2\pi}\int_{-\pi}^{\pi} |f(re^{it})|^2\,dt = \sum_0^\infty |a_n|^2 r^{2n}$$

and the result is immediate from this.

Suppose now that f is in $H^2(\Delta)$. Let $g(e^{i\theta})$ be the L^2 function

$$g(e^{i\theta}) = \sum_0^\infty a_n e^{in\theta}$$

and set

$$f_r(e^{i\theta}) = \sum_0^\infty a_n r^n e^{in\theta}, \qquad 0 < r < 1$$

A computation gives

$$\|f_r - g\|_2^2 = \sum_1^\infty |a_n|^2 (1 - r^{2n})$$

and so $f_r \to g$ in L^2, and a subsequence converges a.e. to g. However, we know that each $f \in H^2(\Delta)$ has radial limits a.e. $d\theta$ from Theorem 1.2.2 so that we actually see that $f_r \to g$ both a.e. and in L^2, as r increases to 1.

Suppose next that $f \in H^p(\Delta)$ and write $f = BF$ where $|B| \leqslant 1$, $F \in H^p(\Delta)$, $F \neq 0$ on Δ. Then $F^{p/2}$ is in $H^2(\Delta)$ and so has radial limits a.e. $d\theta$ and the radial limits define a function in L^2. Further, B is bounded and thus B has radial limits a.e. $d\theta$ and these limits define a function B^* which is in L^∞ and is bounded by 1. Hence, f has radial limits a.e. $d\theta$ and the limits define a function f^* which is in L^p.

The function B^* actually has modulus 1 a.e. $d\theta$. To see this, let B_n be defined by (3.4) and note that $g_n = B/B_n$ is analytic on Δ and bounded by 1 for each n. Thus,

$$\prod_{n+1}^{\infty} |z_j|^2 = |g_n(0)|^2 = \left|\frac{1}{2\pi}\int_{-\pi}^{\pi} g_n(re^{it})\,dt\right|^2$$

$$\leqslant \frac{1}{2\pi}\int_{-\pi}^{\pi} |g_n(re^{it})|^2\,dt$$

$$\leqslant \frac{1}{2\pi}\int_{-\pi}^{\pi} |g_n^*(e^{it})|^2\,dt$$

$$= \frac{1}{2\pi}\int_{-\pi}^{\pi} |B^*(e^{it})|^2\,dt \leqslant 1$$

However, as $n \to \infty$ we know that $\prod_{n+1}^{\infty}|z_j|$ converges to 1. Hence, the L^2 norm of B^* is 1 which is also its sup norm. Thus, $|B^*| = 1$ a.e. $d\theta$.

We already know that $F_r^{p/2}$ converges in L^2 to $(F^*)^{p/2}$ where F^* is the radial limit of F. Hence,

$$\limsup\{M_p(f; r) : r \to 1\} \leqslant \limsup\{M_p(F; r) : r \to 1\}$$

$$= \|F^*\|_p = \|f^*\|_p$$

where the second equality follows from the fact that $f^* = F^*B^*$ and $|B^*| = 1$ a.e. $d\theta$. However, by Fatou's lemma

$$\liminf\{M_p(f; r) : r \to 1\} \geqslant \|f^*\|_p$$

since $|f_r|^p \to |f^*|^p$ a.e. Thus,

$$\lim\{M_p(f; r) : r \to 1\} = \|f^*\|_p$$

We now apply problem 15 of this chapter and conclude that $f(re^{it}) \to f^*(e^{it})$ in L^p. We summarize all of this.

Theorem 3.3. *Let $f \in H^p(\Delta)$ $f \not\equiv 0$, $0 < p < \infty$. Then*

$$\lim_{r \to 1} f(re^{it}) = f^*(e^{it}) \text{ exists a.e. } dt \tag{3.6a}$$

$$f^* \in L^p \tag{3.6b}$$

$$\int_{-\pi}^{\pi} |f(re^{it}) - f^*(e^{it})|^p\,dt \to 0 \quad \text{as } r \to 1 \tag{3.6c}$$

$$\log|f(re^{i\theta})| \leqslant \frac{1}{2\pi}\int_{-\pi}^{\pi} P(r, \theta - t)\log|f^*(e^{it})|\,dt \tag{3.6d}$$

Proof. All that remains to be proved is (3.6d). There is no loss in assuming $f(re^{i\theta}) \neq 0$. For $\rho < 1$ we have

$$\log|f(\rho re^{i\theta})| \leqslant \log|F(\rho re^{i\theta})| = \frac{1}{2\pi}\int_{-\pi}^{\pi} P(r, \theta - t)\log|F(\rho e^{it})|\, dt$$

since $\log|F(\rho z)|$ is harmonic on the closed disc. Hence, for $\varepsilon > 0$,

$$\log|f(\rho re^{i\theta})| \leqslant \frac{1}{2\pi}\int_{-\pi}^{\pi} P(r, \theta - t)\log\big(|F(\rho e^{it})| + \varepsilon\big)\, dt$$

Let ρ increase to 1 and apply Fatou's lemma in the form

$$\limsup \int u_n \leqslant \int \limsup u_n$$

Then let ε decrease to 0; this gives (3.6d) upon recalling that $|f^*| = |F^*|$ a.e. on $\mathbf{T}$.

The next item is the factorization theorem and to state this we must define the various factors.

A *singular inner function*, $S(z)$, is a holomorphic function on Δ of the form

$$S(z) = \exp\left[-\int_{\mathbf{T}} \frac{e^{i\theta} + z}{e^{i\theta} - z}\, d\mu(\theta)\right] \tag{3.7a}$$

where μ is a non-negative measure on $\mathbf{T}$ singular with respect to Lebesgue measure (3.7b)

An *outer function* $F(z)$ in $H^p(\Delta)$ is a function on Δ of the form

$$F(z) = \exp\left\{\frac{1}{2\pi}\int_{-\pi}^{\pi} \frac{e^{i\theta} + z}{e^{i\theta} - z} u(\theta)\, d\theta\right\} \tag{3.8a}$$

$$\text{where} \quad u \in L^1, \qquad e^u \in L^p, \qquad u \text{ real-valued} \tag{3.8b}$$

Finally, a *Blaschke product* is a function B of the form

$$B(z) = \prod_1^{\infty}\left(\frac{-\bar{z}_j}{|z_j|}\right)\left(\frac{z - z_j}{1 - \bar{z}_j z}\right) \tag{3.9a}$$

$$\text{where} \quad \sum_1^{\infty}(1 - |z_j|) < \infty \tag{3.9b}$$

(Appropriate modifications are needed if there are only a finite number of z_j.)

Theorem 3.4. *If $f \in H^p(\Delta)$, then*

$$f = \lambda BSF \tag{3.10}$$

where λ is a unimodular constant, B is a Blaschke product, S is a singular inner function, and F is an outer function in $H^p(\Delta)$. This factorization is unique.

Proof. We begin by writing

$$f = BG$$

where G is in $H^p(\Delta)$, $G \neq 0$ on Δ, $\|G\|_p = \|f\|_p$; this is just (3.3) with G in place of F. Let F be the outer function in $H^p(\Delta)$ given by

$$F(z) = \exp\left\{\frac{1}{2\pi}\int_{-\pi}^{\pi}\frac{e^{i\theta}+z}{e^{i\theta}-z}\log|f^*(e^{it})|\,dt\right\}$$

Then by (3.6d) applied to G rather than f, we have

$$\log|G(re^{i\theta})| \leqslant \log|F(re^{i\theta})|, \qquad 0 < r < 1, \qquad 0 \in [0, 2\pi]$$

Hence, the holomorphic function $W = G/F$ is bounded by 1 in Δ and does not vanish. Thus,

$$-\log|W(re^{i\theta})| = \int_{\mathbf{T}} P(r, \theta - t)\,d\mu(t)$$

for some non-negative measure μ on $\mathbf{T}$; see Corollary 1.2.4. However, $|G^*| = |F^*| = |f^*|$ a.e. $d\theta$ on $\mathbf{T}$ so that $|W^*| = 1$ a.e. $d\theta$. But we know from Theorem 1.2.2. that

$$0 = \lim\{-\log|W(re^{i\theta})|: r \to 1\}$$

$$= \frac{d\mu}{d\theta} \quad \text{a.e. } d\theta$$

Thus, μ is singular with respect to Lebesgue measure. Let S be the singular inner function defined by μ; then $|W| = |S|$ on Δ and so $W = \lambda S$ for some unimodular constant λ and this gives (3.10).

To see the uniqueness suppose

$$f = \lambda_1 B_1 S_1 F_1$$

Clearly we must have $B = B_1$ since the other factors are zero-free. Furthermore,

$$|F_1^*| = |f^*| = |F^*| \quad \text{a.e. } d\theta$$

so that $F = cF_1$ where c is a unimodular constant. But $F(0)$, $F_1(0)$ are both positive so $c = 1$. Next, $S(0)$ and $S_1(0)$ are both positive so $\lambda = \lambda_1$ and the proof is finished.

We conclude this section with a characterization of $H^p(\Delta)$ functions from their boundary values and a short discussion of the Banach space structure of $H^p(\Delta)$.

Theorem 3.5. *Let $1 \leqslant p \leqslant \infty$ and suppose $u \in L^p(\mathbf{T}, d\theta)$. Then there is an $f \in H^p(\Delta)$ with $f^* = u$ a.e. $d\theta$ if and only if*

$$0 = \int_{-\pi}^{\pi} u(e^{it}) e^{int}\, dt, \qquad n = 1, 2, 3, \dots \tag{3.11}$$

Proof. Suppose (3.11) holds and set

$$f(re^{i\theta}) = \frac{1}{2\pi} \int_{-\pi}^{\pi} P(r, \theta - t) u(e^{it})\, dt$$

Then f is harmonic in Δ and $\lim\{f(re^{i\theta}): r \to 1\} = u(e^{i\theta})$ a.e. $d\theta$ by Theorem 1.2.2. However,

$$\begin{aligned} f(re^{i\theta}) &= \frac{1}{2\pi} \int_{-\pi}^{\pi} u(e^{it}) \left[1 + 2\sum_{1}^{\infty} r^n \cos n(\theta - t)\right] dt \\ &= \frac{1}{2\pi} \int_{-\pi}^{\pi} u(e^{it}) \left[1 + \sum_{1}^{\infty} r^n \left(e^{in\theta} e^{-int} + e^{-in\theta} e^{int}\right)\right] dt \\ &= \sum_{0}^{\infty} c_n r^n e^{in\theta} \end{aligned}$$

where

$$c_n = \frac{1}{2\pi} \int_{-\pi}^{\pi} u(e^{it}) e^{-int}\, dt, \qquad n = 0, 1, 2, \dots$$

Hence, f is actually analytic in Δ. Finally, for $1 \leqslant p < \infty$

$$\frac{1}{2\pi} \int_{-\pi}^{\pi} |f(re^{i\theta})|^p\, d\theta \leqslant \frac{1}{2\pi} \int_{-\pi}^{\pi} |u(e^{it})|^p\, dt$$

so that $f \in H^p(\Delta)$.

Conversely, if $f \in H^p(\Delta)$ and $f^* = u$ a.e. on $\mathbf{T}$, then $f_r \to u$ in L^1 as $r \to 1$ and so for $n = 1, 2, \dots$

$$0 = \frac{1}{2\pi} \int_{-\pi}^{\pi} f_r(e^{it}) e^{int}\, dt \to \frac{1}{2\pi} \int_{-\pi}^{\pi} u(e^{it}) e^{int}\, dt$$

Corollary 3.6. *For $1 \leqslant p < \infty$, $H^p(\Delta)$ is a separable Banach space and for $1 \leqslant p \leqslant \infty$, $H^p(\Delta)$ is (isometrically isomorphic to) a dual space.*

Proof. The mapping $f \mapsto f^*$ is an isometry of $H^p(\Delta)$ onto the closed subspace of L^p consisting of functions which satisfy (3.11). Thus, the first assertion holds. The second assertion for $1 < p \leqslant \infty$ is easy: H^p "is" the dual space of the quotient L^q/A_q where $1/p + 1/q = 1$ and A_q is the closed linear span of $\{e^{int}: n = 1, 2, \dots\}$. For $H^1(\Delta)$, this is a little more complicated and requires the F. and M. Riesz theorem which is to be found in Chapter 4, exercise 1; namely, if μ is a measure on $\mathbf{T}$ and

$$0 = \int_{\mathbf{T}} e^{int}\, d\mu(t), \qquad n = 1, 2, \dots \tag{3.12}$$

then $d\mu = g^*\, dt$ where $g \in H^1(\Delta)$. Assuming this (its proof is in no way dependent on what we're doing) we see that H^1 "is" the dual space of $C(\mathbf{T})/\mathbf{A}_0$, where $\mathbf{A}_0$ is the closed linear span of $\{e^{int}: n = 1, 2, \dots\}$ in $C(\mathbf{T})$.

The space $\mathbf{A}$ is defined to be the uniformly closed linear span of the polynomials in z and is called the *disc algebra*. It is almost immediate that $f \in \mathbf{A}$ if and only if f is continuous on $\Delta \cup \mathbf{T}$ and holomorphic on Δ. Further, for $1 \leqslant p < \infty$, $\mathbf{A}$ is dense in $H^p(\Delta)$ [this is a consequence of (3.6c)] and $\mathbf{A}$ is boundedly pointwise dense in $H^\infty(\Delta)$; that is, if $h \in H^\infty(\Delta)$ then the functions $h_r(z) = h(rz)$, $0 < r < 1$, converge pointwise to h in Δ and satisfy $\|h_r\|_\infty \leqslant \|h\|_\infty$. Hence, the space $\mathbf{A}$ is weak-* dense in H^∞, when H^∞ is viewed as a weak-* closed subspace of L^∞.

Proposition 3.7. Let $1 \leqslant p \leqslant \infty$. A sequence $\{f_n\}$ of elements of H^p converges weakly to $f \in H^p$ (or weak-* if $p = \infty$) if and only if their norms are uniformly bounded and $f_n(z) \to f(z)$ for each $z \in \Delta$.

Proof. Suppose the norms are bounded and the sequence converges pointwise in Δ. Then the convergence is actually uniform on compact subsets on Δ. If g is any weak-cluster point of $\{f_n\}$ in $L^p(\mathbf{T}, d\theta)$, then

$$f(re^{i\theta}) = \frac{1}{2\pi}\int_{-\pi}^{\pi} P(r, \theta - t) g(e^{it})\, dt$$

so that $f^* = g$ a.e. $d\theta$ and hence $\{f_n\}$ converges weakly to f^*. The converse is very easy and so is omitted.

3.4. $H^p/\mathfrak{G}$ AND $H^p(\Omega)$

There is a lovely connection between $H^p(\Omega)$ and a subspace of $H^p = H^p(\Delta)$ involving the uniformizer discussed in Chapter 2.

Let Ω be a domain in $\mathbb{C}$ and suppose $H^p(\Omega)$ contains some nonconstant function. Temporarily let X stand for either the open unit disc Δ or the whole complex plane $\mathbb{C}$, depending which is the universal covering surface of Ω. Let $T: X \to \Omega$ be the uniformizer of Ω. If $f \in H^p(\Omega)$, then $f \circ T$ is in $H^p(X)$. For if u_f is the least harmonic majorant of $|f|^p$, then $u_f \circ T$ is a harmonic majorant of $|f \circ T|^p$. Since $H^p(\mathbb{C})$ is trivial, we see that X must be the unit disc. Let $\mathfrak{G}$ be the group of linear fractional transformations of Δ onto Δ which fix T.

Proposition 4.1. $u_f \circ T$ is the least harmonic majorant of $|f \circ T|^p$. If $T(0) = z_0$, then the lifting $f \mapsto f \circ T$ is an isometry of $H^p(\Omega)$ onto the closed subspace of $H^p(\Delta)$ defined by

$$H^p/\mathfrak{G} = \{f \in H^p : f \circ h = f \text{ for all } h \in \mathfrak{G}\} \tag{4.1}$$

Proof. Again we concentrate on the case $p < \infty$. Let v be the least harmonic majorant of $|f \circ T|^p$. If $h \in \mathfrak{G}$ then $v \circ h$ majorizes $|f \circ T \circ h|^p$. But the latter is exactly $|f \circ T|^p$. Hence, $v \circ h \geqslant v$. Replace h by h^{-1} and see that $v \circ h^{-1} \geqslant v$, or equivalently $v \geqslant v \circ h$ which implies equality holds. Hence, there is a function u on Ω with $u \circ T = v$. Since T is locally one-to-one we see that u is harmonic on Ω and since $v \geqslant |f \circ T|^p$ we also see that $u \geqslant |f|^p$. Thus, $u_f \leqslant u$ so that $u_f \circ T \leqslant u \circ T = v$ and so $v = u_f \circ T$. Since $T(0) = z_0$, we have

$$\|f\|_p^p = u_f(z_0) = (u_f \circ T)(0) = \|f \circ T\|_p^p$$

Finally, it is clear that $f \circ T \in H^p/\mathfrak{G}$ if $f \in H^p(\Omega)$. If g is an element of $H^p/\mathfrak{G}$, then g is invariant under $\mathfrak{G}$ so that $g = f \circ T$ for some continuous function f on Ω. f must be analytic on Ω since T is locally one-to-one. Further, the least harmonic majorant v of $|g|^p$ is invariant under $\mathfrak{G}$ as we showed above and so has the form $u \circ T$ where u is harmonic on Ω. Thus, u majorizes $|f|^p$ on Ω and so $g = f \circ T$ where $f \in H^p(\Omega)$.

Corollary 4.2. *Let Ω be a domain. Then*

(*a*) *$H^p(\Omega)$ is separable if $1 \leqslant p < \infty$*
(*b*) *$H^p(\Omega)$ is uniformly convex if $1 < p < \infty$*
(*c*) *For $1 < p \leqslant \infty$, $H^p(\Omega)$ is a dual space*
(*d*) *For $1 < p < \infty$, a sequence $\{f_n\}$ converges weakly to f in $H^p(\Omega)$ if and only if the norms are uniformly bounded and $f_n \to f$ uniformly on compact subsets of Ω*

Proof. These assertions all follow from the corresponding statements for $H^p(\Delta)$ since $H^p(\Omega)$ can be realized as a closed subspace of $H^p(\Delta)$.

3.5. NULL SETS AND ESSENTIAL BOUNDARY POINTS FOR $H^\infty(\Omega)$

This section, and the next two as well, are devoted to the study of certain aspects of the space $H^\infty(\Omega)$ and, in particular, to the issues of how Ω effects $H^\infty(\Omega)$ and how knowledge of $H^\infty(\Omega)$ determines Ω.

Definition. A compact subset $\mathbf{E}$ of $\mathbb{C}$ is a (Painleve) null set if the only functions which are holomorphic and bounded on $\mathbf{S}^2 \setminus \mathbf{E}$ are the constants.

The Riemann mapping theorem shows that if a compact set $\mathbf{E}$ contains a closed connected subset with more than one point, then $\mathbf{E}$ is not a null set. Hence, a null set is totally disconnected. The exact characterization of null sets is not known but if $\mathbf{E}$ is a subset of a line then such a characterization is easy.

Theorem 5.1. *Suppose* $\mathbf{E}$ *is a compact subset of the real line* $\mathbb{R}$. *Then* $\mathbf{E}$ *is a null set if and only if* $\mathbf{E}$ *has zero length.*

Proof. Suppose $\mathbf{E}$ has zero length. Given a small positive number ε let $\mathcal{U}$ be the union of a finite number of discs which cover $\mathbf{E}$ and whose radii sum to less than ε. Let R be a large positive number so big that $\mathbf{E}$ lies within the disc $\{z : |z| < R\}$. Suppose f is analytic on $\mathbf{S}^2 \setminus \mathbf{E}$ and is bounded by 1 there. For a z with $|z| < R$, $z \notin \mathbf{E}$, we have

$$f(z) = \frac{1}{2\pi i}\int_{|\zeta|=R} \frac{f(\zeta)}{\zeta - z}\,d\zeta + \frac{1}{2\pi i}\int_{\partial\mathcal{U}} \frac{f(\zeta)}{\zeta - z}\,d\zeta \tag{5.1}$$

The second integral in (5.1) is bounded above by ε multiplied by some constant depending on the distance from z to $\partial\mathcal{U}$, and hence this term may be made arbitrarily small. Thus, only the first integral in (5.1) is needed to represent $f(z)$. But the first integral is an analytic function of z within $|z| = R$ so that f is actually analytic on all of $\mathbf{S}^2$ and hence constant.

Conversely, if $\mathbf{E} \subset \mathbb{R}$ has positive length then the function

$$g(z) = \int_{\mathbf{E}} \frac{dt}{t - z} \tag{5.2}$$

is holomorphic on $\mathbf{S}^2 \setminus \mathbf{E}$, nonconstant, and its imaginary part is bounded. Hence, $f(z) = \exp[ig(z)]$ is a nontrivial bounded holomorphic function on $\mathbf{S}^2 \setminus \mathbf{E}$; see problem 7 at the end of the chapter.

The concept of a null set is actually a local one although it does not at first seem so.

Proposition 5.2. Let $\mathbf{E}$ be compact. $\mathbf{E}$ is a null set if and only if whenever Ω is a domain containing $\mathbf{E}$ then each bounded holomorphic function on $\Omega \setminus \mathbf{E}$ extends to be holomorphic on Ω.

Proof. One direction is immediate so that we need only prove the "only if" assertion. Let $\mathbf{E}$ be a null set, then, and let f be bounded and holomorphic on $\Omega \setminus \mathbf{E}$. Suppose $z \in \Omega \setminus \mathbf{E}$ and δ is the minimum of the distance from z to $\mathbf{E}$ and the distance from z to $\partial\Omega$. Let γ_1 and γ_2, respectively, denote piecewise smooth curves in $\Omega \setminus \mathbf{E}$ with γ_1 within distance $\delta/2$ of $\mathbf{E}$ and γ_2 within distance $\delta/2$ of $\partial\Omega$ (there is no loss in assuming Ω to be bounded). Then Cauchy's formula yields

$$f(z) = \frac{1}{2\pi i}\int_{\gamma_1} \frac{f(\zeta)}{\zeta - z}\, d\zeta + \frac{1}{2\pi i}\int_{\gamma_2} \frac{f(\zeta)}{\zeta - z}\, d\zeta \tag{5.3}$$

Let f_j be the value of the integral over γ_j in (5.3), $j = 1, 2$. Then f_1 is independent of γ_1 so long as γ_1 is near $\mathbf{E}$ and hence f_1 is actually holomorphic on $\mathbf{S}^2 \setminus \mathbf{E}$. Further, from (5.3) we see that $f_1 = f - f_2$; but both f and f_2 are bounded near $\mathbf{E}$ so that f_1 is bounded near $\mathbf{E}$ and so is bounded on all of $\mathbf{S}^2 \setminus \mathbf{E}$. Thus, f_1 is constant and, in fact, zero since $f_1(\infty) = 0$. This means $f = f_2$ and so f is holomorphic across $\mathbf{E}$ since f_2 clearly is.

Definition. A point $x \in \partial\Omega$ is *essential* if there is some $f \in H^\infty(\Omega)$ and some $\varepsilon > 0$ such that f does not extend to be holomorphic in the disc $\{z : |z - x| < \varepsilon\}$. A boundary point which is not essential is called *removable*, the nomenclature being fully explained by the next proposition.

Proposition 5.3. Let Ω be a domain, $\infty \in \Omega$, and let $\mathbf{E} = \mathbf{S}^2 \setminus \Omega$. A point $x \in \mathbf{E}$ is a removable boundary point of Ω if and only if there is some $\varepsilon > 0$ such that $\mathbf{E} \cap \{z : |z - x| \leqslant \varepsilon\}$ is a null set.

Proof. Suppose first that $\mathbf{E} \cap \{z : |z - x| \leqslant \varepsilon\} = \mathbf{E}_\varepsilon$ is a null set for some $\varepsilon > 0$, and let f be bounded and holomorphic in Ω. Then f is bounded and holomorphic in the set $\{z : |z - x| < \varepsilon\} \setminus \mathbf{E}_\varepsilon$ and, because $\mathbf{E}_\varepsilon$ is a null set, we may invoke Proposition 5.2 to conclude that f is actually holomorphic in $\{z : |z - x| < \varepsilon\}$ as well. Thus, x is removable.

To see the converse suppose $\mathbf{E}_\varepsilon = \{z : |z - x| \leqslant \varepsilon\} \cap \mathbf{E}$ is not a null set for any $\varepsilon > 0$. Let $\mathfrak{U}$ be a neighborhood of x. Suppose for every $y \in \mathfrak{U}$, $y \neq x$, there were a $\delta > 0$ such that $\mathbf{E} \cap \{z : |z - y| \leqslant \delta\}$ is a null set. Then each f bounded and holomorphic on $\mathfrak{U} \setminus \mathbf{E}$ extends to be bounded and holomorphic on $\mathfrak{U} \setminus \{x\}$ and hence also to $\mathfrak{U}$ so that $\mathbf{E}_\varepsilon$ is a null set if ε is so small that $\mathfrak{U}$ contains $\{z : |z - x| \leqslant \varepsilon\}$. This contradiction shows that there are points $\{x_n\}$ with $x_n \to x$ and (small) positive numbers $\{\delta_n\}$ such that the disc $\{z : |z - x_n| \leqslant \delta_n\}$ are pairwise disjoint and such that $\mathbf{E}_n = \mathbf{E} \cap \{z : |z - x_n| \leqslant \delta_n\}$ is not a null set. Let $\Omega_n = \mathbf{S}^2 \setminus \mathbf{E}_n$; then Ω_n is a domain and Ω is a subset of Ω_n for all n. Let f_n be bounded, nonconstant, and holomorphic in Ω_n with $\|f_n\| \leqslant 2^{-n}$ there. Put $f = \Sigma f_n$. Then f is holomorphic and bounded on Ω. Further, $f - f_n$ is holomorphic over $\mathbf{E}_n$ and f_n is not. Thus, f can not be holomorphic in any neighborhood of $\mathbf{E}_n$ and consequently f is not holomorphic in any neighborhood of x, either. We conclude that x is an essential boundary point.

Remark. It is worth emphasizing that Proposition 5.3 shows that if $x \in \partial\Omega$ is removable then there is a fixed disc $\{z : |z - x| < \varepsilon\}$ into which *all* elements of $f \in H^\infty(\Omega)$ extend. It is also worth making explicit the observation that if x is removable then the supremum of each f bounded and holomorphic on Ω is no bigger in $\Omega \cup \{z : |z - x| < \varepsilon\}$ than it is in Ω. This is straightforward and its proof is left to the exercises.

Definition. A domain Ω is *maximal* if each point in $\partial\Omega$ is essential.

Theorem 5.4. *Let Ω be a domain and let B^* consist of all essential boundary points of Ω. Then B^* is perfect. Further, if Ω^* is that component of the complement of B^* which contains Ω, then Ω^* is maximal, Ω is dense in Ω^*, and Ω^* is the smallest maximal domain containing Ω.*

Proof. We, of course, assume that B^* is not empty. According to Proposition 5.3 if $x \in \partial\Omega$ is removable then there is an open disc centered at x which does not meet B^*. The union of these discs and Ω is precisely Ω^* and Ω must be dense in Ω^* since null sets contain no interior. Further, each f in $H^\infty(\Omega)$ extends to an element of $H^\infty(\Omega^*)$, so all that remains is to show Ω^* is maximal. Let y be a boundary point of Ω^*. Then y must not only be in boundary of Ω but it obviously must be in B^*, as well. Thus, some element f of $H^\infty(\Omega)$, and hence of $H^\infty(\Omega^*)$ does not extend to be holomorphic across y so that y is essential.

Proposition 5.5. (a) Let $\mathcal{D}$ be a dense, open subset of a domain Ω. If ϕ is analytic in Ω and one-to-one in $\mathcal{D}$, then ϕ is one-to-one in Ω.

(b) Suppose ϕ is a one-to-one holomorphic function from a domain Ω onto a domain $\mathcal{D}$. If $\mathcal{D}$ is maximal, then so is Ω.

Proof. Suppose z_1 and z_2 are points of Ω with $\phi(z_1) = \phi(z_2)$. Let $\mathcal{V}_1$ and $\mathcal{V}_2$ be disjoint discs in Ω centered at z_1 and z_2, respectively. Then $\phi(\mathcal{D} \cap \mathcal{V}_j)$ is an open set, dense in $\phi(\mathcal{V}_j)$ for $j = 1, 2$ and hence $\phi(\mathcal{D} \cap \mathcal{V}_1)$ must meet $\phi(\mathcal{D} \cap \mathcal{V}_2)$, a contradiction.

To prove (b), suppose to the contrary that Ω is not maximal. Then there is a point $x \in \partial\Omega$ and a (small) disc $\mathcal{U}$ such that each $f \in H^\infty(\Omega)$ extends to be bounded and holomorphic in $\Omega' = \Omega \cup \mathcal{U}$. In particular, ϕ extends to Ω' since ϕ is bounded near x. The extension of ϕ is still one-to-one by part (a). Let $\mathcal{D}' = \phi(\Omega')$ so that $\mathcal{D}'$ is a domain properly containing $\mathcal{D}$. Let $f \in H^\infty(\mathcal{D})$ and put $g = f \circ \phi$. Then $g \in H^\infty(\Omega)$ and so g extends to Ω'; thus, $g \circ \phi^{-1}$ extends f to $\mathcal{D}'$, which contradicts the maximality of $\mathcal{D}$.

3.6. $H^\infty(\Omega)$ DETERMINES Ω^*

In this section we shall see that the algebraic structure of $H^\infty(\Omega)$ as a ring over the reals determines the maximal domain Ω^* of Ω, at least up to conformal equivalence.

Theorem 6.1. *Let Ω and $\mathfrak{D}$ be two domains and suppose ϕ is a one-to-one holomorphic function mapping $\mathfrak{D}^*$ onto Ω^*. Then the rule*

$$\psi f = f \circ \phi, \qquad f \in H^{\infty}(\Omega) \tag{6.1}$$

defines a real algebra isomorphism ψ of $H^{\infty}(\Omega)$ onto $H^{\infty}(\mathfrak{D})$ with $\psi(i) = i$. Conversely, if ψ is a real algebra isomorphism of $H^{\infty}(\Omega)$ onto $H^{\infty}(\mathfrak{D})$ with $\psi(i) = i$, then there is a one-to-one holomorphic function ϕ from $\mathfrak{D}^*$ onto Ω^* such that (6.1) *holds.*

Proof. Of course if ψ is given by (6.1) then it is easy to see that ψ is an algebra isomorphism so we concentrate on the second part of the theorem, namely, given ψ we must produce ϕ. We begin by assuming that both Ω and $\mathfrak{D}$ are maximal since this in no way affects $H^{\infty}(\Omega)$, $H^{\infty}(\mathfrak{D})$, or ψ. We may also assume ∞ is not in either Ω or $\mathfrak{D}$.

We first show that $\psi(c) = c$ for all constants c. This is clearly true if $c = r_1 + ir_2$ where r_1, r_2 are rational numbers; such a constant c is a complex rational and is denoted by r. If h is a bounded holomorphic function on some domain we let $\mathrm{c}(h)$ stand for the closure of the range of f. For a nonconstant $f \in H^{\infty}(\Omega)$ and a complex rational r we have

$$\begin{aligned} r \notin \mathrm{c}(f) \quad & \text{iff } f - r \text{ is invertible in } H^{\infty}(\Omega) \\ & \text{iff } \psi(f) - r \text{ is invertible in } H^{\infty}(\mathfrak{D}) \\ & \text{iff } r \notin \mathrm{c}(\psi(f)) \end{aligned}$$

Hence, if f is nonconstant, then $\mathrm{c}(f) = \mathrm{c}(\psi(f))$. Suppose that $c \in \mathbb{C}$ and f is a nonconstant element of $H^{\infty}(\Omega)$ with $0 \in \mathrm{c}(f)$. Put $f_n = c + f/n$. Then

$$c = \bigcap_{n=1}^{\infty} \mathrm{c}(f_n) = \bigcap_{n=1}^{\infty} \mathrm{c}(\psi(f_n)) = \psi(c)$$

Next we extend ψ from $H^{\infty}(\Omega)$ to the quotient field of $H^{\infty}(\Omega)$ by setting

$$\psi\left(\frac{f}{g}\right) = \frac{\psi(f)}{\psi(g)}, \qquad f, g \in H^{\infty}(\Omega)$$

The extension is well defined since ψ is an algebra isomorphism. Let $I(z) = z$, $z \in \Omega$. If $f \in H^{\infty}(\Omega)$ is nonconstant and $a \in \Omega$, then

$$g(z) = \frac{f(z) - f(a)}{z - a}$$

lies in $H^{\infty}(\Omega)$ and hence

$$I = a + \frac{f - f(a)}{g}$$

is in the field of quotients. Set $\phi = \psi(I)$; we know that ϕ is the quotient of two elements of $H^\infty(\mathcal{D})$. We must show that ϕ is actually a one-to-one holomorphic map of $\mathcal{D}$ onto Ω.

For $c \in \mathbf{C}$ we have

$$\begin{aligned} c \notin \mathrm{c}(I) \quad &\text{iff } \frac{1}{I-c} \text{ is in } H^\infty(\Omega) \\ &\text{iff } \frac{1}{\phi - c} \text{ is in } H^\infty(\mathcal{D}) \\ &\text{iff } c \notin \mathrm{c}(\phi) \end{aligned}$$

Thus, $\mathrm{c}(\phi) = \mathrm{c}(I) = \mathrm{cl}(\Omega)$ so that those points of Ω that are in the range of ϕ are dense in Ω.

Let $a \in \Omega$ and let $\mathfrak{M}_a$ be the maximal ideal in $H^\infty(\Omega)$ consisting of those elements of $H^\infty(\Omega)$ which vanish at a. $\psi(\mathfrak{M}_a)$ is a maximal ideal in $H^\infty(D)$. Further, if $f \in \mathfrak{M}_a$, then $f/(I-a)$ is in $H^\infty(\Omega)$ and consequently $\psi(f)/(\phi - a)$ is in $H^\infty(\mathcal{D})$. We see that $\psi(f)$ vanishes at any and all points $b \in \mathcal{D}$ at which $\phi(b) = a$. Since $\psi(\mathfrak{M}_a)$ is maximal there can be at most one such point b. Hence, ϕ is one-to-one on the set

$$\mathcal{D}' = \{z \in \mathcal{D} : \phi(z) \in \Omega\}$$

But $\mathcal{D}'$ is dense and open in $\mathcal{D}$ so that ϕ is actually one-to-one in $\mathcal{D}$ by Proposition 5.5(a). Thus, ϕ is a conformal mapping on $\mathcal{D}$ into Ω and we need only show that ϕ is onto.

Let $\Lambda = \phi(\mathcal{D})$ so that Λ is a maximal domain by Proposition 5.5(b). Suppose first that $f \in H^\infty(\Omega)$ and that $a \in \Lambda$ and $\phi(b) = a$, $b \in \mathcal{D}$. Then $f - f(a) \in \mathfrak{M}_a$ so that $\psi(f) - f(a) \in \psi(\mathfrak{M}_a)$; but the latter consists of those elements of $H^\infty(\mathcal{D})$ which vanish at b. Hence, $f(a) = \psi(f)(b)$; equivalently,

$$f(z) = \psi(f)(\phi^{-1}(z)), \qquad z \in \Lambda, \qquad f \in H^\infty(\Omega)$$

If $g \in H^\infty(\Lambda)$, then $h(z) = g(\phi(z))$ lies in $H^\infty(\mathcal{D})$ and so there is an $f \in H^\infty(\Omega)$ with $h = \psi(f)$. Thus, for $a \in \Lambda$, $a = \phi(b)$, we have

$$g(a) = h(b) = \psi(f)(b) = \psi(f)(\phi^{-1}(a)) = f(a)$$

This says exactly that each element of $H^\infty(\Lambda)$ can be extended to an element of $H^\infty(\Omega)$. However, both Λ and Ω are maximal so that $\Lambda = \Omega$ and we are finished.

Theorem 6.1 says that the algebraic structure of $H^\infty(\Omega)$ determines Ω. There is one case in which this determination can be made "explicit."

Let Ω be the annulus

$$\Omega = \{z : 1 < |z| < \rho\}$$

and suppose f is a holomorphic function mapping Ω into itself. If γ is any smooth curve in Ω with winding number $+1$ about $z = 0$, then

$$\frac{1}{2\pi i}\int_\gamma \frac{f'(z)}{f(z)}\,dz = q \tag{6.2}$$

is an integer which is independent of γ. Let us suppose that q is *not* zero and see what follows. By replacing f with ρ/f, if necessary, we may assume that $q > 0$. By hypothesis, the integral of

$$\frac{q}{z} - \frac{f'(z)}{f(z)}$$

around any smooth simple closed curve in Ω must vanish and hence we can find a single-valued branch, $F(z)$, of $\log(z^q/f(z))$ in Ω. Let

$$u(re^{i\theta}) = \operatorname{Re} F(re^{i\theta}) = q\log r - \log|f(re^{i\theta})|$$

The integrals

$$I(r) = \int_{-\pi}^{\pi} u(re^{i\theta})\,d\theta$$

are independent of r by Cauchy's theorem and, further,

$$q\log(r/\rho) \leqslant u(re^{i\theta}) \leqslant q\log r, \qquad \text{all } \theta, \qquad 1 < r < \rho$$

since f maps Ω into itself. Thus,

$$\limsup\{I(r) : r \to 1\} \leqslant 0, \qquad \liminf\{I(r) : r \to \rho\} \geqslant 0$$

Hence, $I(r) \equiv 0$. Now set

$$\begin{aligned} J(r) &= \int_{-\pi}^{\pi} |u(re^{i\theta})|\,d\theta \\ &= \int_{u\geqslant 0} + \int_{u\leqslant 0} \\ &= -2\int_{u<0} u(re^{i\theta})\,d\theta \leqslant 4\pi q\log(\rho/r) \end{aligned}$$

and, likewise,

$$J(r) = 2\int_{u>0} u(re^{i\theta})\,d\theta \leqslant 4\pi q\log r$$

Since $|u(z)|$ is subharmonic it is immediate that

$$J(r) \leqslant \max\{J(r_1), J(r_2)\}, \qquad 1 < r_1 \leqslant r \leqslant r_2 < \rho$$

Let r_1 decrease to 1 and r_2 increase to ρ. Then

$$0 \leqslant J(r) \leqslant \max\{0, 0\}$$

Hence, $J(r) \equiv 0$ and so u vanishes identically. This means that

$$f(z) = \lambda z^q \quad \text{where } |\lambda| = 1 \tag{6.3}$$

However, we must have $\rho > |f(re^{i\theta})| = r^q$ for all $r \in (1, \rho)$ so that q can only be 1. We have thus proved.

Theorem 6.2. *Let Ω be the annulus $\{z : 1 < |z| < \rho\}$, let f be a holomorphic function mapping Ω into Ω, and let q be defined by* (6.2). *Then* $q = 0, 1,$ *or* -1. *In the last two cases, either* $f(z) = \lambda z$, $|\lambda| = 1$, or $f(z) = \lambda\rho/z$, $|\lambda| = 1$.

Theorem 6.2 is all that is needed to characterize ρ from within $H^\infty(\Omega)$. Specifically, let U consist of those elements of $H^\infty(\Omega)$ for which $1/f(z)$ is also in $H^\infty(\Omega)$ and let R consist of those $f \in H^\infty(\Omega)$ which have a holomorphic nth root for all $n = 1, 2, \ldots$. Then

$$\rho = \inf\{\|f\|\|f^{-1}\| : f \in U, f \notin R\} \tag{6.4}$$

To see this, note that $f(z) = z$ certainly has $\|f\|\|f^{-1}\| = \rho$. On the other hand, if $f \in U$, $f \notin R$, satisfies $\|f\|\|f^{-1}\| < \rho$, then we can multiply f by a constant to assume that $\|f\| < \rho$ and $\|f^{-1}\| < 1$. Thus, f maps Ω into itself. If q is given by (6.2) then q cannot be zero since then f would have a holomorphic logarithm and hence $f \in R$. Thus, Theorem 6.2 implies that f is a constant multiple of z or $1/z$ and so $\|f\|\|f^{-1}\| = \rho$.

3.7. WEAK PEAK POINTS FOR $H^\infty(\Omega)$

We now shall prove a theorem of Beck (1964a) concerning the boundary behavior of H^∞ functions at essential boundary points. We begin with a definition.

Definition. Let Ω be a domain and let $x \in \partial\Omega$. x is a *weak peak point* for $H^\infty(\Omega)$ if there is an $f \in H^\infty(\Omega)$ with

$$\limsup\{|f(z)| : z \to x\} = 1 \tag{7.1a}$$

$$\limsup\{|f(z)| : z \to \zeta\} < 1 \quad \text{if} \quad \zeta \in \partial\Omega, \qquad \zeta \neq x \tag{7.1b}$$

Theorem 7.1. *If $x \in \partial\Omega$ is essential, then x is a weak peak point for $H^\infty(\Omega)$.*

Proof. The proof must proceed in several steps. We first show that the set of weak peak points in any domain is closed. Let $x_1, x_2, \ldots$ be points in $\partial\Omega$ each of which is a weak peak point for $H^\infty(\Omega)$ and suppose $x_j \to x \in \partial\Omega$. Let $\mathcal{U}_1, \mathcal{U}_2, \ldots$ be pairwise disjoint (small) discs with $\mathcal{U}_j$ centered at $x_j, j = 1, 2, \ldots$ and $\operatorname{diam}(\mathcal{U}_j) \to 0$ as $j \to \infty$. If f_j is the function given by (7.1) for x_j, then $|f_j|$ is bounded below 1 off $\mathcal{U}_j$ so an appropriately high power of f_j is smaller in modulus than 4^{-j} off $\mathcal{U}_j$ and still satisfies (7.1). There is no loss in letting f_j denote such a function. We now define a sequence $\{b_j\}$ of complex numbers so that we will obtain

$$|b_j| \leqslant 2, \qquad j = 1, 2, \ldots \tag{7.2a}$$

$$\sum_1^\infty b_j f_j \quad \text{is in } H^\infty(\Omega) \tag{7.2b}$$

$$\left|\sum_1^m b_j f_j\right| \leqslant 2 - \frac{1}{m} \quad \text{on } \mathcal{U}_m, \qquad m = 1, 2, \ldots \tag{7.2c}$$

To find such numbers, start by setting $b_1 = 1$. If $b_1, \ldots, b_{m-1}$ have been chosen, we find b_m as follows. Set, for $|\zeta| \leqslant 2$,

$$u_m(\zeta) = \sup\left\{\left|\sum_1^{m-1} b_j f_j(z) + \zeta f_m(z)\right| : z \in \mathcal{U}_m\right\}$$

Then $u_m(0) \leqslant \Sigma_1^{m-1}|b_j||f_j(z)| < 2\Sigma_1^{m-1}4^{-j} < \frac{2}{3}$. Choose a point $y_m \in \mathcal{U}_m$ with $|f_m(y_m)| > 1 - 1/2m$; define

$$\mu_m = \sum_1^{m-1} b_j f_j(y_m)$$

and

$$c_m = 2\frac{\mu_m}{|\mu_m|}\frac{|f_m(y_m)|}{f_m(y_m)}$$

Then

$$u_m(c_m) \geqslant \left|\sum_1^{m-1} b_j f_j(y_m) + c_m f_m(y_m)\right|$$

$$\geqslant |c_m f_m(y_m)| > 2 - \frac{1}{m}$$

Because u_m is continuous on the disc $|\zeta| \leqslant 2$ and satisfies $u_m(0) < \frac{2}{3} < 2 - 1/m < u_m(c_m)$, there is a point b_m with $|b_m| \leqslant 2$ and $u_m(b_m) = 2 - 1/m$. Hence, b_m is chosen with the knowledge of $b_1, \dots, b_{m-1}$ and (7.2) holds. Further, for $z \in \mathfrak{U}_m$ we have

$$\left|\sum_1^\infty b_j f_j(z)\right| \leqslant \left|\sum_1^m b_j f_j(z)\right| + \sum_{m+1}^\infty |b_j||f_j(z)|$$

$$< 2 - \frac{1}{m} + \sum_{m+1}^\infty 2 \cdot 4^{-j}$$

$$= 2 - \frac{1}{m} + \left(\frac{2}{3}\right)4^{-m}$$

In a similar fashion,

$$\left|\sum_1^\infty b_j f_j(y_m)\right| \geqslant \left|\sum_1^m b_j f_j(y_m)\right| - \sum_{m+1}^\infty |b_j||f_j(y_m)|$$

$$\geqslant 2 - \frac{1}{m} - \left(\frac{2}{3}\right)4^{-m}$$

so that

$$\sup\left\{\left|\sum_1^\infty b_j f_j(z)\right| : z \in \mathfrak{U}_m\right\}$$

lies in the interval from $2 - 1/m - (\frac{2}{3})4^{-m}$ to $2 - 1/m + (\frac{2}{3})4^{-m}$. Next, for $z \in \Omega$ but z not in any $\mathfrak{U}_m$ we estimate

$$\left|\sum_1^\infty b_j f_j(z)\right| \leqslant 2\sum_1^\infty 4^{-j} = \frac{2}{3}$$

The series defining $f(z) = \frac{1}{2}\sum_1^\infty b_j f_j(z)$ converges uniformly on compact subsets of Ω and f satisfies $|f(z)| < 1$, $z \in \Omega$. Finally, $|f(z)|$ is near 1 in the disc $\mathfrak{U}_m$ when m is big and only then. This f then satisfies (7.1) at the point x, which shows that x is a weak peak point for $H^\infty(\Omega)$. This completes the proof that the weak peak points in $\partial\Omega$ are a closed set.

The second step of the proof is to show

$$\text{if there is a continuum in } \Omega^C \text{ containing } x \in \partial\Omega, \text{ then } x \text{ is a weak peak point for } H^\infty(\Omega) \tag{7.3}$$

Let **C** be a (nontrivial) continuum in Ω^c containing x and let $\mathfrak{D}$ be that component of the complement of $\mathbf{C}$ which contains Ω. Then the Dirichlet problem is solvable in $\mathfrak{D}$ by Corollary 1.4.5. Let u be a continuous real-valued function on $\partial\mathfrak{D}$ with $u(x) = 0$ and $0 > u(y) \geqslant -1$ for all $y \in \partial\mathfrak{D}$, $y \neq x$. Again denote by u the harmonic extension to $\mathfrak{D}$ and let $^*u$ be the harmonic conjugate of u on $\mathfrak{D}$; note that $\mathfrak{D}$ is simply connected so $^*u$ is well defined. Then $f = \exp[u + i\,{}^*u]$ is in $H^\infty(\mathfrak{D})$ and f satisfies (7.1) at x.

It follows from (7.3) and the first step of the proof that if x is in a nontrivial component of $\partial\Omega$ or is a limit point of such points, then x is a weak peak point for $H^\infty(\Omega)$.

The third and final step in the proof is this:

$$\text{if } x \in \partial\Omega \text{ is essential and if } x \text{ has a neighborhood } \mathfrak{U} \text{ such that } \mathfrak{U} \cap \partial\Omega \text{ is totally disconnected, then } x \text{ is a weak peak point for } H^\infty(\Omega). \tag{7.4}$$

The proof of (7.4) is very much like the proof of the first step of the theorem. We know from Theorem 5.4 that the essential boundary points form a perfect set, so let $x_1, x_2, \dots$ be a sequence of essential boundary points in $\partial\Omega$ with $x_j \to x$. Let $\mathfrak{U}_1, \mathfrak{U}_2, \dots$ be disjoint open discs, $\mathfrak{U}_j$ centered at x_j, with $\operatorname{diam}(\mathfrak{U}_j) \to 0$ as $j \to \infty$. Let $\mathbf{V}_j$ be a closed disc centered at x_j with radius half that of $\mathfrak{U}_j$ for $j = 1, 2, \dots$. For each j, there is a bounded holomorphic function f_j on the set $\mathfrak{D}_j = \mathbf{S}^2 \setminus (\mathbf{V}_j \cap \partial\Omega)$ satisfying $\|f_j\|_{\mathfrak{D}_j} = 1$ and $\sup\{|f_j(z)| : z \notin \mathfrak{U}_j\} \leqslant 4^{-j}$. Since Ω^c is nowhere dense in $\mathfrak{U}_j$ we also have $\|f_j\|_\Omega = 1$ and $\sup\{|f_j(z)| : z \notin \mathfrak{U}_j\} \leqslant 4^{-j}$. Now a repetition of the construction which leads to (7.2) produces a function $f \in H^\infty(\Omega)$ which peaks at x. Thus, (7.4) is proved and that completes the proof of Theorem 7.1.

Corollary 7.2. *A point $x \in \partial\Omega$ is a weak peak point for $H^\infty(\Omega)$ if and only if x is an essential boundary point.*

Proof. Suppose x is a weak peak point for $H^\infty(\Omega)$; let f satisfy (7.1). If x were removable, then f would be holomorphic in a domain Ω' containing both Ω and x and $\sup\{|f(z)| : z \in \Omega'\} = 1$. But then $|f|$ would have an interior maximum at x and hence would be constant. However, f is not constant.

ADDITIONAL READINGS AND NOTES

The idea of H^p spaces on general domains was put forth by Parreau (1951) and, independently, by Rudin (1955b). The fundamentals of $H^p(\Delta)$, and a great deal more, can be found in the books of Duren (1970) and Hoffman (1962). D. Hejhal (1975) and Hasumi (1978) have shown that there are domains Ω in $\mathbf{C}$ for which $H^p(\Omega)$ is trivial if $p > p_0$ but $H^{p_0}(\Omega)$ is not trivial; in particular, $H^\infty(\Omega)$ can be trivial even though $\cap_{p>0} H^p(\Omega)$ contains nonconstant functions. The connection between $H^p(\Omega)$ and $H^p/\mathfrak{G}$ was noted in

Rudin (1955b); Forelli exploited it to give a proof of the corona theorem for finitely connected Ω, as we will see in Chapter 6. The problem of characterizing removable sets for bounded analytic functions (or other classes of holomorphic functions) is an old and honored one. The paper of Ahlfors and Beurling (1950) has a lot of substantive material on this topic. One of their constructions produces a domain Ω for which $H^\infty(\Omega)$ is nontrivial but Ω supports no bounded schlicht functions. Also see Zalcman (1968b). Much of what is in Sections 5 and 6 is from Rudin (1955a); in particular, Theorem 6.1 is his. The concept of a weak peak point for $H^\infty(\Omega)$ was introduced by Beck (1964a) although with slightly different notation. Beck also solved the "rings on rings" problem (1964b), a sequel to Beck (1964a) pagewise. The presentation here is from Richards (1968); the lovely proof of Theorem 6.2 is from Reich (1966). The notion of a peak point for a space of analytic functions is well developed. Let **K** be a compact set in $\mathbb{C}$ and **A(K)** those continuous functions on **K** which are analytic on INT **K** (if **K** has interior). A point $x \in \mathbf{K}$ is a peak point for **A(K)** if there is an $f \in \mathbf{A(K)}$ with $f(x) = 1$ but $|f(z)| < 1$ for all $z \in \mathbf{K}$, $z \neq x$. Let **R(K)** be the uniform closure on **K** of the rational functions with no poles on **K**; clearly $\mathbf{R(K)} \subseteq \mathbf{A(K)}$. The questions as to whether $\mathbf{R(K)} = \mathbf{A(K)}$ or which points in **K** are peak points for **R(K)** or **A(K)** have received considerable attention; see the monograph of Zalcman (1968a), and the books of Gamelin (1969) and Stout (1971) for many results in these areas.

EXERCISES

1. If Ω and Λ are two domains with $\Omega \subset \Lambda$ and if f is holomorphic on Λ, show that the $H^p(\Lambda)$ norm of f exceeds the $H^p(\Omega)$ norm. Can equality hold if f is not constant?
2. If Ω^* is the maximal domain containing Ω, then $\sup\{|f(z)| : z \in \Omega\} = \sup\{|f(z)| : z \in \Omega^*\}$ for all $f \in H^\infty(\Omega)$.
3. If Ω and Λ are as in problem 1 and if for each $f \in H^\infty(\Lambda)$ we have $\sup\{|f(z)| : z \in \Lambda\} = \sup\{|f(z)| : z \in \Omega\}$, then $\Lambda \subset \Omega^*$.
4. Prove directly that $H^p(\mathbb{C})$ is trivial for $0 < p \leqslant \infty$.
5. Let **E** be a compact subset of $\mathbb{R}$ of zero length and set $\Omega = \mathbf{E}^c$. Show that $H^p(\Omega)$ is trivial for $1 \leqslant p < \infty$.
6. Let **E** be a compact set in $\mathbb{C}$ of positive area. Define

$$F(z) = \int_{\mathbf{E}} \frac{d\sigma\, d\tau}{z - \zeta}, \qquad \zeta = \sigma + i\tau$$

   Show F is continuous on the sphere, analytic off **E**, and nonconstant.
7. Show that the function g given by (5.2) is nonconstant and has bounded imaginary part.

8. Describe those algebra isomorphisms ψ of $H^\infty(\Omega)$ onto $H^\infty(D)$ for which $\psi(i) = -i$.

9. Show that $(1 - z)^{-1}$ is in $H^p(\Delta)$ for all $p < 1$ but is not in $H^1(\Delta)$.

10. Let B be a Blaschke product on Δ with zeros $\{z_j\}$ and let $\mathbf{E}$ be the set of limit points in $\mathbf{T}$ of $\{z_j\}$, if any. Show B is holomorphic off the set $\mathbf{Z} = \mathbf{E} \cup \{\bar{z}_j^{-1} : j = 1, 2, \dots\}$ and B is *not* holomorphic in a neighborhood of any point of $\mathbf{Z}$.

11. Let S be a singular inner function on Δ with associated singular measure μ; see (3.7a). Let $\mathbf{C}$ be the closed support of μ in $\mathbf{T}$. Show S is holomorphic on the complement of $\mathbf{C}$ and is not holomorphic in a neighborhood of any point of $\mathbf{C}$.

12. Suppose $f \in H^\infty(\Delta)$, $g \in H^1(\Delta)$ and that there is an arc $\alpha = \{e^{it} : a < t < b\}$ with these two properties:

1. $f^*(e^{it})g^*(e^{it}) \geqslant 0$ a.e. dt on α
2. $|f^*(e^{it})| = 1$ a.e. dt on α

Prove that both f and g are analytic across α. Use problems 10 and 11.

13. Suppose $f \in H^\infty(\Delta)$ and $\lim\{f(re^{i\theta_0}) : r \to 1\} = L$ for some θ_0. Let U_δ be the triangle whose vertices are located at $\exp[i\theta_0]$, $\exp[i(\theta_0 + \delta)]$, and $\exp[i(\theta_0 - \delta)]$, $\delta > 0$. Show that

$$\lim\{f(z);\, z \in U_\delta,\, z \to e^{i\theta_0}\} = L$$

HINT: a translation and rotation allows us to assume that f is in $H^\infty(\Omega)$ where $\Omega = \{z : |z - 1| < 1\}$ and that

$$\lim\{f(x) : x \downarrow 0\} = L$$

Let

$$f_n(z) = f\left(\frac{z}{n}\right), \qquad z \in \Omega$$

Use Proposition 2.4 to conclude $\{f_n\}$ is a normal family in Ω; any subsequential limit must be identically equal to L by our assumption. [This nice proof of nontangential convergence is from Duren (1970, p. 6).]

14. If $\{f_n\}$ is a sequence of elements of $H^1(\Delta)$ with $f_n \to f$ weakly in H^1 and $\|f_n\|_1 \to \|f\|_1$, then show $\|f_n - f\|_1 \to 0$. See Zalcman (1968c).

15. Let $0 < p < \infty$. If $\{u_n\}$ is a sequence of L^p functions and if there is a $u \in L^p$ with $u_n \to u$ a.e. and $\|u_n\|_p \to \|u\|_p$, then $\{u_n\}$ converges to u in L^p.

16. Show directly from the definition that $H^p(\Omega) \subset H^q(\Omega)$ if $p > q$.

17. Let f be an outer function in $H^2(\Delta)$. Show that the subspace $FH^\infty(\Delta)$ is dense in $H^2(\Delta)$. Conversely, if $f \in H^2(\Delta)$ and if $fH^\infty(\Delta)$ is dense in $H^2(\Delta)$, show that f is outer.

18. Let Ω_1 and Ω_2 be domains in $\mathbb{C}$ with $\Omega_j = \text{INT}(\text{CL}(\Omega_j))$ for $j = 1, 2$. Define $\mathbf{A}(\Omega)$ to be those continuous functions on $\text{CL}(\Omega)$ which are also holomorphic on Ω. Find all the algebra isomorphisms of $\mathbf{A}(\Omega_1)$ onto $\mathbf{A}(\Omega_2)$.

19. Let Ω be bounded by a finite number of disjoint analytic simple closed curves. If $x \in \partial\Omega$, show that x is a peak point for $\mathbf{A}(\Omega)$; that is, there is an $f \in \mathbf{A}(\Omega)$ with $f(x) = 1$ and $|f(y)| < 1$ for all $y \in \partial\Omega$, $y \neq x$.

20. If $h \in H^1(\Delta)$, show that $h = g_1 g_2$ where g_1 and g_2 are both in $H^2(\Delta)$ and $|h^*| = |g_1^*|^2 = |g_2^*|^2$ a.e. on $\mathbf{T}$.

21. Let E be a compact subset of an analytic curve γ in the plane. Show E is a null set for H^∞ if and only if E has arc length zero. This result extends to the case when γ is smooth or even rectifiable but the proof is considerably more difficult.

4

DOMAINS OF FINITE CONNECTIVITY

4.1. INTRODUCTION

If the topological boundary of Ω is simple, then much more specific information about the H^p spaces of Ω can be obtained. This is most evident in the case when Ω is the open unit disc Δ as we saw in Section 3 of Chapter 3. We shall see in this chapter that much of what is known for the unit disc extends to a domain whose boundary consists of finitely many disjoint analytic simple closed curves, sometimes, however, with appropriate modifications.

If Ω is a domain on the sphere whose complement relative to the sphere consists of exactly $m + 1$ (closed) components, each of which is nontrivial, then $m + 1$ applications of the Riemann mapping theorem produces a one-to-one holomorphic map of Ω onto a bounded domain whose boundary consists of $m + 1$ disjoint analytic simple closed curves. Since such a holomorphic map gives an isometry of the corresponding H^p spaces, we may assume initially that Ω is the second kind of domain. Specifically,

$$\Gamma = \partial\Omega = \Gamma_0 \cup \cdots \cup \Gamma_m \tag{1.1}$$

where Γ_j is an analytic simple closed curve and $\Gamma_j \cap \Gamma_k = \varnothing$ if $j \neq k$. We may further suppose that Γ_0 is the boundary of the unbounded component of the complement of Ω. Let

$$\mathcal{U}_0 = \text{bounded component of } \mathbf{S}^2 \setminus \Gamma_0 \tag{1.2}$$

and

$$\mathcal{U}_j = \text{unbounded component of } \mathbf{S}^2 \setminus \Gamma_j, \qquad j = 1, \ldots, m \tag{1.3}$$

(We shall assume $m \geqslant 1$ since the case $m = 0$ is just the case of the unit disc Δ which is covered in Section 3 of Chapter 3.)

We let $\mathbf{R}(\Omega)$ denote those rational functions whose poles are off $\Omega \cup \Gamma$ and $\operatorname{Re}\mathbf{R}(\Omega)$ denote the real parts of functions in $\mathbf{R}(\Omega)$. Let $\mathbf{A}(\Omega)$ be those functions which are continuous on $\Omega \cup \Gamma$ and holomorphic in Ω.

4.2. THE DEFECT OF Re R(Ω) IN $\mathbf{C}_r(\Gamma)$

We begin with a theorem which measures how much $\operatorname{Re}\mathbf{R}(\Omega)$ misses being dense in $\mathbf{C}_r(\Gamma)$, the space of real-valued continuous functions on Γ. To this end, let a_j be a point in the bounded component of the complement of Γ_j for $j = 1, \dots, m$ and set

$$L_j(z) = \log|z - a_j|, \qquad j = 1, \dots, m \tag{2.1}$$

Theorem 2.1. *The real linear span of* $\operatorname{Re}\mathbf{R}(\Omega)$ *and the functions* $L_1, \dots, L_m$ *is dense in* $\mathbf{C}_r(\Gamma)$.

Proof. Let v be a real-valued continuous function on Γ and suppose initially that v is harmonic on some neighborhood of Γ. Let V denote the harmonic extension to Ω of v. Then V is actually harmonic in some neighborhood of $\Omega \cup \Gamma$ since each Γ_j is analytic. (The function $V - v$ is harmonic in Ω near Γ and vanishes on Γ; hence by the reflection principle it is harmonic over Γ and thus so is V since v is.)

Fix j and let b be a point of Γ_j. In a small neighborhood $\mathcal{O}$ of b_j, V has a harmonic conjugate uniquely determined by the requirement that $V + i{}^*V$ is analytic in $\mathcal{O}$ and ${}^*V(b) = 0$. Analytic continuation of the function element $V + i{}^*V$ along Γ_j in the positive direction is possible and eventually produces in $\mathcal{O}$ the function element $V + i{}^*V + ic_j$ where c_j is real; c_j is the period of ${}^*V$ along Γ_j. Suppose now that is done for $j = 1, \dots, m$. Let

$$V_1 = V - \frac{1}{2\pi}\sum_{j=1}^{m} c_j L_j$$

The period of ${}^*L_k$ along Γ_j is $2\pi\delta_{jk}$ so that ${}^*V_1$ has zero period along Γ_j for $j = 1, \dots, m$. Further, V_1 is harmonic in a neighborhood of $\Omega \cup \Gamma$. Thus, V_1 has a single-valued conjugate ${}^*V_1$ on all of this neighborhood. By Runge's theorem there is a rational function f with

$$\sup\{|f(z) - (V_1(z) + i{}^*V_1(z))| : z \in \Gamma\} < \varepsilon$$

This immediately implies that

$$\left|V(z) - \frac{1}{2\pi}\sum_{1}^{n} c_j L_j(z) - \operatorname{Re} f(z)\right| < \varepsilon, \qquad z \in \Gamma$$

which is the desired conclusion since $V = v$ on Γ. Hence, the theorem is proved in the case when v is harmonic in a neighborhood of Γ. However, it is easy to show that such functions v are uniformly dense in $\mathbf{C}_r(\Gamma)$. First note that if $\mathbf{T}$ is the unit circle, then any continuous real-valued function v on $\mathbf{T}$ can be uniformly approximated on $\mathbf{T}$ by a trigonometric polynomial p of the form $p(e^{i\theta}) = \sum_{-N}^{N} a_j e^{ij\theta}$, $\overline{a_j} = a_{-j}$, $j = 0, \ldots, N$, which is harmonic in a neighborhood of $\mathbf{T}$ because $p(z) = a_0 + 2\operatorname{Re}\sum_1^N a_j z^j$ if $|z| = 1$. Since each Γ_j is analytic the Riemann map of $\mathcal{U}_j$ onto Δ extends to be analytic in a neighborhood of $\mathcal{U}_j \cup \Gamma_j$ and maps Γ_j to $\mathbf{T}$ homeomorphically. Thus, if v is continuous on Γ_j then v can be uniformly approximated on Γ_j by functions harmonic in some neighborhood of Γ_j.

Corollary 2.2. *If u is continuous on Γ, then u may be approximated uniformly on Γ by functions harmonic in a neighborhood of $\Omega \cup \Gamma$.*

Theorem 2.3. *There are m linearly independent real measures $\nu_1, \ldots, \nu_m$ on Γ orthogonal to* Re $\mathbf{R}(\Omega)$ *and of the form*

$$d\nu_j = Q_j \, d\omega_{z_0}, \qquad 1 \leqslant j \leqslant m \tag{2.2}$$

where Q_j is C^∞ on Γ; Q_j is non-negative on Γ_j and nonpositive on Γ_k, $k \neq j$. The $\{\nu_j\}$ form a basis for the real measures on Γ which are orthogonal to $\mathbf{R}(\Omega)$.

Proof. Let h_j be the harmonic function on Ω whose boundary values are 1 on Γ_j and 0 on Γ_k, $k \neq j$. Then h_j is actually harmonic in a neighborhood of $\Omega \cup \Gamma$ by the reflection principle.

Let u be harmonic and real-valued on a neighborhood of $\Omega \cup \Gamma$. Then by Green's theorem

$$\int_\Gamma \left\{ u \frac{\partial h_j}{\partial n} - h_j \frac{\partial u}{\partial n} \right\} ds = \iint_\Omega \left(u \Delta h_j - h_j \Delta u \right) dx\, dy$$
$$= 0$$

Hence,

$$\int_\Gamma u \frac{\partial h_j}{\partial n} ds = \int_{\Gamma_j} \frac{\partial u}{\partial n} ds \tag{2.3}$$

Let $x \in \Gamma_j$ be fixed. In a small neighborhood of x, u has a unique harmonic conjugate $^*u$ with $^*u(x) = 0$. Analytic continuation of $u + i\,^*u$ along Γ_j will result in the function element $u + i{*u} + ic_j$ where c_j is a real scalar. We can apply the Cauchy–Riemann equations to find

$$\int_{\Gamma_j} \frac{\partial u}{\partial n}\, ds = \int_{\Gamma_j} \frac{\partial(*u)}{\partial \tau}\, ds = c_j \tag{2.4}$$

where τ is the tangential derivative in the positive direction along Γ_j. Indeed, the relation (2.4) is valid if Γ_j is replaced by a smooth curve γ within Ω but close to Γ_j. The number c_j is the period of the harmonic conjugate of u about Γ_j. Hence,

$$\int_{\Gamma} u \frac{\partial h_j}{\partial n}\, ds = \text{period of harmonic conjugate of } u \text{ about } \Gamma_j \tag{2.5}$$

Since each function u continuous on Γ can be uniformly approximated by functions harmonic in a neighborhood of $\Omega \cup \Gamma$ by Corollary 2.2 we see that (2.5) holds for all $u \in \mathbf{C}_r(\Gamma)$. Note that if $u = \operatorname{Re} f$ where $f \in \mathbf{R}(\Omega)$, then the harmonic conjugate of u is $\operatorname{Im} f$ and this is single-valued. Thus the real measure

$$d\nu_j = \left(\frac{\partial h_j}{\partial n}\right) ds \tag{2.6}$$

is orthogonal to $\mathbf{R}(\Omega)$. Further, if we take

$$u(z) = \log|z - a|$$

where $a \notin \operatorname{CL}\mathcal{U}_k$, then

$$\int_{\Gamma} u\, d\nu_j = 2\pi\delta_{jk}$$

which shows that $\nu_1, \dots, \nu_m$ are linearly independent.

Finally, since the measures ds and $d\omega_{z_0}$ are related by $ds = \rho\, d\omega_{z_0}$ where ρ is a positive continuous function (see 1.6.5), we can write

$$d\nu_j = \left(\frac{\partial h_j}{\partial n}\right) ds = Q_j\, d\omega_{z_0}$$

The sign of Q_j is that of $\partial h_j/\partial n$ and this is easy to find since $0 < h_j < 1$ in Ω and $h_j = 1$ on Γ_j and $h_j = 0$ on Γ_k if $k \neq j$.

Let $\mathbf{S}$ consist of all real measures on Γ which are orthogonal to $\operatorname{Re}\mathbf{R}(\Omega)$ (equivalently, orthogonal to $\mathbf{R}(\Omega)$). $\mathbf{S}$ is a closed subspace of the real measures on Γ. Theorem 2.1 asserts that the closure of $\operatorname{Re}\mathbf{R}(\Omega)$ in $\mathbf{C}_r(\Gamma)$ has codimension exactly m. Since $\nu_1, \dots, \nu_m$ are linearly independent elements of $\mathbf{S}$, they must be a basis.

Proposition 2.4. There are functions $W_1, \ldots, W_m$ analytic on a neighborhood of $\Omega \cup \Gamma$ such that

$$dv_j = W_j\, dz, \qquad j = 1, \ldots, m \tag{2.7}$$

Hence, no linear combination of $Q_1, \ldots, Q_m$ has more than a finite number of zeros on Γ.

Proof. Let W_j be the (complex) derivative of $h_j + i\,{}^*h_j$ where h_j is as in the proof of Theorem 2.3 and ${}^*h_j$ is its harmonic conjugate. Then W_j is analytic and single-valued on $\Omega \cup \Gamma$. Further, a computation like that in the proof of Proposition 1.6.5 yields

$$iW_j(\zeta)\, d\zeta = \left(\frac{\partial h_j}{\partial n}\right) ds$$

which is (2.7) up to a factor of i.

Remark. Actually this argument (like that in Proposition 1.6.6) shows that W_j and hence Q_j has no zeros at all on Γ.

4.3. MEASURES ORTHOGONAL TO R(Ω)

We shall show in this section that each measure on Γ which is orthogonal to $\mathbf{R}(\Omega)$ is actually absolutely continuous with respect to ω (equivalently, to ds). More detailed information on the Radon-Nikodym derivative will be available in Section 5. We shall also show that each closed set $\mathbf{E}$ in Γ of arc length zero is a peak interpolation set for $\mathbf{R}(\Omega)$. The proof of Theorem 3.1 which follows is more from functional analysis than from function theory or potential theory.

In what follows, $\mathbf{X}$ is a compact Hausdorff space, $\mathbf{C(X)}$ is the space of continuous complex-valued functions on $\mathbf{X}$, $\mathbf{A}$ is a closed subspace of $\mathbf{C(X)}$, and $\mathbf{E}$ is a closed set in $\mathbf{X}$ with this property:

if μ is a measure on $\mathbf{X}$ orthogonal to $\mathbf{A}$, then μ has no mass on $\mathbf{E}$. (3.1)

Theorem 3.1. *Let hypothesis* (3.1) *hold and let p be a positive continuous function on* $\mathbf{X}$. *If u is a continuous function on* $\mathbf{E}$ *with $|u(x)| \leqslant p(x)$ for all $x \in \mathbf{E}$ then there is an element f of* $\mathbf{A}$ *with*

$$f = u \qquad \text{on } \mathbf{E} \tag{3.2a}$$

$$|f(x)| \leqslant p(x), \qquad \text{all } x \in \mathbf{X} \tag{3.2b}$$

Proof. The proof procceds in several steps. Note that (3.1) immediately implies that $\mathbf{A}_{\mathbf{E}}$, the restriction of **A** to **E**, is dense in **C(E)**. We shall show first that $\mathbf{U}_{\mathbf{E}}$, the restriction to **E** of the unit ball of **A**, is dense in the closed unit ball of **C(E)**. If this fails, then there is a measure ν on **E** and a function $u_0 \in \mathbf{C(E)}$ with norm one such that

$$\operatorname{Re} \int f \, d\nu \leqslant 1, \qquad f \in \mathbf{A}, \qquad \|f\| \leqslant 1 \tag{3.3a}$$

$$\operatorname{Re} \int u_0 \, d\nu > 1 \tag{3.3b}$$

Now (3.3a) implies that the linear functional

$$f \mapsto \int f \, d\nu$$

has norm on **A** no more than 1 and so by the Hahn-Banach theorem can be extended to a linear functional on **C(X)** of norm no more than 1. Thus, there is a measure μ of norm at most 1 with

$$\int f \, d\mu = \int f \, d\nu, \qquad f \in \mathbf{A}$$

Hence, $\mu - \nu$ is orthogonal to **A** and so (3.1) implies $\mu - \nu$ has no mass on **E**. But ν is supported on **E**; hence, $\mu = \nu$ on **E**. Consequently, for the u_0 above we find

$$\begin{aligned} 1 < \operatorname{Re} \int_{\mathbf{E}} u_0 \, d\nu &\leqslant \left| \int_{\mathbf{E}} u_0 \, d\nu \right| \\ &= \left| \int_{\mathbf{E}} u_0 \, d\mu \right| \leqslant \|u_0\|_{\mathbf{E}} \|\mu\| \\ &\leqslant 1 \end{aligned}$$

This contradiction shows that the closure of $\mathbf{U}_{\mathbf{E}}$ is all of the unit ball of **C(E)**.

Next, let $\varepsilon > 0$ be given and let $u \in \mathbf{C(E)}$, $\|u\|_{\mathbf{E}} \leqslant 1$. By the foregoing there is an element f_1 of **A** with

$$\|u - f_1\|_{\mathbf{E}} < \frac{\varepsilon}{2}, \qquad \|f_1\|_{\mathbf{X}} \leqslant 1$$

Then there is, in a similar fashion, an element $f_2 \in \mathbf{A}$ with

$$\|u - f_1 - f_2\|_{\mathbf{E}} < \frac{\varepsilon}{4}, \qquad \|f_2\|_{\mathbf{X}} < \frac{\varepsilon}{2}$$

Continuing in this way we obtain a sequence $\{f_n\}$ with

$$\left\| u - \sum_{k=1}^{n} f_k \right\|_{\mathbf{E}} < \frac{\varepsilon}{2^k}, \qquad \|f_n\|_{\mathbf{X}} < \frac{\varepsilon}{2^{n-1}}, \qquad n = 2, 3, \ldots$$

Put $f = \Sigma_1^\infty f_k$; then $f \in \mathbf{A}$, and

$$f = u \quad \text{on } \mathbf{E} \tag{3.4a}$$

$$\|f\|_{\mathbf{X}} < 1 + \varepsilon \tag{3.4b}$$

Now let p be a positive function on $\mathbf{X}$. Let $\mathbf{A}/p$ be the subspace of functions of the form f/p, $f \in \mathbf{A}$. A measure μ is orthogonal to $\mathbf{A}/p$ precisely when the measure μ/p is orthogonal to $\mathbf{A}$. Thus, by the results above which culminated in (3.4) we find that if $u \in \mathbf{C(E)}$ and $|u(x)| \leqslant p(x)$, $x \in \mathbf{E}$, then there is a $g \in \mathbf{A}$ with

$$g = u \qquad \text{on } \mathbf{E} \tag{3.5a}$$

$$|g(x)| \leqslant p(x)(1 + \varepsilon), \qquad x \in \mathbf{X} \tag{3.5b}$$

We next show that (3.2) holds in the special case when $p \equiv 1$. This will be adequate to prove Theorem 3.1 by the device previously employed; namely, replacing $\mathbf{A}$ by $\mathbf{A}/p$.

Let $u \in \mathbf{C(E)}$, $\|u\|_{\mathbf{E}} \leqslant 1$. There is an element $g_1 \in \mathbf{A}$ with $g_1 = u$ on $\mathbf{E}$ and $\|g_1\| \leqslant \frac{5}{4}$. Let $g_1, \ldots, g_{n-1}$ be chosen with $g_j = u$ on $\mathbf{E}$, $1 \leqslant j \leqslant n-1$. Let $\mathfrak{U}_n$ be those points $x \in \mathbf{X}$ with $|g_j(x)| < 1 + 2^{-n-1}$ for $1 \leqslant j \leqslant n-1$. By (3.5) there is a $g_n \in \mathbf{A}$ with

$$g_n = u \quad \text{on } \mathbf{E} \tag{3.6a}$$

$$\|g_n\| \leqslant 1 + 2^{-n-1} \tag{3.6b}$$

$$|g_n(x)| \leqslant \tfrac{1}{2} \quad \text{if} \quad x \notin \mathfrak{U}_n \tag{3.6c}$$

Let this be done for $n = 1, 2, \ldots$ and set $g = \Sigma_1^\infty g_n/2^n$. Clearly, $g \in \mathbf{A}$ and $g = u$ on $\mathbf{E}$. We need only show that $|g(x)| \leqslant 1$ for all $x \in \mathbf{X}$. If $x \in \mathfrak{U}_n$ for all n, then $|g_j(x)| \leqslant 1 + 2^{-n-1}$ for all n so that $|g_j(x)| \leqslant 1$ for each j and so $|g(x)| \leqslant 1$ also. If $x \in \mathfrak{U}_n$ but $x \notin \mathfrak{U}_{n+1}$, then $|g_j(x)| \leqslant 1 + 2^{-n-1}$ for $j = 1, \ldots, n$ and $|g_j(x)| \leqslant \frac{1}{2}$ if $j > n$. Hence, we can estimate g in this way:

$$\begin{aligned} |g(x)| &\leqslant \sum_1^n |g_j(x)| 2^{-j} + \sum_{n+1}^\infty |g_j(x)| 2^{-j} \\ &\leqslant (1 + 2^{-n-1}) \sum_1^n 2^{-j} + \frac{1}{2} \sum_{n+1}^\infty 2^{-j} \\ &= 1 - 2^{-2n-1} \end{aligned}$$

This completes the proof of Theorem 3.1.

We now must show that the critical hypothesis (3.1) holds in some reasonable situations.

Proposition 3.2. Let **E** be a closed set in the unit circle **T** of Lebesgue measure zero. Then there is a sequence $\{f_n\}$ of entire functions such that

$$|f_n| \leqslant 1 \quad \text{on } \mathbf{T} \tag{3.7a}$$

$$f_n \to 1 \quad \text{a.e. } d\theta \tag{3.7b}$$

$$f_n \to 0 \quad \text{on } \mathbf{E} \tag{3.7c}$$

$$f_n \to 1 \text{ uniformly on compact sets in } \Delta \tag{3.7d}$$

Proof. Let $\mathcal{G}_n$ be an open set containing **E** such that the measure of $\mathcal{G}_n \setminus \mathbf{E}$ is no more than n^{-4}. There is a real-valued function $v_n \in C(\mathbf{T})$ such that

$$-n \leqslant v_n \leqslant 0, \qquad v_n = 0 \quad \text{on } \mathbf{T} \setminus \mathcal{G}_n, \qquad v_n = -n \quad \text{on } \mathbf{E}$$

By Fejer's theorem, there is a real-valued trigonometric polynomial u_n with $-n \leqslant u_n \leqslant 0$ on **T** and $|u_n - v_n| \leqslant 1/n$ on all of **T**. Hence

$$\frac{-1}{n} \leqslant u_n \leqslant 0 \quad \text{on } \mathbf{T} \setminus \mathcal{G}_n \quad \text{and} \quad -n \leqslant u_n \leqslant -n + \frac{1}{n} \quad \text{on } \mathbf{E}$$

Let $^*u_n$ be the harmonic conjugate of u_n so that

$$\|^*u_n\|_{L^2}^2 \leqslant \|u_n\|_{L^2}^2 = \int_{\mathcal{G}_n} |u_n|^2 + \int_{\mathbf{T} \setminus \mathcal{G}_n} |u_n|^2$$

$$\leqslant n^{-2} + n^{-2}$$

and so

$$\|u_n + i\,^*u_n\|_2 \leqslant \frac{2}{n} \to 0 \quad \text{as } n \to \infty.$$

By passing to a subsequence if necessary, we may assume $u_n + i\,^*u_n \to 0$ a.e. $d\theta$. Put

$$f_n = \exp(u_n + i\,^*u_n)$$

Then f_n is actually an entire function, $|f_n| \leqslant 1$ on **T**, $f_n \to 1$ a.e. $d\theta$, and on **E** we have

$$|f_n| = \exp(u_n) \leqslant \exp\left[-n + \frac{1}{n}\right] \to 0, \quad \text{as } n \to \infty$$

Finally, (3.7d) follows from (3.7a) and the fact that

$$1 = \lim \frac{1}{2\pi} \int_{\mathbf{T}} f_n(e^{it})\, dt = \lim f_n(0)$$

Theorem 3.3. *Let* **E** *be a closed set in* Γ *of harmonic measure* 0. *If* μ *is a measure on* Γ *which is orthogonal to* $\mathbf{R}(\Omega)$, *then* μ *has no mass on* **E**. *Consequently,* μ *is absolutely continuous with respect to harmonic measure.*

Proof. If μ had some mass on **E**, then there would be a closed subset **F** of **E**, with $\mathbf{F} \subset \Gamma_j$ for some j, and a nonzero complex number λ with

$$\int_{\mathbf{F}} d\mu = \lambda$$

The curve Γ_j is analytic. Let ϕ be the Riemann mapping of $\mathcal{U}_j$ onto Δ; then ϕ extends to be holomorphic in a neighborhood of $\mathcal{U}_j \cup \Gamma_j$ and maps Γ_j homeomorphically onto **T**. Let $\mathbf{F}' = \phi(\mathbf{F})$; then $\mathbf{F}'$ is a closed set in **T** of length zero and so there is a sequence $\{f_n\}$ as in Proposition 3.2 for the set $\mathbf{F}'$. Put $g_n = f_n \circ \phi^{-1}$; then g_n is analytic in a neighborhood of $\mathcal{U}_j \cup \Gamma_j$, $g_n \to 0$ on **F** uniformly, $|g_n| \leqslant 1$ on $\Omega \cup \Gamma$, $g_n \to 1$ a.e. ω on Γ_j and $g_n \to 1$ uniformly on Γ_k for all $k \neq j$. Hence, μ is orthogonal to g_n and we find

$$\begin{aligned} 0 = \lim \int_{\Gamma} g_n\, d\mu &= \int_{\Gamma \setminus \mathbf{F}} d\mu \\ &= \int_{\Gamma} d\mu - \int_{\mathbf{F}} d\mu \\ &= -\lambda \end{aligned}$$

a contradiction.

Corollary 3.4. *Let* **E** *be a closed set of harmonic measure zero (equivalently, of arc length zero) in* Γ. *Let* p *be a positive continuous function on* Γ. *If* $u \in \mathbf{C}(\mathbf{E})$ *and* $|u(x)| \leqslant p(x)$ *for* $x \in \mathbf{E}$, *then there is an element* $f \in \mathbf{R}(\Omega)$ *with*

$$f = u \qquad \text{on } \mathbf{E} \tag{3.8a}$$

$$|f(x)| \leqslant p(x), \qquad x \in \Gamma \tag{3.8b}$$

4.4. $H^p(\Omega)$

We now investigate various aspects of the behavior of an $H^p(\Omega)$ function at Γ. Here we shall find there is much in common with what is true for the case of the open unit disc; Section 3 of Chapter 3.

Proposition 4.1. Let $\mathfrak{U}_0, \ldots, \mathfrak{U}_m$ be the domains defined by (1.2) and (1.3). If $f \in H^p(\Omega)$, then

$$f = f_0 + \cdots + f_m \quad \text{on } \Omega \tag{4.1}$$

where $f_j \in H^p(\mathfrak{U}_j)$ for $0 \leqslant j \leqslant m$. Further, for each $j = 1, \ldots, m$ the map $f \mapsto f_j$ is a bounded linear projection of $H^p(\Omega)$ onto those $H^p(\mathfrak{U}_j)$ functions which vanish at ∞; for $j = 0$, the map $f \mapsto f_0$ maps $H^p(\Omega)$ onto $H^p(\mathfrak{U}_0)$.

Proof. The proof involves Cauchy's formula. If $z \in \Omega$ let $C_0, \ldots, C_m$ be smooth simple closed curves so near $\Gamma_0, \ldots, \Gamma_m$, respectively, that z is exterior to $C_1, \ldots, C_m$ and interior to C_0. Set

$$f_k(z) = \frac{1}{2\pi i} \int_{C_k} \frac{f(w)}{w - z} dw \qquad k = 0, \ldots, m \tag{4.2}$$

Clearly f_k is independent of the choice of C_k, is holomorphic in $\mathfrak{U}_k$ for $k = 0, \ldots, m$, and $f_k(\infty) = 0$ for $k = 1, \ldots, m$. Further, (4.1) certainly holds. Note that if f already lies in $H^p(\mathfrak{U}_j)$, then $f_k = 0$ for all $k \neq j$ so that $f = f_j$. Note as well that the rule (4.2) defining f_k is linear. All that remains is to show that $f_k \in H^p(\mathfrak{U}_k)$. For $j \neq k$, we know that f_j is bounded in a neighborhood of Γ_k, so we find from (4.1) that f_k is in $H^p(\mathcal{O})$ for some neighborhood $\mathcal{O}$ of Γ_k since H^p is a linear space. However, f_k is clearly bounded off any neighborhood of Γ_k and hence $f_k \in H^p(\mathfrak{U}_k)$.

Proposition 4.2. $\mathbf{R}(\Omega)$ is dense in $H^p(\Omega)$ if $1 \leqslant p < \infty$ and boundedly pointwise dense in $H^\infty(\Omega)$; $\mathbf{R}(\Omega)$ is uniformly dense in $\mathbf{A}(\Omega)$.

Proof. We make use of (4.1). Suppose first that $1 \leqslant p < \infty$. Since the $H^p(\mathfrak{U}_j)$ norm is larger than the $H^p(\Omega)$ norm, it suffices to prove that f_j is the limit in $H^p(\mathfrak{U}_j)$ of a sequence of functions holomorphic in a neighborhood of $\mathfrak{U}_j \cup \Gamma_j$. Let ϕ be the Riemann mapping of $\mathfrak{U}_j$ onto Δ; then ϕ extends to be holomorphic and one-to-one in some neighborhood of $\mathfrak{U}_j \cup \Gamma_j$ since Γ_j is analytic. Further, $g_j = f_j \circ \phi^{-1}$ is in $H^p(\Delta)$ and thus there is a function G analytic on a neighborhood of $\Delta \cup \mathbf{T}$ with $\|G - g_j\| < \varepsilon$ in $H^p(\Delta)$. Hence, $\|f_j - G \circ \phi\| < \varepsilon$ in $H^p(\mathfrak{U}_j)$ and $G \circ \phi$ is analytic in a neighborhood of $\mathfrak{U}_j \cup \Gamma_j$. All that remains is to apply Runge's theorem and approximate $G \circ \phi$ uniformly on $\mathfrak{U}_j \cup \Gamma_j$ by an element of $\mathbf{R}(\Omega)$.

The argument for $p = \infty$ is virtually the same. The function g_j is in $H^\infty(\Delta)$ and so there are rational functions G_{jn}, $n = 1, 2, \ldots$ with no poles in $\Delta \cup \mathbf{T}$ such that

$$\|G_{jn}\|_{\mathbf{T}} \leqslant \|g_j\|_{\mathbf{T}}$$

$$G_{jn}(z) \to g_j(z) \quad \text{as } n \to \infty, \qquad z \in \Delta$$

Thus, the functions $F_{jn} = G_{jn} \circ \phi$ are holomorphic in a neighborhood of $\mathcal{U}_j \cup \Gamma_j$, are uniformly bounded on Γ_j, and converge at each point of $\mathcal{U}_j$ to f_j as $n \to \infty$. Runge's theorem again allows F_{jn} to be replaced by a rational function with no poles on $\mathcal{U}_j \cup \Gamma_j$.

In the formula (4.1) we see that if $f \in \mathbf{A}(\Omega)$ then $f_j \in \mathbf{A}(\mathcal{U}_j \cup \Gamma_j)$, $j = 0,\ldots, m$. An argument just like that above shows that there is a sequence $\{p_{jn}\}_{n=1}^{\infty}$ of polynomials with $p_{jn}(1/(z - a_j)) \to f_j$ uniformly on Γ_j, and hence uniformly on Γ, for $1 \leqslant j \leqslant m$, and a sequence $\{p_{0n}\}$ of polynomials such that $p_{0n}(z) \to f_0(z)$ uniformly on Γ_0 and hence uniformly on Γ. Then

$$q_n(z) = \sum_{j=1}^{m} p_{jn}\left(\frac{1}{z - a_j}\right) + p_{0n}(z),$$

is an element of $\mathbf{R}(\Omega)$ and $q_n \to f$ uniformly on Γ.

Proposition 4.3. If $u \in L^1(\Gamma, ds)$ and

$$0 = \int_\Gamma \frac{u(\zeta)}{\zeta - z}\, d\zeta, \qquad z \notin \Gamma \tag{4.3}$$

then $u = 0$ a.e. ds.

Proof. Write

$$g_j(z) = \int_{\Gamma_j} \frac{u(\zeta)}{\zeta - z}\, d\zeta, \qquad z \notin \Gamma_j, \qquad j = 0,\ldots, m$$

Then g is holomorphic off Γ_j, $g_j(\infty) = 0$, and from (4.3) we find $g_0 + \cdots + g_m \equiv 0$ off Γ. But then each g_j is actually analytic over Γ_j since g_k is analytic over Γ_j for all $k \neq j$. Hence, $g_j \equiv 0$. Now we concentrate on one particular g_j, say g_0. By Runge's theorem and the fact that $g_0 \equiv 0$ we know that

$$\int_{\Gamma_0} u(\zeta) h(\zeta)\, d\zeta = 0$$

if h is analytic in some neighborhood of Γ_0. Since Γ_0 is analytic,

$$\int_{\mathbf{T}} u(\phi(e^{it})) H(e^{it})\, dt = 0 \tag{4.4}$$

if H is analytic in some neighborhood of $\mathbf{T}$; here ϕ is a function holomorphic and one-to-one in some neighborhood of $\mathbf{T}$, mapping $\mathbf{T}$ onto Γ_0. But (4.4) clearly implies $u \circ \phi = 0$ a.e. dt since we can take $H(e^{it}) = e^{int}$ for $n = 0, \pm 1, \pm 2, \ldots$. Hence, $u = 0$ a.e. ds on Γ_0. In a similar way we can show $u = 0$ on each Γ_j.

Let z_0 be the point in Ω at which the $H^p(\Omega)$ norm is determined and let ω be harmonic measure on Γ for z_0. We are now ready for the major result on the boundary behavior of an $H^p(\Omega)$ function.

Theorem 4.4. *Each $f \in H^p(\Omega)$ has boundary values f^* almost everywhere $(d\omega)$ on Γ and $f^* \in L^p(\Gamma, \omega)$. Further, we have both*

$$f(z) = \frac{1}{2\pi i}\int_\Gamma \frac{f^*(w)}{w - z}\,dw, \qquad z \in \Omega \tag{4.5a}$$

$$0 = \int_\Gamma \frac{f^*(w)}{w - z}\,dw, \qquad z \notin \Omega \cup \Gamma \tag{4.5b}$$

and

$$f(z) = \int_\Gamma f^*(\zeta)\,d\omega_z(\zeta), \qquad z \in \Omega \tag{4.6}$$

Finally, the mapping $f \mapsto f^*$ is an isometry of $H^p(\Omega)$ onto a closed subspace of $L^p(\Gamma, \omega)$.

Proof. Because of (4.1) it clearly suffices to prove that f_j has boundary values a.e. ds on Γ and that these boundary values lie in $L^p(\Gamma, \omega)$. There is nothing to show if we look at Γ_k when $k \neq j$ since f_j is actually analytic on Γ_k. Let ϕ be the Riemann mapping of $\mathcal{U}_j$ onto Δ. Then ϕ extends to be analytic and conformal on a neighborhood of $\mathcal{U}_j \cup \Gamma_j$ since Γ_j is analytic. Further, $g_j = f_j \circ \phi^{-1}$ lies in $H^p(\Delta)$ and so g_j has boundary values a.e. $d\theta$ on $\mathbf{T}$ and $g_j^* \in L^p(\mathbf{T}, d\theta)$. Clearly then $f_j = g_j \circ \phi$ has boundary values a.e. ds on Γ_j and $f_j^* = g_j^* \circ \phi$ a.e. so that f_j^* is in $L^p(\Gamma_j, ds)$ and hence $f_j^* \in L^p(\Gamma, \omega)$.

If $z \in \Omega$, then

$$\begin{aligned} f_j(z) &= g_j(\phi(z)) \\ &= \frac{1}{2\pi i}\int_{|\xi|=1} \frac{g_j^*(\xi)}{\xi - \phi(z)}\,d\xi \end{aligned}$$

Write $\xi = \phi(\zeta)$ so that

$$\begin{aligned} f_j(z) &= \frac{1}{2\pi i}\int_{\Gamma_j} \frac{f_j^*(\zeta)}{\phi(\zeta) - \phi(z)}\phi'(\zeta)\,d\zeta \\ &= \frac{1}{2\pi i}\int_{\Gamma_j} \frac{f_j^*(\zeta)}{\zeta - z}[1 + (\zeta - z)S(\zeta)]\,d\zeta \end{aligned}$$

where S is holomorphic in a neighborhood of $\Omega \cup \Gamma$ and $S(z) = 0$. It is immediate that

$$0 = \int_{\Gamma_j} f_j^*(\zeta) S(\zeta)\, d\zeta$$

so that (4.5) follows for f_j since $\int_{\Gamma_k}[f_j(\zeta)/(\zeta - z)]\, d\zeta = 0$ if $k \neq j$. Next, to see (4.6) recall from Proposition 1.6.5 that

$$d\omega_z(\zeta) = \frac{i}{2\pi} Q_z'(\zeta)\, d\zeta$$

where $Q_z(\zeta) = g(\zeta; z) + ih(\zeta; z)$. Hence,

$$Q_z'(\zeta) = \frac{1}{z - \zeta} + S(\zeta) \tag{4.7}$$

where S is holomorphic in a neighborhood of $\Omega \cup \Gamma$. Consequently,

$$\begin{aligned}\int_\Gamma f^*(\zeta)\, d\omega_z(\zeta) &= \frac{-i}{2\pi}\int_\Gamma \frac{f^*(\zeta)}{\zeta - z}\, d\zeta + \frac{i}{2\pi}\int f^*(\zeta) S(\zeta)\, d\zeta \\ &= f(z)\end{aligned}$$

by (4.5). To show that the map is an isometry let $f \in \mathbf{R}(\Omega)$ and let $u(z)$ be the harmonic function on Ω given by

$$u(z) = \int_\Gamma |f(\zeta)|^p\, d\omega_z(\zeta) \tag{4.8}$$

Then $|f(z)|^p \leqslant u(z)$ by (4.6) But further, if v is any harmonic majorant of $|f|^p$ we have $u(x) = |f(x)|^p \leqslant \liminf\{v(z) : z \to x\}$ so that the harmonic function $v - u$ is non-negative at Γ and hence on all of Ω. Thus, the function u given by (4.8) is the least harmonic majorant of $|f|^p$, if $f \in \mathbf{R}(\Omega)$ and so in this case we have $\|f\|_{L^p(\Gamma, \omega)} = \|f\|_{H^p(\Omega)}$. Next, let $f \in H^p(\Omega)$ and let f_n be elements of $\mathbf{R}(\Omega)$ converging on $H^p(\Omega)$ to f. Then f_n converges to f uniformly on compact subsets of Ω. Further,

$$\|f_n - f_m\|_{H^p(\Omega)} = \|f_n - f_m\|_{L^p(\Gamma, \omega)}$$

by the foregoing, so that $\{f_n\}$ is a Cauchy sequence in $L^p(\Gamma, \omega)$. If $f_n \to g$ in $L^p(\Gamma, \omega)$, then clearly

$$f(z) = \int_\Gamma g(\zeta)\, d\omega_z(\zeta), \qquad z \in \Omega$$

since all the harmonic measures are boundedly mutually absolutely continuous. Further, we have $f_n \to g$ in $L^p(\Gamma, ds)$ so that

$$2\pi i f(z) = \int_\Gamma \frac{g(\zeta)}{\zeta - z} d\zeta, \qquad z \in \Omega$$

and

$$0 = \int_\Gamma \frac{g(\zeta)}{\zeta - z} d\zeta = \int_\Gamma \frac{f^*(\zeta)}{\zeta - z} d\zeta \quad \text{if} \quad z \notin \Omega \cup \Gamma$$

Proposition 4.3 now implies that $g = f^*$ a.e. $d\omega$. Consequently, $f_n \to f^*$ in $L^p(\Gamma, \omega)$ and we have

$$\|f^*\|_{L^p(\Gamma, \omega)} = \lim \|f_n\|_{L^p(\Gamma, \omega)} = \lim \|f_n\|_{H^p(\Omega)}$$
$$= \|f\|_{H^p(\Omega)}$$

This completes the proof if $1 \leqslant p < \infty$. If $f \in H^\infty(\Omega)$, then by (4.6) $|f(z)| \leqslant \|f^*\|_{L^\infty(\Gamma, \omega)}$ so that $\|f\|_{H^\infty(\Omega)} \leqslant \|f^*\|_{L^\infty(\Gamma, \omega)}$. On the other hand,

$$\|f\|_{H^\infty(\Omega)} \geqslant \limsup_{p \to \infty} \|f\|_{H^p(\Omega)} = \limsup_{p \to \infty} \|f^*\|_{L^p(\Gamma, \omega)}$$
$$= \|f^*\|_{L^\infty(\Gamma, \omega)}$$

Corollary 4.5. *If $f \in H^p(\Omega)$, $1 \leqslant p < \infty$, then*

$$u_f(z) = \int_\Gamma |f^*(\zeta)|^p \, d\omega_z(\zeta), \qquad z \in \Omega$$

is the least harmonic majorant of $|f(z)|^p$ on Ω.

Corollary 4.6. *If $f \in H^1(\Omega)$, $f \not\equiv 0$, then $\log|f^*(\zeta)|$ is in $L^1(\Gamma, \omega)$ and*

$$\log|f(z)| \leqslant \int_\Gamma \log|f^*(\zeta)| \, d\omega_z(\zeta), \qquad z \in \Omega$$

Proof. Suppose first that $f \in \mathbf{R}(\Omega)$. The subharmonic function $\log|f(z)|$ and the harmonic function

$$w_\varepsilon(z) = \int_\Gamma \log(|f(\zeta)| + \varepsilon) \, d\omega_z(\zeta)$$

satisfy

$$\limsup\{\log|f(z)| - w_\varepsilon(z) : z \to x, x \in \Gamma\} \leqslant 0$$

and hence

$$\log|f(z)| \leqslant w_\varepsilon(z) = \int_\Gamma \log(|f(\zeta)| + \varepsilon)\, d\omega_z(\zeta)$$

for each $\varepsilon > 0$; let ε decrease to 0. In general, let $\{f_n\}$ be a sequence of elements of $\mathbf{R}(\Omega)$ with $f_n \to f$ both in $L^1(\Gamma, \omega)$ and almost everywhere $(d\omega)$. Then $f_n \to f$ uniformly on compact sets in Ω and for $\varepsilon > 0$ we have

$$\begin{aligned}\log|f(z)| &= \lim \log|f_n(z)| \\ &\leqslant \limsup \int_\Gamma \log|f_n(\zeta)|\, d\omega_z(\zeta) \\ &\leqslant \limsup \int_\Gamma \log(|f_n(\zeta)| + \varepsilon)\, d\omega_z(\zeta) \\ &\leqslant \int_\Gamma \log(|f^*(\zeta)| + \varepsilon)\, d\omega_z(\zeta)\end{aligned}$$

Now just let ε decrease to 0.

We see from Theorem 4.4 that $H^p(\Omega)$ is isometrically isomorphic to a closed subspace of $L^p(\Gamma, \omega)$. What is needed is a way to identify which $L^p(\Gamma, \omega)$ functions are boundary values of $H^p(\Omega)$ functions and the next theorem provides exactly that.

Theorem 4.7. *Let $f \in L^p(\Gamma, \omega)$, $1 \leqslant p \leqslant \infty$. There is an $F \in H^p(\Omega)$ with $F^* = f$ a.e. ω if and only if*

$$0 = \int_\Gamma \frac{f(\zeta)}{\zeta - w}\, d\zeta \quad \text{for all } w \notin \Omega \cup \Gamma \tag{4.9}$$

Proof. Suppose (4.9) holds. Set

$$F(z) = \int_\Gamma f(\zeta)\, d\omega_z(\zeta), \qquad z \in \Omega$$

Then F is certainly harmonic on Ω, and

$$|F(z)|^p \leqslant \int_\Gamma |f(\zeta)|^p\, d\omega_z(\zeta), \qquad 1 \leqslant p < \infty$$

and

$$|F(z)| \leqslant \|f\|_\infty \quad \text{if} \quad p = \infty$$

so that $|F|^p$ has a harmonic majorant. Further, by (4.7)

$$d\omega_z(\zeta) = \frac{1}{2\pi i}\frac{d\zeta}{\zeta - z} + S(\zeta)\, d\zeta$$

where S is holomorphic in a neighborhood of $\Omega \cup \Gamma$. Thus,

$$F(z) = \frac{1}{2\pi i}\int_\Gamma \frac{f(\zeta)}{\zeta - z}\, d\zeta, \qquad z \in \Omega$$

so that F is actually analytic in Ω and hence $F \in H^p(\Omega)$. Finally, by Theorem 4.4 we know that F^* exists a.e. and

$$F(z) = \frac{1}{2\pi i}\int_\Gamma \frac{F^*(\zeta)}{\zeta - z}\, d\zeta, \qquad z \in \Omega$$

$$0 = \int_\Gamma \frac{F^*(\zeta)}{\zeta - z}\, d\zeta, \qquad z \notin \Omega \cup \Gamma$$

These facts and (4.9) again imply $F^* = f$ a.e. ds.
The converse is just (4.5b).

We shall also have need of a characterization of F^*, $F \in H^p(\Omega)$, in terms of the measure ω. Recall that if $g(z; z_0)$ is the Green's function for z_0 then the critical points of $g(z; z_0)$ are those points, exactly m in number, at which the (complex) derivative of $g(z; z_0) + ih(z; z_0)$ vanishes. Let

$$z_1^*, \ldots, z_m^* \text{ be the critical points of } g(z; z_0) \tag{4.10}$$

and

$$P(z) = \prod_1^m (z - z_j^*) \tag{4.11}$$

Theorem 4.8. *Let $f \in L^p(\Gamma, \omega)$. Then*

$$0 = \int_\Gamma f(\zeta)h^*(\zeta)\, d\omega(\zeta), \qquad \text{all } h \in H^\infty(\Omega), \qquad h(z_0) = 0 \tag{4.12}$$

if and only if there is an $F \in H^p(\Omega)$ such that

$$f = \frac{F^*}{P} \quad \text{a.e. } d\omega \text{ on } \Gamma \tag{4.13}$$

Proof. Suppose (4.13) holds. If $h \in H^\infty(\Omega)$ and $h(z_0) = 0$, then

$$-2\pi i\int_\Gamma fh^*\, d\omega = \int_\Gamma \frac{F^*(\zeta)h^*(\zeta)}{P(\zeta)} Q'(\zeta)(\zeta - z_0)\frac{d\zeta}{\zeta - z_0}$$

where $Q(z) = g(z; z_0) + ih(z; z_0)$. However, the function

$$W(z) = \frac{Q'(z)(z - z_0)}{P(z)}$$

is analytic and single-valued on a neighborhood of $\Omega \cup \Gamma$ so that

$$-\int_\Gamma fh^* \, d\omega = \frac{1}{2\pi i}\int_\Gamma F^*(\zeta)h^*(\zeta)W(\zeta)\frac{d\zeta}{\zeta - z_0}$$

$$= F(z_0)W(z_0)h(z_0) = 0$$

Conversely, if (4.12) holds, then again write $d\omega = (i/2\pi)Q'(\zeta)\,d\zeta$ so that for all $h \in H^\infty(\Omega)$ we have

$$0 = \int_\Gamma f^*(\zeta)h^*(\zeta)(\zeta - z_0)Q'(\zeta)\,d\zeta$$

$$= \int_\Gamma f^*(\zeta)h^*(\zeta)S(\zeta)P(\zeta)\,d\zeta$$

where S is holomorphic on a neighborhood of $\Omega \cup \Gamma$ and zero-free there. The proof is completed by invoking Theorem 4.7.

4.5. *N* AGAIN

Let $Q_1, \ldots, Q_m$ be the m functions given in (2.2) and let **N** be their complex linear span. The letter **N** has also been employed in Section 3 of Chapter 2; this multiple use of **N** is deliberate and the connection between the two spaces will be made explicit later in this section. Let $H^p(\Gamma)$ denote the closed subspace of $L^p(\Gamma, \omega)$ consisting of boundary values of $H^p(\Omega)$ functions and let $H_0^p(\Gamma)$ be those $H^p(\Gamma)$ functions f with $f(z_0) = 0 = \int_\Gamma f\,d\omega$.

We begin by showing that for $f \in H^p(\Gamma)$ and $u \in \mathbf{N}$,

$$\int_\Gamma fu\,d\omega = 0 \tag{5.1}$$

This is straightforward. There is no loss in assuming $u = Q_j$ for some j; then (5.1) holds for $f \in \mathbf{R}(\Omega)$ and hence for $f \in H^p(\Gamma)$ by Proposition 4.2.

Now let $\overline{H_0^p}(\Gamma)$ denote the complex conjugates of the elements of $H_0^p(\Gamma)$. We now show that

$$H^p(\Gamma) + \overline{H_0^p}(\Gamma) + \mathbf{N} \text{ is dense in } L^p(\Gamma, \omega), \qquad 1 \leqslant p < \infty \tag{5.2}$$

For if $g \in L^{p'}(\Gamma, \omega)$ and g is orthogonal to $H^p(\Gamma)$ and to the complex conjugate of $H^p(\Gamma)$, then the real part of g is also orthogonal to $H^p(\Gamma)$, as is the imaginary part. Likewise, both the real and the imaginary parts of g are still orthogonal to **N**, since **N** has a basis of real functions. So it suffices to consider the case when g is real. But if g is real, then by Theorem 2.3

$$g = \sum_{1}^{m} a_j Q_j \in \mathbf{N}$$

where $a_1, \ldots, a_m$ are real. Since g is also orthogonal to **N**, we must have $g = 0$.

In the special case when $p = 2$ each of the terms in (5.2) is closed and orthogonal to the others so that

$$H^2(\Gamma) \oplus \overline{H_0^2(\Gamma)} \oplus \mathbf{N} = L^2(\Gamma, \omega) \tag{5.3}$$

Of course, the proof just given shows that

$$H^\infty(\Gamma) + \overline{H_0^\infty(\Gamma)} + \mathbf{N} \text{ is weak-* dense in } L^\infty(\Gamma, \omega) \tag{5.4}$$

Let $P(z) = \prod_1^m (z - z_j^*)$ be the function given in (4.11); $z_1^*, \ldots, z_m^*$ are the critical points of the Green's function with pole at z_0.

Theorem 5.1. $P(\mathbf{N} + H^\infty(\Gamma)) = H^\infty(\Gamma)$.

Proof. Clearly, $P \cdot H^\infty(\Gamma) \subset H^\infty(\Gamma)$. Also if $u \in \mathbf{N}$ then Theorem 4.8 implies $Pu \in H^\infty(\Gamma)$ and hence

$$P(\mathbf{N} + H^\infty(\Gamma)) \subset H^\infty(\Gamma)$$

Now if $F \in H_0^\infty(\Gamma)$, then Theorem 4.8 also implies that $u = F/P$ is orthogonal to $H_0^\infty(\Gamma)$, but we do not immediately get $u \in \mathbf{N}$. Let us show instead that

$$P(\mathbf{N} \oplus H^2(\Gamma)) = H^2(\Gamma) \tag{5.5}$$

Now (5.5) will be true by (5.3) if we can establish that the closed subspace

$$\frac{H^2(\Gamma)}{P} = \left\{ \frac{f}{P} : f \in H^2(\Gamma) \right\}$$

is orthogonal to $\overline{H_0^2(\Gamma)}$. However, Theorem 4.8 says exactly that. Hence, if $f \in H^\infty(\Gamma)$, then there is an $h \in H^2(\Gamma)$ and a $u \in \mathbf{N}$ with

$$f = P(u + h)$$

But then $|h| \leqslant |f|/|P| + |u|$, a quantity which is bounded on Γ and so $h \in H^\infty(\Gamma)$, as desired.

Let us now make specific the connection between the functions $Q_1, \dots, Q_m$ and the space **N** defined in Section 3 of Chapter 2.

Let T be the uniformizer of Ω and $\mathfrak{G}$ the group of linear fractional transformations of Δ to itself that fix T; $\mathfrak{G}$ is isomorphic to $\pi_1(\Omega)$. Let $z_0 = T(0)$ and let $\gamma_1, \dots, \gamma_m$ be smooth simple closed curves in Ω each beginning and ending at z_0 such that for each j, $\mathfrak{U}_j$ lies in the bounded component of the complement of γ_j and $\cup_{k \neq j} \mathfrak{U}_k$ lies in the unbounded component of the complement of γ_j. Let $C_1, \dots, C_m$ be the lifts to Δ of $\gamma_1, \dots, \gamma_m$, respectively, so that $C_1, \dots, C_m$ are curves in Δ (not closed) which begin at 0 and which end at $a_1, \dots, a_m$, respectively. Let $h_1, \dots, h_m$ be the elements of $\mathfrak{G}$ with $h_j(0) = a_j$, $j = 1, \dots, m$; then $h_1, \dots, h_m$ are a basis for $\mathfrak{G}$ since $\gamma_1, \dots, \gamma_m$ are a basis for $\pi_1(\Omega, z_0)$. Let

$$H_j = \mathfrak{E}\left(-{}^*P_{a_j}\right), \qquad 1 \leqslant j \leqslant m \tag{5.6}$$

where $\mathfrak{E}$ is the conditional expectation operator defined in Section 3 of Chapter 2. Finally, let $u = v \circ T$ where v is real and continuous on Γ. Then

$$\int_{\mathbf{T}} H_j u \, d\sigma = -\int_{\mathbf{T}} {}^*P_{a_j} u \, d\sigma, \qquad d\sigma = \frac{1}{2\pi} d\theta$$

$$= \int_{\mathbf{T}} P_{a_j} {}^*u \, d\sigma$$

$$= {}^*\tilde{u}(a_j)$$

However, $\tilde{u} = \tilde{v} \circ T$ where $\tilde{v}$ is a real-valued bounded harmonic function on Ω. Analytic continuation of the function element $\tilde{v} + i\,{}^*\tilde{v}$ along γ_j, beginning at z_0 corresponds to analytic continuation of $\tilde{u} + i\,{}^*\tilde{u}$ along C_j, beginning at 0. Hence,

$$\begin{aligned} \text{period of } {}^*\tilde{v} \text{ along } \gamma_j &= \text{change in } {}^*\tilde{u} \text{ along } C_j \\ &= {}^*\tilde{u}(a_j) - {}^*\tilde{u}(0) \\ &= {}^*\tilde{u}(a_j) \\ &= \int_{\mathbf{T}} H_j u \, d\sigma \end{aligned}$$

But the period of ${}^*\tilde{v}$ along γ_j is also given by $\int_\Gamma v Q_j \, d\omega_{z_0}$ and hence

$$\int_\Gamma v Q_j \, d\omega_{z_0} = \int_{\mathbf{T}} H_j v \circ T \, d\sigma \tag{5.7}$$

Now recall (2.4.1):

$$\int_{\mathbf{T}} u \circ T \, d\sigma = \int_{\Gamma} u \, d\omega, \qquad u \in L^1(\Gamma, \omega)$$

Combined with (5.7) we see that the invariant function $H_j \in L^\infty/\mathfrak{G}$ is exactly the lift to $\mathbf{T}$ of Q_j. In other words, the space $\mathbf{N}$, defined on Γ as the linear span of $Q_1, \dots, Q_m$, when lifted to $\mathbf{T}$ by the uniformizer T produces exactly the space $\mathbf{N}$ on $\mathbf{T}$ defined as the linear span of $H_1, \dots, H_m$.

Let us rephrase Theorem 5.1 in this light. Let $\Phi = P \circ T$; then we have from Theorem 5.1

$$\Phi(H^\infty/\mathfrak{G} + \mathbf{N}) = H^\infty/\mathfrak{G} \tag{5.8}$$

where we are viewing all the functions on $\mathbf{T}$. Formula (5.8) is one element in the proof of the next result.

Theorem 5.2. *There is a bounded linear projection $\mathcal{P}$ from $H^\infty = H^\infty(\Delta)$ onto $H^\infty/\mathfrak{G}$ satisfying*

$$\mathcal{P}(fg) = f\mathcal{P}(g), \qquad f \in H^\infty/\mathfrak{G}, \qquad g \in H^\infty \tag{5.9}$$

Proof. We begin by showing that

$$\mathcal{E}(H^\infty) = H^\infty/\mathfrak{G} + \mathbf{N} \tag{5.10}$$

Here, as usual, H^∞ is $H^\infty(\Delta)$ viewed on $\mathbf{T}$ and $H^\infty/\mathfrak{G}$ consists of those elements of H^∞ which are invariant under $\mathfrak{G}$. To see that $\mathcal{E}(H^\infty) \subset H^\infty/\mathfrak{G} + \mathbf{N}$, let $f \in H^\infty$. Then for any $g \in H_0^2/\mathfrak{G}$ we have

$$\int_{\mathbf{T}} \mathcal{E}(f) g \, d\sigma = \int_{\mathbf{T}} \mathcal{E}(fg) \, d\sigma = \int_{\mathbf{T}} fg \, d\sigma = 0$$

Hence, $\mathcal{E}(f)$ lies in $\mathbf{N} \oplus H^2/\mathfrak{G}$ by (2.3.14). But $\mathcal{E}(f)$ is bounded and so is every element of $\mathbf{N}$ so that $\mathcal{E}(f)$ actually lies in $H^\infty/\mathfrak{G} + \mathbf{N}$. To see the reverse inclusion we only need show that $\mathbf{N} \subset \mathcal{E}(H^\infty)$. Let $h \in \mathfrak{G}$ and put $a = h(0)$. For $u \in L^2/\mathfrak{G}$ we find

$$\begin{aligned} \int_{\mathbf{T}} u\mathcal{E}(P_a) \, d\sigma &= \int_{\mathbf{T}} \mathcal{E}(uP_a) \, d\sigma = \int_{\mathbf{T}} uP_a \, d\sigma \\ &= \tilde{u}(a) = \tilde{u}(h(0)) = \tilde{u}(0) \\ &= \int_{\mathbf{T}} u \, d\sigma \end{aligned}$$

Hence, $\mathcal{E}(P_a) = 1$ so that we can write

$$iv_h = \mathcal{E}(P_a + i\,{}^*P_a - 1) \in \mathcal{E}(H^\infty)$$

This finishes the proof of (5.10). Combining (5.8) and (5.10) we have

$$\Phi\mathcal{E}(H^\infty) = \Phi(H^\infty/\mathfrak{G} + \mathbf{N}) = H^\infty/\mathfrak{G} \tag{5.11}$$

Select an element $g_0 \in H^\infty$ with

$$\Phi\mathcal{E}(g_0) = 1 \tag{5.12}$$

The linear projector $\mathcal{P}$ is defined by

$$\mathcal{P}(f) = \mathcal{E}(f\Phi g_0) = \Phi\mathcal{E}(fg_0), \qquad f \in H^\infty \tag{5.13}$$

Then the range of $\mathcal{P}$ is in $H^\infty/\mathfrak{G}$ by (5.11) and, if $f \in H^\infty/\mathfrak{G}$, we have

$$\mathcal{P}(f) = \Phi\mathcal{E}(fg_0) = f\Phi\mathcal{E}(g_0) = f$$

by (5.12). Finally, if $f \in H^\infty/\mathfrak{G}$ and $g \in H^\infty$ then

$$\mathcal{P}(fg) = \Phi\mathcal{E}(fgg_0) = \Phi f\mathcal{E}(gg_0) = f\mathcal{P}(g)$$

The next result is just an observation about the projection $\mathcal{P}$ given in (5.13); its proof is simple.

Corollary 5.3. *Let $\mathcal{P}$ be defined by* (5.13). *Then $\mathcal{P}$ actually projects H^p onto $H^p/\mathfrak{G}$ for $1 \leqslant p \leqslant \infty$ and*

$$\mathcal{P}(fg) = f\mathcal{P}(g) \quad \text{if } f \in H^r/\mathfrak{G}, \qquad g \in H^s, \qquad \frac{1}{r} + \frac{1}{s} = \frac{1}{p} \tag{5.14}$$

4.6. FUNCTIONS WITH PERIODS

If u is a real-valued harmonic function on Ω, then its harmonic conjugate ${}^*u$ has periods around each Γ_j. There are many times when it is highly advantageous to "correct" these periods; that is, to add some function u_1 to u so that ${}^*(u + u_1)$ is single-valued, and hence $u + u_1 + i\,{}^*(u + u_1)$ is analytic on Ω. There are, of course, several ways to do this. The simplest is the device employed in Section 2: take

$$u_1(z) = \sum_1^m r_j \log|z - a_j| \tag{6.1}$$

for an appropriate choicc of real scalars $r_1,\ldots, r_m$. However, if u is required to have some particular value on a segment in $\partial\Omega$, then the addition of a u_1 of the form (6.1) will surely spoil this. We shall show in this section several ways to correct the periods and not do "too much" damage to the values of u on Γ. We shall also show that the class of multiple-valued analytic functions with single-valued modulus is naturally related to this question and to a class of "automorphic" functions on Δ.

Proposition 6.1. Let E be a measurable set in Γ of positive harmonic measure (equivalently, of positive length) and let Λ consist of all m-tuples of the form

$$\left(\int_\Gamma uQ_1\,d\omega,\ldots,\int_\Gamma uQ_m\,d\omega\right),\quad u\in L^\infty(\Gamma),\quad 0\leqslant u\leqslant 1 \text{ on } E,\quad u=0 \text{ off } E \tag{6.2}$$

where $Q_1,\ldots, Q_m$ are the functions given in (2.2) and $\omega=\omega_{z_0}$. Then Λ has interior in $\mathbb{R}^m$.

Proof. Λ is a (closed) convex set in $\mathbb{R}^m$; if it fails to have interior then it lies in a hyperplane through the origin (since $0\in\Lambda$). That is, there are real scalars $r_1,\ldots, r_m$, not all zero, with

$$0=\sum_1^m r_j\int uQ_j\,d\omega=\int uQ\,d\omega,\qquad 0\leqslant u\leqslant 1\quad\text{on } E$$

where $Q=\Sigma_1^m r_jQ_j$. Thus, Q vanishes on E which contradicts Proposition 2.4.

Corollary 6.2. *If E has interior and Λ is formed with C^∞ functions which vanish off E, the conclusion of Proposition* 6.1 *holds.*

It is a consequence of Proposition 6.1 and Corollary 6.2 that there is a constant M depending only on Ω and E with this property:

> Given $(r_1,\ldots, r_m)\in\mathbb{R}^m$ there is a $u\in L^\infty(E)$ (or $u\in \mathbf{C}^\infty(\Gamma)$ and vanishing off E if E has interior) such that $0\leqslant u\leqslant M$ on E and (6.3)

$$\int_\Gamma uQ_j\,d\omega\equiv r_j\,(\mathrm{mod}\,2\pi)\quad\text{for } 1\leqslant j\leqslant m$$

We shall have much use for (6.3) in the succeeding sections.

If μ is a real measure on Γ we define the harmonic extension of μ to Ω by the rule

$$\tilde{\mu}(z)=\int_\Gamma H_z(x)\,d\mu(x) \tag{6.4}$$

where H_z is the Radon-Nikodym derivative of ω_z with respect to ω. This definition is consistent with the usual one in the case when $d\mu = u\, d\omega$, $u \in \mathbf{C}_r(\Gamma)$. We can find the periods of $*\tilde{\mu}$ about Γ_j, $1 \leqslant j \leqslant m$, by the simple device

$$\text{period of } *\tilde{\mu} \text{ about } \Gamma_j = \int Q_j\, d\mu, \qquad 1 \leqslant j \leqslant m$$

where Q_j is given by (2.2). This follows by choosing real-valued continuous functions u_n such that the measures $\{u_n\, d\omega\}$ converge weak-* to $d\mu$. If we specialize to the case when μ is a single point mass at x, then we have the m-tuple of periods

$$\pi(x) = (Q_1(x), \ldots, Q_m(x))$$

Let E be a set in Γ with no interior, Λ be the convex hull of the points $\{\pi(x) : x \in \Gamma \setminus E\}$. We show now that Λ contains a neighborhood of the origin. If this is not so, then there is a nonzero vector $r = (r_1, \ldots, r_m) \in \mathbb{R}^m$ with

$$0 \leqslant r \cdot \pi(x), \qquad x \in \Gamma \setminus E$$

Equivalently,

$$0 \leqslant \sum_1^m r_j Q_j(x), \qquad x \in \Gamma \setminus E$$

and hence $\Sigma\, r_j Q_j$ is non-negative on all of Γ since E has no interior. However, $\int_\Gamma Q_j\, d\omega = 0$ for all j so that we must have $\Sigma\, r_j Q_j \equiv 0$. But this contradicts the linear independence of $Q_1, \ldots, Q_m$. Hence, we conclude that 0 is interior to Λ and hence interior to the convex hull of $m + 1$ points of Λ by Carathéodory's theorem. We have thus proved this result.

Proposition 6.3. Let E be a set in Γ with no interior. Then there are $m + 1$ points $x_0, \ldots, x_m$ in $\Gamma \setminus E$ such that each m-tuple $(r_1, \ldots, r_m)$ of real numbers has the form

$$r_j = \sum_{k=0}^m c_k Q_j(x_k), \qquad j = 1, \ldots, m \tag{6.5}$$

for some choice of non-negative scalars $c_0, \ldots, c_m$. There is a constant M with the property that if $\Sigma_1^m r_j^2 \leqslant 1$, then the $c_0, \ldots, c_m$ can be chosen with $\Sigma_0^m c_j^2 \leqslant M$.

Let us now spend a moment investigating the harmonic extension of a single point mass to Ω.

Proposition 6.4. Let $x \in \Gamma$ and let $u(z)$ be the harmonic extension to Ω of a point mass at x; that is, $u(z) = H_z(x)$ where H_z is the Radon-Nikodym derivative of ω_z with respect to ω. Then u is a positive harmonic function on Ω, and u is harmonic over $\Gamma \setminus \{x\}$ and vanishes identically on $\Gamma \setminus \{x\}$. Further, in a sufficiently small neighborhood $\mathcal{O}$ of x there is a nonvanishing analytic function g such that $u(z) = \mathrm{Re}(g(x)/(z - x))$ for $z \in \mathcal{O} \cap \Omega$.

Proof. Clearly u is a positive harmonic function on Ω. Let $y \in \Gamma$, $y \neq x$, and let $\mathcal{D}$ be a small disc centered at y. Let $\{I_n\}$ be a nested sequence of closed segments in Γ centered at x with $\omega(I_n) = 1/n$ and let u_n be the harmonic extension to Ω of the function which is n on I_n and 0 on $\Gamma \setminus I_n$. Then $\{u_n\}$ converges uniformly on compact subsets of Ω to u; each u_n is harmonic in $\mathcal{D}$ by the reflection principle. Suppose, with no loss of generality, that $x \in \Gamma_0$. Let v_n be the harmonic function on $\mathcal{U}_0$ which is zero on all of Γ_0 except on I_n where it is identically n. Then the maximum principle implies that on Ω we must have $v_n > u_n$. However, as $n \to \infty$ we know that $\{v_n\}$ converges uniformly on compact subsets of $\mathcal{U}_0 \cup \Gamma_0 \setminus \{x\}$ to a positive harmonic function v and $v = 0$ on $\Gamma_0 \setminus \{x\}$. Thus, $\{u_n\}$ remains uniformly bounded on $\mathcal{D}$ and so at least a subsequence converges uniformly on compact subsets of $\mathcal{D}$ to a harmonic function which must be u in $\mathcal{D} \cap \Omega$ and which must vanish on $\mathcal{D} \cap \Gamma$. Thus, the first and second assertions of Proposition 6.4 are proved.

To prove the last assertion let $\mathcal{V}$ be a domain in Ω bounded by a piecewise analytic simple closed curve such that a segment α of Γ centered at x lies in $\partial\mathcal{V}$. Let $*u$ be the harmonic conjugate of u in $\mathcal{V}$. Let ϕ be the Riemann mapping of Δ onto $\mathcal{V}$ with $\phi(1) = x$. Now $h = \exp[-u - i*u]$ is a zero-free holomorphic function $\mathcal{V}$ bounded by 1 and h is analytic over $\alpha \setminus \{x\}$ and has unit modulus there. Thus, $f = h \circ \phi$ is a zero-free holomorphic function on Δ which is bounded by one on Δ, holomorphic across some arc of $\mathbf{T}$ centered at 1 except possibly at the point 1 itself, and has unit modulus on this arc, again except possibly at 1. If f were holomorphic at 1 then its value there would have modulus one and we could conclude u was continuous at x and vanished there. Thus, u would be identically zero, a contradiction. Hence, f is not holomorphic at 1 and so the factorization theorem for $H^\infty(\Delta)$ implies that f has a singular factor with mass at the point 1:

$$f(z) = H(z)\exp\left[a\frac{z+1}{z-1}\right]$$

where $a > 0$ and H is holomorphic over $\mathbf{T}$ in a neighborhood of 1 and has unit modulus at 1. Unwinding the composition with ϕ we see that the final conclusion holds.

We employ the results of Propositions 6.3 and 6.4 to prove a theorem about inner functions in $\mathbf{A}(\Omega)$.

Definition. A function ϕ in $\mathbf{A}(\Omega)$ is *inner* if $|\phi| \equiv 1$ on $\Gamma = \partial\Omega$.

Theorem 6.5. *Let $x_1, \ldots, x_N$ be distinct points in Γ and let $\lambda_1, \ldots, \lambda_N$ be points of* $\mathbf{T}$. *Then there is an inner function ϕ in* $\mathbf{A}(\Omega)$ *with $\phi(x_j) = \lambda_j$, $j = 1, \ldots, N$.*

Proof. Let δ_j be the unit point mass at x_j, $j = 2, \ldots, N$; by Proposition 6.3 there are $m + 1$ points $z_0, \ldots, z_m$ in $\Gamma \setminus \{x_1, \ldots, x_N\}$ and non-negative scalars $c_0, \ldots, c_m$ such that

$$\sum_{j=2}^{N} Q_k(x_j) + \sum_{j=0}^{m} c_j Q_k(z_j) = 0, \qquad k = 1, \ldots, m$$

That is, the measure which places mass 1 at each of $x_2, \ldots, x_N$ and mass $c_0, \ldots, c_m$ at $z_0, \ldots, z_m$, respectively, when extended to a harmonic function u on Ω has the property that $^*u$ is single-valued. Set $g = u + i\,^*u$ and note that g is analytic on Ω, g maps Ω into the right half-plane, g is analytic over $\Gamma \setminus \{x_2, \ldots, x_N, z_0, \ldots, z_m\}$ and is pure imaginary there and, finally, from the last assertion of Proposition 6.4, g has a pole of order 1 at $x_2, \ldots, x_N$ and at each z_j for which $c_j \neq 0$. Define $h = (1 - g)(1 + g)^{-1}$. Then h is indeed an element of $\mathbf{A}(\Omega)$, h is inner, and $h(x_j) = -1$ for $j = 2, \ldots, N$ while $h(x_1)$ is not -1. Composing h with an appropriate linear fractional transformation of Δ onto Δ we can find an inner function ϕ_1 in $\mathbf{A}(\Omega)$ with $\phi_1(x_1) = \lambda_1$, $\phi_1(x_j) = 1$, $j = 2, \ldots, N$. Repeat this construction for $x_2, \ldots, x_N$, obtaining inner functions $\phi_1, \phi_2, \ldots, \phi_N$ in $\mathbf{A}(\Omega)$ with

$$\phi_k(x_j) = \begin{cases} \lambda_k & j = k \\ 1 & j \neq k \end{cases}, \qquad k = 1, \ldots, N.$$

Then $\phi = \phi_1 \cdots \phi_N$ is the desired inner function in $\mathbf{A}(\Omega)$.

There is an important class of functions (albeit, multiple-valued) on Ω which is directly involved with the issue of periods. An illustration will help to set the ideas. If u is a real-valued harmonic function on Ω with harmonic conjugate $^*u$, then we have already noted in Section 3 of Chapter 2 that $u + i\,^*u$ gives rise to a homomorphism of $\pi_1(\Omega, z_0)$ into $\mathbb{R}$ by means of analytic continuation of $u + i\,^*u$ along Γ_k, $k = 1, \ldots, m$. Equivalently, $f = \exp[u + i\,^*u]$ is a (multiple-valued) holomorphic function on Ω, $|f|$ is single-valued, and f is associated with a homomorphism of $\pi_1(\Omega, z_0)$ into $\mathbf{T}$ by computing the change in arg f along each Γ_k. By employing the uniformizer T we see that $F = f \circ T$ is single-valued and holomorphic on Δ and that $F \circ h = \chi(h)F$ for each $h \in \mathfrak{G}$, where $\chi(h)$ is a unimodular constant (that is, a point of $\mathbf{T}$) with the property that $\chi(h_1 h_2) = \chi(h_1)\chi(h_2)$; that is, χ is a character of $\mathfrak{G}$ in the sense of topological groups. These considerations lead us to the following definitions and theorem.

Definition. Let Ω be a planar region with Δ as its universal covering surface. The class $MH^p(\Omega)$ consists of those functions f on Ω for which $|f|$ is single-valued, $|f|^p$ has a harmonic majorant if $0 < p < \infty$, or is bounded if $p = \infty$, and f is locally analytic in the sense that each point $w \in \Omega$ has a neighborhood $\mathfrak{U}$ and a single-valued holomorphic function $g_{\mathfrak{U}}$ on $\mathfrak{U}$ with $|g_{\mathfrak{U}}| = |f|$ on $\mathfrak{U}$.

Definition. Let T be the uniformizer of Ω and $\mathfrak{G}$ the associated group of linear fractional transformations. A function $F \in H^p(\Delta)$ is *modulus automorphic with respect to* $\mathfrak{G}$ if there is a homomorphism χ of $\mathfrak{G}$ into $\mathbf{T}$ such that $F \circ h = \chi(h)F$ for each $h \in \mathfrak{G}$.

Theorem 6.6. *A function f is in $MH^p(\Omega)$ if and only if there is a modulus automorphic $F \in H^p(\Delta)$ with*

$$F = f \circ T \tag{6.6}$$

Further, two elements f_1, f_2 of $MH^p(\Omega)$ satisfy $|f_1| \equiv |f_2|$ if and only if the corresponding F_1, F_2 satisfy $F_1 = \lambda F_2$ for some constant λ, $|\lambda| = 1$.

Proof. Let $f \in MH^p(\Omega)$ and let g be analytic near z_0 with $|g| = |f|$. Consider $g \circ T$ near $z = 0$ in Δ. Analytic continuation of this function element along any path in Δ is possible so that there is, by the monodromy theorem, a holomorphic function F on Δ with $|F| = |f \circ T|$. Thus, $F \in H^p(\Delta)$ since $|f|^p$ has a harmonic majorant (or is bounded if $p = \infty$). If γ is an element of $\pi_1(\Omega, z_0)$ and h is the corresponding element of $\mathfrak{G}$ [that is, $h(0) = a$ where a is the endpoint of the lifted curve $T^{-1}(\gamma)$] then analytic continuation of F along γ results in the value $F(z)\chi(\gamma)$, $z \in \mathfrak{U}$, where $|\chi(\gamma)| = 1$. Thus, $F(h(\zeta)) = \chi(\gamma)F(\zeta)$, ζ near 0, so that $F \circ h = \chi(\gamma)F$. The rule $\gamma \mapsto \chi(\gamma)$ is a homomorphism of $\pi_1(\Omega, z_0)$ into $\mathbf{T}$ and so F is modulus automorphic.

Conversely, if F is a given modulus automorphic function in $H^p(\Delta)$, then certainly there is a non-negative *single*-valued function u on Ω with $|F| = u \circ T$ since $|F|$ is invariant under $\mathfrak{G}$. If $w \in \Omega$, then w has a small neighborhood $\mathfrak{U}$ on which T is invertible so that $g(z) = F \circ T^{-1}(z)$, $z \in \mathfrak{U}$, is holomorphic and $|g \circ T| = |F|$; this means that $f = F \circ T^{-1}$ is in $MH^p(\Omega)$.

Suppose now that f_1, f_2 are in $MH^p(\Omega)$ and $|f_1| \equiv |f_2|$. Then $|F_1| \equiv |F_2|$ on Δ and so $F_1 = \lambda F_2$ for some unimodular constant λ. The converse is obvious.

Proposition 6.7. Suppose Ω is bounded by $m + 1$ disjoint analytic simple closed curves and $\Gamma = \partial\Omega$. Let $f \in MH^p(\Omega)$, $f \not\equiv 0$. Then

$$|f(z)| \text{ has boundary values a.e. } \omega \text{ on } \Gamma \tag{6.7a}$$

$$|f^*| \in L^p(\Gamma, \omega) \tag{6.7b}$$

$$\log|f^*| \text{ is in } L^1(\Gamma, \omega) \tag{6.7c}$$

$$\log|f(z)| \leqslant \int_\Gamma \log|f^*(\zeta)|\, d\omega_z(\zeta), \qquad z \in \Omega \tag{6.7d}$$

Proof. Let I be a small interval in Γ and let v be a smooth function supported on I such that

$$g = f \cdot \exp[\tilde{v} + i\,{}^*\tilde{v}]$$

is single-valued in Ω; such a function exists by Corollary 6.2. Then g has boundary values a.e. ω on Γ, $|g^*| \in L^p(\Gamma, \omega)$, and $\log|g^*|$ lies in $L^1(\Gamma, \omega)$. Hence, (6.7a)–(6.7c) hold. Suppose now that $|f|$ is continuous on $\Omega \cup \Gamma$. Then

$$u_\varepsilon(z) = \int_\Gamma \log(|f(\zeta)| + \varepsilon)\, d\omega_z(\zeta)$$

is harmonic on Ω, continuous on $\Omega \cup \Gamma$, and satisfies $u_\varepsilon(x) \geqslant \limsup\{\log|f(z)|: z \to x\}$ for all $x \in \Gamma$. Hence, $u_\varepsilon(z) \geqslant \log|f(z)|$; now let ε decrease to 0. In general, we can approximate g in $H^p(\Omega)$ by rational functions and hence $|f^*|$ in L^p by $MH^p(\Omega)$-functions f_n for which $|f_n|$ is continuous on $\Omega \cup \Gamma$. The inequality (6.7d) then follows by taking limits.

4.7. THE FACTORIZATION OF $H^p(\Omega)$ FUNCTIONS

In this section we show that the canonical factorization of a function in $H^p(\Delta)$ into the product of a Blaschke product, a singular inner function, and an outer function, carries over to $H^p(\Omega)$ and even to $MH^p(\Omega)$ if we are somewhat more liberal in our interpretation of what these specialized factors should be. We give the definitions a little later and begin with a theorem on the location of zeros of a (nontrivial) $H^p(\Omega)$ function.

Proposition 7.1. Let Ω be bounded by $m + 1$ disjoint analytic simple closed curves.

(a) Let $z_1, z_2, \ldots$ be points in Ω, not necessarily distinct, with no limit point in Ω. There is an $f \in H^\infty(\Omega)$ vanishing at each z_j and no where else on Ω if and only if

$$\sum_1^\infty g(\zeta; z_j) < \infty, \qquad \text{each } \zeta \in \Omega \tag{7.1}$$

(b) If $h \in H^p(\Omega)$, $h \not\equiv 0$, and if h has zeros at $z_1, z_2, \ldots$ in Ω, then (7.1) holds.

Proof. Suppose (7.1) holds for the points $\{z_j\}$. Let

$$u_n(\zeta) = \sum_{j=1}^n g(\zeta; z_j)$$

and let v_n be a non-negative smooth function supported on some fixed interval in Γ such that the periods of $*(u_n + v_n)$ about $\Gamma_1, \ldots, \Gamma_m$ are all integer multiples of 2π and, further, such that $0 \leqslant v_n \leqslant M$ for all n; such functions exist by (6.3). Let

$$C_n(z) = \exp\left[-(u_n + v_n) - i^*(u_n + v_n)\right] \tag{7.2}$$

Then C_n is a single-valued holomorphic function on Ω which vanishes exactly at $z_1, \ldots, z_n$ and which is continuous to Γ. If $\zeta \in \Omega$ and ζ is not one of the points $\{z_1, z_2, \ldots\}$, then

$$1 \geqslant |C_n(\zeta)| \geqslant \exp[-M] \exp\left[-\sum_1^\infty g(\zeta; z_j)\right]$$

for all n and so at least a subsequence of $\{C_n\}$ converges uniformly on compact subsets of Ω to an H^∞ function which vanishes precisely at the points $\{z_1, z_2, \ldots\}$.

We now prove (b) since this includes the converse of (a). Let C_n be the functions from (7.2) and let $f_n = f/C_n$. On Γ we have $\exp[-M] \leqslant |C_n(x)| \leqslant 1$, $x \in \Gamma$. Thus,

$$\|f\|_p \leqslant \|f_n\|_p \leqslant e^M \|f\|_p$$

and so

$$|f(\zeta)| \exp\left[\sum_1^n g(\zeta; z_j) - \tilde{v}_n(\zeta)\right] = |f_n(\zeta)|$$

$$\leqslant C\|f_n\|_p \leqslant C'\|f\|_p$$

which gives the desired conclusion, since $0 \leqslant \tilde{v}_n(\zeta) \leqslant M$.

Definition. If the points $\{z_j\}$ satisfy (7.1) then we define

$$B(z) = \exp\left[-\sum_1^\infty g(z; z_j) - i^*\left(\sum_1^\infty g(z; z_j)\right)\right]$$

and call B the *Blaschke product* for the points $\{z_j\}$.

Note that $|B(z)|$ is single-valued on Ω although $B(z)$ itself may not be. However, each point in Ω has a neighborhood $\mathcal{U}$ and a single-valued holomorphic function g on $\mathcal{U}$ with $|g| = |B|$ on $\mathcal{U}$. Thus, B is an example of an element of $MH^\infty(\Omega)$.

To state the factorization theorem we introduce the two other types of factors needed. A function $F \in MH^p(\Omega)$ is *outer* if

$$\log|F(z_0)| = \int_\Gamma \log|F^*|\, d\omega$$

A function $G \in MH^\infty(\Omega)$ is *inner* if $|G^*| = 1$ a.e. ω and G is *singular* if it is inner and $G \neq 0$ on Ω.

Theorem 7.3. *Let $f \in MH^p(\Omega)$. Then*

$$|f| = |B||S||F|$$

where B is a Blaschke product, S is a singular inner function in $MH^\infty(\Omega)$, and F is an outer function in $MH^p(\Omega)$. The factors are uniquely determined up to multiplying by constants of modulus one.

Proof. Let $z_1, z_2, \ldots$ be the zeros of f. A glance at the proof of Proposition 7.1(b) shows that (7.1) holds; let B be the Blaschke product for the points $\{z_j\}$ and let $g = f/B$. We need to know that $g \in MH^p(\Omega)$. Let

$$B_n(z) = \exp\left[-\sum_1^n g(z; z_j) - i^*\left(\sum_1^n g(z; z_j)\right)\right]$$

The least harmonic majorant of $|f/B_n|^p$ is exactly the least harmonic majorant for $|f|^p$ since $|B_n| = 1$ continuously on Γ. Letting $n \to \infty$ we see that the least harmonic majorant of $|g|^p$ is also the least harmonic majorant of $|f|^p$ and so $g \in MH^p(\Omega)$. Hence, the function $|g|$ has boundary values a.e. ω, $|g^*| \in L^p(\Gamma, \omega)$, and $u = \log|g^*|$ lies in $L^1(\Gamma, \omega)$. Let

$$G = \exp[\tilde{u} + i^*\tilde{u}]$$

Then $G \in MH^p(\Omega)$ since

$$|G(z)|^p = \exp[p\tilde{u}(z)]$$

$$\leqslant \int_\Gamma |g(\zeta)|^p\, d\omega_z(\zeta), \qquad z \in \Omega$$

Further, G is actually outer since

$$\log|G(z_0)| = \int_\Gamma u\, d\omega_{z_0} = \int_\Gamma \log|G^*(\zeta)|\, d\omega(\zeta)$$

Note that $|g(z)| \leqslant |G(z)|$ in Ω by (6.7d). To finish, set $S = g/G$. Then $|S| \leqslant 1$ by the foregoing, $S \in MH^\infty(\Omega)$, and $|S^*| = 1$ a.e. ω since $|g^*| = |G^*|$ a.e. ω on Γ. Thus, S is a zero-free inner function and so is singular.

Uniqueness is direct. If

$$|f| = |B_1||S_1||F_1| \quad \text{on } \Omega$$

where B_1 is a Blaschke product, S_1 is singular, and F_1 is outer, then clearly $|B| = |B_1|$ since B_1 must incorporate all the zeros of f. However, if we put $g_1 = f/B_1$, then we know $|F_1| = |g_1|$ a.e. ω on Γ and $|g_1| = |g| = |F|$ a.e. ω, as well. Thus, $|f_1| = |F|$ a.e. ω on Γ and so $|F| = |F_1|$ on Ω. This leaves $|S_1| = |S|$ on Ω. The maximum principle then implies $S_1 = \lambda_1 S$, $F_1 = \lambda_2 F$ for unimodular constants λ_1, λ_2.

ADDITIONAL READINGS AND NOTES

The defect of $\operatorname{Re}\mathbf{R}(\Omega)$ in $\mathbf{C}_r(\Gamma)$, the characterization of measures orthogonal to $\mathbf{R}(\Omega)$, and the existence of peak-interpolation sets in Γ are all easily handled in the setting of this chapter. If the boundary is less smooth and/or more complicated topologically, these same problems become more difficult. If $\mathbf{K}$ is a compact set whose complement has only a finite number of components, then much the same conclusions hold [in our setting, $\mathbf{K} = \text{CL}(\Omega)$]. The reader is referred to the book of Stout (1971) for the details. The whole idea of $\operatorname{Re}\mathbf{R}(\Omega)$ having finite defect in $\mathbf{C}_r(\Gamma)$ has been enormously generalized and abstracted, as has the idea of Hardy spaces; see Stout's book mentioned above and that of Gamelin (1969) and the references therein. Some aspects of the geometry of unit ball of the space of real annihilating measures are discussed in Ahern (1969). Proposition 3.2, which provides an easy route to the F. and M. Riesz theorem, I learned in a course given by Forelli. The basic properties of $H^p(\Omega)$ covered in Section 4 are set forth in Rudin (1955b). The material of Section 5 is largely from Forelli (1966), although he works almost entirely on Δ; in particular, he proved Theorem 5.2 and then used it to prove the Corona theorem on a finitely connected Ω. Also see Earle and Marden (1969). Proposition 6.1 is from a paper of Gamelin and Voichick (1968) and Proposition 6.3 is from Ahlfors (1950). The factorization theorem of Section 7 is from Voichick and Zalcman (1965). Voichick was the first to extend to domains of the type discussed here the invariant subspace and closed ideal theory developed in the unit disc; Theorem 6.6 is basically his, as well; see Voichick (1964). Theorem 6.5 is proved in Stout (1966) but in the context of a Riemann surface.

The proof here can be adapted to work in that setting and provides an alternative approach; also see Abrahamse and Fisher (1980).

EXERCISES

In all the exercises Ω is a domain bounded by a finite number of disjoint, analytic simple closed curves.

1. If μ is a measure on Γ with

$$0 = \int_\Gamma \frac{d\mu(\zeta)}{\zeta - z}, \qquad \text{all } z \notin \Omega \cup \Gamma$$

show that there is a $g \in H^1(\Omega)$ with $d\mu = g^* \, dz$.

2. If $\{f_n\}$ is a sequence of functions in the unit ball of $H^1(\Omega)$ show there is an f in the unit ball of $H^1(\Omega)$ and a subsequence $\{f_{n_j}\}$ of $\{f_n\}$ such that $f_{n_j} \to f$ uniformly on compact subsets and the measures $\{f_{n_j}\, d\omega\}$ converge to the measure $f\, d\omega$ weak-* in the space of measures on Γ.
3. Let **E** be a compact, proper subset of Γ. Show that the restriction to **E** of **R**(Ω) is dense in **C(E)** but that this restriction is not equal to **C(E)** if the harmonic measure of **E** is positive.
4. $H^2(\Omega)$ is a Hilbert space. Find the inner product and then an orthonormal basis for $H^2(\Omega)$ when $\Omega = \{z : 1 < |z| < \rho\}$.
5. Let $h(z; \zeta)$ be the harmonic conjugate of $g(z; \zeta)$. Show that the period of $h(z; \zeta)$ along Γ_k is exactly $2\pi\omega_\zeta(\Gamma_k)$, $k = 1, \ldots, m$.
6. Let F be holomorphic and nonconstant on a neighborhood of $\Omega \cup \Gamma$ and satisfy $|F(z)| = 1$ if $\zeta \in \Gamma$. Show F has at least $m + 1$ zeros in Ω.
7. Let F be as in problem 6 and let $z_0, \ldots, z_r$, $r \geqslant m$, be the zeros of F in Ω. Show that

$$F(z) = \lambda \exp\left\{ -\sum_0^r g(z; z_j) - i \sum_0^r h(z; z_j) \right\}$$

for some unimodular constant λ.

8. Let u be a positive harmonic function on Ω. Show there is a non-negative measure μ on Γ with

$$u(z) = \tilde{\mu}(z) = \int_\Gamma H_z(\zeta)\, d\mu(\zeta); \quad \text{see (6.4)}$$

9. Let μ be a real measure on Γ with Lebesgue decomposition relative to $\omega : d\mu = v\, d\omega + d\beta$, β singular with respect to ω. Let $\tilde{\mu}$ be the harmonic

extension of μ to Ω. Use the technique of Proposition 6.4 to show that

$$\lim\{\tilde{\mu}(z): z \to \zeta\} = v(\zeta) \quad \text{a.e. } \omega, \qquad \zeta \in \Gamma$$

where the convergence of z to ζ is normal to Γ at ζ. (Of course, this is known for the open unit disc Δ; see Section 2 of Chapter 1.)

10. Suppose S is analytic, zero-free, and bounded by one on Ω and that $|S^*| = 1$ a.e. ω on Γ. Show that

$$S(z) = c\exp[-\tilde{\mu} - i\,{}^*\tilde{\mu}], \qquad |c| = 1$$

for some non-negative measure μ on Γ, μ singular with respect to ω.

11. Let Λ consist of all positive harmonic functions u on Ω with $u(z_0) = 1$. Show that Λ is convex and compact in the topology of uniform convergence on compact sets in Ω. Find the extreme points of Λ; see Heins (1950).

12. Let $f \in H^\infty(\Omega)$, $\|f\|_\infty = 1$. Show f is an extreme point of the unit ball of $H^\infty(\Omega)$ if and only if $\log|f^*|$ is *not* in $L^1(\Gamma, \omega)$. Prove the same result with $\mathbf{A}(\Omega)$ in place of $H^\infty(\Omega)$; see Gamelin and Voichick (1968).

13. Let μ be a non-negative measure on $\Gamma = \partial\Omega$ such that

a. $$\int_\Gamma f\,d\mu = f(z_0), \qquad f \in \mathbf{A}(\Omega)$$

b. $$\int_\Gamma \log|f|\,d\mu \geqslant \log|f(z_0)|, \qquad f \in \mathbf{A}(\Omega)$$

Show that $\mu = \omega$, harmonic measure for z_0. HINT: begin by showing that μ is absolutely continuous with respect to ω and then that (a) and (b) hold as well for $f \in H^\infty(\Omega)$.

14. If $F \in \mathbf{A}(\Omega)$ is inner, show that the range of F on Ω is all of Δ and each point in Δ has exactly the same number of inverse images in Ω (counting multiplicities, of course).

5

BLASCHKE PRODUCTS, INNER FUNCTIONS, AND EXTREMAL PROBLEMS

One of the most recurrent and productive themes in complex analysis is the solution of extremal problems. In this chapter we describe several such problems, some old and classical and others new. Their solutions share the common characteristic of being one (in modulus) as much as possible.

5.1. THE AHLFORS FUNCTION

Let Ω be a domain for which $H^\infty(\Omega)$ consists of more than constants. Let $p \in \Omega$ and put

$$\gamma = \sup\{|h'(p)| : h \in H^\infty(\Omega), \|h\|_\infty \leqslant 1\} \tag{1.1}$$

A function $h \in H^\infty(\Omega)$, $\|h\|_\infty = 1$, for which $h'(p) = \gamma$ will be called extremal. Clearly at least one extremal exists (apply a normal families argument).

Theorem 1.1. *There is precisely one extremal and it vanishes at p.*

Proof. Let F be an extremal. We first show that $F(p) = 0$. To see this, let

$$g(z) = [F(z) - F(p)]\left[1 - \overline{F(p)}\,F(z)\right]^{-1}, \qquad z \in \Omega$$

A computation gives

$$g'(p) = \gamma\left(1 - |F(p)|^2\right)^{-1}$$

Since $\|g\|_\infty \leqslant 1$, we must have $F(p) = 0$.

Next, if both F_1 and F_2 are extremals, then so is $\frac{1}{2}(F_1 + F_2)$. The proof will thus be complete if we show that any extremal F is an extreme point of the unit ball of H^∞. Suppose, then, that $g \in H^\infty(\Omega)$ and $\|F \pm g\|_\infty \leqslant 1$. Consequently,

$$|F|^2 \pm 2\,\mathrm{Re}\, F\bar{g} + |g|^2 \leqslant 1$$

so that

$$|F|^2 + |g|^2 \leqslant 1$$

Hence,

$$|g|^2 \leqslant 1 - |F|^2 = (1 + |F|)(1 - |F|) \leqslant 2(1 - |F|)$$

Define $h = \frac{1}{2}(g^2)$; then $h \in H^\infty$ and

$$|F| + |h| \leqslant 1 \quad \text{on } \Omega. \tag{1.2}$$

We shall show that (1.2) implies that h vanishes identically; hence, so does g, and F will be proved to be an extreme point of the unit ball of $H^\infty(\Omega)$.

First note that $h(p) = 0$. For otherwise, there is a choice of λ, $|\lambda| = 1$, which gives the function $F + \lambda Fh$ a larger derivative at p than F, a contradiction since this function is bounded by 1 in Ω:

$$|F + \lambda Fh| \leqslant |F| + |F||h| \leqslant |F| + |h| \leqslant 1$$

If h is not identically zero, let r be the order of its zero at p, $r \geqslant 1$. Let ε be a small complex number and consider

$$F(z) + \varepsilon h(z)(z - p)^{-r+1} = G(z).$$

Then $|G'(p)| = |\gamma + \varepsilon h^{(r)}(p)/r!| > \gamma$, for an appropriate choice of the argument of ε. Furthermore, once out of a neighborhood of p the quantity $|\varepsilon||z - p|^{1-r}$ is less than 1 so that $|G| \leqslant 1$ by the maximum principle. This contradiction establishes that $h \equiv 0$ and the theorem follows.

The unique function specified by Theorem 1.1 will be called the *Ahlfors function* for p and Ω. It has a number of interesting properties, which we will see in the remainder of this section.

Definition. A subset S of the domain Ω is *dominating* if

$$\sup_{z \in S} |f(z)| = \sup_{z \in \Omega} |f(z)| \quad \text{for all } f \in H^\infty(\Omega)$$

Proposition 1.2. There is a countable dominating set in Ω with no limit point in Ω.

Proof. Let $\{\mathfrak{F}_j\}$ be a sequence of compact subsets of Ω with $\mathfrak{F}_j \subset \text{INT}(\mathfrak{F}_{j+1})$ and $\cup_j \mathfrak{F}_j = \Omega$. Let $\mathfrak{C}_n = \mathfrak{F}_n \setminus \text{INT}\, \mathfrak{F}_{n-1}$ for $n = 2, 3, \ldots$ so that $\mathfrak{C}_n$ is compact. Further

$$\max_{\mathfrak{C}_n} |f(z)| = \max_{\mathfrak{F}_n} |f(z)| \to \|f\|_\infty, \qquad \text{as } n \to \infty$$

for each $f \in H^\infty(\Omega)$. The unit ball of $H^\infty(\Omega)$ is equicontinuous on each $\mathfrak{C}_n$. Let δ_n be chosen so

$$|f(w) - f(z)| < \left(\tfrac{1}{2}\right)^n \quad \text{if} \quad |w - z| < \delta_n, \qquad w, z \in \mathfrak{C}_n$$

and $\|f\|_\infty \leqslant 1$. Let $\mathfrak{E}_n$ be a finite set in $\mathfrak{C}_n$ such that each point of $\mathfrak{C}_n$ is no more than $\frac{1}{2}\delta_n$ from some point of $\mathfrak{E}_n$. Then

$$\max_{\mathfrak{C}_n} |f(z)| \leqslant \max_{\mathfrak{E}_n} |f(z)| + \left(\tfrac{1}{2}\right)^n$$

Set $\mathfrak{S} = \cup_{n=1}^{\infty} \mathfrak{E}_n$; then $\mathfrak{S}$ is the desired set.

We shall make use of dominating sets to prove the next result.

Theorem 1.3. *Let F be the Ahlfors function for Ω and $p \in \Omega$. For each $h \in H^\infty$ we have*

$$\|Fh\|_\infty = \|h\|_\infty \tag{1.3}$$

Proof. Let us define Ω' to consist of those points $a \in \Omega$ for which there is some $h \in H^\infty(\Omega)$ with $|h(a)| > 1$ yet $\|Fh\| \leqslant 1$. Suppose $p \in \Omega'$; let $F_1 = hF$ so that $\|F_1\| \leqslant 1$. However,

$$|F_1'(p)| = |F'(p)||h(p)| > |F'(p)|$$

a contradiction. Hence, $p \notin \Omega'$. We shall show that Ω' is both open and closed in Ω. It will follow that Ω' is empty and this will prove the theorem.

Clearly Ω' is open. To show Ω' is closed, we show that $\Omega \setminus \Omega'$ is open. Let $q \in \Omega \setminus \Omega'$; then $\|Fh\| \leqslant 1$ implies that $|h(q)| \leqslant 1$ if $h \in H^\infty(\Omega)$. Let $\{z_j\}$ be a countable dominating sequence in Ω with no limit point in Ω; there is no loss in assuming no z_j is the point q. At this point we assume some familiarity with the maximal ideal space of a Banach algebra. Let $\mathbf{X}$ be the maximal ideal space of ℓ^∞; each function u which is bounded on a neighborhood of $\{z_j\}$ gives rise to an element $\hat{u}$ on $\mathbf{X}$; in particular, this is true for functions in $H^\infty(\Omega)$.

Our assumption that $q \notin \Omega'$ implies that the linear functional from FH^∞ into $\mathbb{C}$ given by

$$Fh \mapsto h(q)$$

has norm 1. Hence, by the Hahn-Banach theorem there is a measure λ on $\mathbf{X}$ of norm 1 with

$$\int \hat{F}\hat{h}\,d\lambda = h(q), \qquad h \in H^\infty(\Omega)$$

Thus,

$$1 = \int \hat{1}\hat{F}\,d\lambda \leqslant \int |\hat{F}||d\lambda| \leqslant \|\lambda\| \leqslant 1$$

so that the measure $d\mu = \hat{F}\,d\lambda$ is non-negative and has mass 1 and μ is supported on the set where $|\hat{F}| = 1$. For $\zeta \in \Omega$ and ζ near q let

$$s_\zeta(z) = (z - q)/(z - \zeta), \qquad z \in \Omega$$

and

$$u(\zeta) = \int \hat{s}_\zeta\,d\mu$$

Clearly u is continuous for ζ near q and since $u(q) = 1$ we know $u(\zeta) \neq 0$ if ζ is near q. Let ζ be a point near q at which $u(\zeta) \neq 0$. Let $h \in H^\infty(\Omega)$ and put

$$g(z) = \frac{h(z) - h(\zeta)}{z - \zeta}(z - q)$$

Then

$$0 = \int \hat{g}\,d\mu \quad \text{since } g(q) = 0.$$

Hence,

$$h(\zeta)u(\zeta) = \int \hat{h}\hat{s}_\zeta\,d\mu$$

which implies

$$|h(\zeta)| \leqslant C \sup\{|\hat{h}(m)| : m \in \mathbf{A}\}, \qquad \mathbf{A} = \text{support } \mu$$

where C is a constant independent of h. Now replace h by h^n, extract nth roots, and then let $n \to \infty$. We find

$$|h(\zeta)| \leqslant \|\hat{h}\|_{\mathbf{A}} = \sup\{|\hat{h}(m)| : m \in \mathbf{A}\}$$

Whence, if $\|hF\| \leqslant 1$, then $|\hat{h}| \leqslant 1$ on $\mathbf{A}$ and so $|h(\zeta)| \leqslant 1$. This shows $\zeta \notin \Omega'$ and completes the proof that $\Omega \setminus \Omega'$ is open.

There is a nice connection between the Ahlfors function and essential boundary points.

Theorem 1.4. *Let F be the Ahlfors function for Ω and $p \in \Omega$. Then a point $x \in \partial\Omega$ is essential if and only if*

$$\limsup\{|F(z)| : z \in \Omega,\ z \to x\} = 1 \tag{1.4}$$

Proof. Suppose first that x is essential but (1.4) fails, say $\limsup\{|F(z)| : z \to x\} = 1 - \delta$, $\delta > 0$. By Theorem 3.7.1 there is a function $h \in H^\infty(\Omega)$ with

$$\limsup\{|h(z)| : z \to x\} = 1$$

$$\limsup\{|h(z)| : z \to y\} < 1, \qquad y \in \partial\Omega, \qquad y \neq x$$

Consider Fh. At x we have

$$\limsup\{|F(z)h(z)| : z \to x\} \leqslant 1 - \delta$$

while at $y \neq x$, $y \in \partial\Omega$, we have

$$\limsup\{|F(z)h(z)| : z \to y\} < 1$$

since $|F| \leqslant 1$ in Ω. Hence, $\|Fh\|_\infty < 1 = \|h\|_\infty$, a contradiction to Theorem 1.3.

Conversely, suppose x is removable. Then there is a domain Ω^* containing Ω and x and all functions in $H^\infty(\Omega)$ extend to be in $H^\infty(\Omega^*)$ without increasing their norm. In particular this holds for F, so that $1 = \|F\|_{H^\infty(\Omega^*)}$. If (1.4) held, then $|F|$ would have an interior maximum on Ω^* and hence would be constant, a contradiction.

The final result on the Ahlfors function in this generality is the following, which describes a property of its range.

Theorem 1.5. *Let Ω be maximal, let F be the Ahlfors function for Ω and $p \in \Omega$, and let E consist of those points in Δ which are not in the range of F. Then $E \cap \{|z| \leqslant r\}$ is a null set for all $r < 1$.*

Proof. Suppose the conclusion fails for some $r < 1$. Let $D = \Delta \setminus E_r$ and let ϕ be the Ahlfors function for D and 0. Then $\phi'(0) > 1$ for otherwise $\phi'(0) = 1$ and by uniqueness we would have $\phi(z) = z$, contradicting Theorem 1.4. But then

$$\begin{aligned}(\phi \circ F)'(p) &= \phi'(F(p))F'(p) \\ &= \phi'(0)F'(p) \\ &> F'(p)\end{aligned}$$

a contradiction. This establishes the theorem.

EXAMPLE.

There is one interesting case in which the Ahlfors function can be given explicitly, although as far as we are concerned the demonstration that we have the correct function will be only half complete.

Let **E** be a compact set in the real line $\mathbb{R}$ of positive Lebesgue measure and define a function h by

$$h(z) = \frac{1}{2}\int_{\mathbf{E}} \frac{dt}{z-t}, \qquad z \notin \mathbf{E}$$

Then h maps $\Omega = S^2 \setminus \mathbf{E}$ into the horizontal strip $|\text{Im } z| < \pi/2$. Hence,

$$F(z) = \frac{e^{h(z)} - 1}{e^{h(z)} + 1}$$

maps Ω into Δ and $F(\infty) = 0$. Further,

$$F'(\infty) = \tfrac{1}{4}L \quad \text{where} \quad L = \int_{\mathbf{E}} dt$$

Thus, $\frac{1}{4}L \leqslant \gamma(\Omega)$. Now Pommerenke (1960) has shown that, in fact, $\frac{1}{4}L = \gamma(\Omega)$, and so F must be the Ahlfors function for Ω and the point ∞.

When Ω has a nice boundary the Ahlfors function is particularly well behaved.

Theorem 1.6. *Let Ω be bounded by $m + 1$ disjoint analytic simple closed curves $\Gamma_0, \ldots, \Gamma_m$ and let F be the Ahlfors function for Ω and $p \in \Omega$. Then*

F maps Ω onto Δ exactly $m + 1$ times (1.5a)

F extends analytically over each Γ_j and maps each Γ_j homeomorphically on **T** (1.5b)

F' is not zero on any Γ_j. (1.5c)

Proof. Let $\Gamma = \partial\Omega$ and let $\mathbf{A}(\Omega)$ denote those functions which are continuous on $\Omega \cup \Gamma$ and holomorphic on Ω. Set

$$\alpha = \sup\{|f'(p)| : f \in \mathbf{A}(\Omega), \|f\| \leqslant 1\} \tag{1.6}$$

There is a measure μ on $\partial\Omega$ of total variation α with

$$\int f\, d\mu = f'(p), \qquad f \in \mathbf{A}(\Omega) \tag{1.7}$$

Let $g(z; p)$ be the Green's function on Ω with pole at p, let h be its harmonic conjugate, and put $Q = g + ih$. Recall from Propositions 1.6.5 and 1.6.6 that

$$iQ'(z)\,dz = d\omega_p(z), \qquad z \in \Gamma$$

We also know that

$$Q'(z) = -(z-p)^{-1} + w(z)$$

where w is holomorphic in a neighborhood of $\Omega \cup \partial\Omega$. Thus,

$$\frac{i}{2\pi}\int_\Gamma f(z)Q'(z)\,dz = f(p), \qquad f \in \mathbf{A}(\Omega)$$

Consequently,

$$0 = \int_\Gamma f(z)(z-p)\left\{\frac{i}{2\pi}\frac{Q'(z)}{z-p}dz - d\mu(z)\right\}, \qquad f \in \mathbf{A}(\Omega)$$

so that Theorem 4.3.3 implies that the measure in the curly brackets is absolutely continuous with respect to dz and so the same is true for $d\mu$. Moreover, because Q' does not vanish on Γ we may write

$$d\mu(z) = r(z)\frac{i}{2\pi}Q'(z)\,dz, \qquad z \in \partial\Omega$$

where r is in $L^1(\Gamma, ds)$. Hence,

$$0 = \int_\Gamma f(z)(z-p)\left\{\frac{1}{z-p} - r(z)\right\}\frac{i}{2\pi}Q'(z)\,dz, \qquad f \in \mathbf{A}(\Omega)$$

We now may invoke Theorem 4.4.7 and conclude that

$$(1-(z-p)r(z))Q'(z) = h(z), \qquad z \in \Omega \cup \partial\Omega$$

where $h \in H^1(\Omega)$. Equivalently,

$$r(z) = \frac{1}{z-p} + \sum_1^m c_j(z-t_j)^{-1} + g(z), \qquad z \in \Omega \cup \partial\Omega \tag{1.8}$$

where $g \in H^1(\Omega)$, $c_1, \ldots, c_m$ are scalars and $t_1, \ldots, t_m$ are the critical points of Q, that is, the zeros of Q'. Thus if F is the Ahlfors function for p, then

$$\begin{aligned}\gamma = F'(p) &= \int_\Gamma F(z)r(z)\frac{i}{2\pi}Q'(z)\,dz\\ &\leqslant \int_\Gamma |F(z)||r(z)|\frac{i}{2\pi}Q'(z)\,dz\\ &\leqslant \|F\|_\infty \alpha = \alpha \leqslant \gamma\end{aligned}$$

Hence,

$$F(z)r(z) \geqslant 0 \quad \text{a.e. } |dz| \text{ on } \Gamma \tag{1.9a}$$

$$|F(z)| = 1 \quad \text{a.e. where } r \neq 0 \tag{1.9b}$$

However, each point $z \in \Gamma$ has a neighborhood $\mathcal{U}$ such that $\mathcal{U} \cap \Omega$ is a conformal to the open unit disc Δ and r is in $H^1(\mathcal{U} \cap \Omega)$. Thus, by problem 12, Chapter 3, we conclude that both F and r may be analytically continued over each point of Γ and certainly $|F| = 1$ on all of Γ. This means that the meromorphic function $r(z)F(z)$ is real and positive on Γ and so has as many poles as zeros in Ω. However, it has precisely $m + 1$ poles and so F has at most $m + 1$ zeros in Ω. On the other hand, the following argument shows arg F is increasing (locally) on Γ. Put $u = \log|F|$ near a point $x \in \Gamma$. Then $^*u = \arg F$ and the Cauchy–Riemann equations imply

$$0 \leqslant \frac{\partial u}{\partial n} = \frac{\partial^* u}{\partial \tau} \quad \text{on } \Gamma$$

which is what we claimed. Since arg F must increase on each Γ_j by an integer multiple of 2π, we see F must have at least $m + 1$ zeros in Ω. Thus, F has exactly $m + 1$ zeros in Ω and arg F increases by exactly 2π on each Γ_j. This completes the proof.

5.2. BLASCHKE PRODUCTS

This section and the next three as well are all directly concerned with Blaschke products, so we begin with a result which identifies Blaschke products from among the elements of the unit ball of $H^\infty(\Delta)$. The result which follows shows that Blaschke products are dense in the set of inner functions.

Proposition 2.1. Let $f \in H^\infty(\Delta)$, $\|f\| \leqslant 1$. Then f is a Blaschke product if and only if

$$\lim_{r \to 1} \left\{ \int_0^{2\pi} \log|f(re^{i\theta})| \, d\theta \right\} = 0 \tag{2.1}$$

Proof. Suppose f is a Blaschke product; if f has only a finite number of zeros then (2.1) is immediate so we may suppose that f has zeros at $z_1, z_2, \ldots$ where $\Sigma_1^\infty(1 - |z_j|) < \infty$. The function $\log|f(z)|$ is subharmonic on Δ so that the quantity

$$\int_0^{2\pi} \log|f(re^{i\theta})| \, d\theta$$

increases with r. Clearly it is always less than zero. Given $\varepsilon > 0$ choose N so big that

$$-\varepsilon < \log \prod_{N+1}^{\infty} |z_j| \tag{2.2}$$

Let $B_N(z) = \Pi_1^N (z - z_j)(1 - \bar{z}_j z)^{-1}$ and $g = f/B_N$. We certainly know that $|B_N(z)| \to 1$ continuously as $|z| \to 1$ and so

$$\int_0^{2\pi} \log|g(re^{i\theta})|\, d\theta \leqslant \int_0^{2\pi} \log|f(re^{i\theta})|\, d\theta + \varepsilon$$

for r near to 1. But

$$\begin{aligned} -\varepsilon < \log \prod_{N+1}^{\infty} |z_j| &= \log|g(0)| \\ &\leqslant \frac{1}{2\pi} \int_0^{2\pi} \log|g(re^{i\theta})|\, d\theta \\ &\leqslant \frac{1}{2\pi} \int_0^{2\pi} \log|f(re^{i\theta})|\, d\theta + \varepsilon \end{aligned}$$

and we're done with half the proof.

Conversely, suppose (2.1) holds. Write $f = \lambda BSF$ where B is a Blaschke product, S is singular, F is outer and λ is a unimodular constant. Then

$$\frac{1}{2\pi} \int_0^{2\pi} \log|f(re^{i\theta})|\, d\theta = \frac{1}{2\pi} \int_0^{2\pi} \log|B(re^{i\theta})|\, d\theta + \log|S(0)| + \log|F(0)| \tag{2.3}$$

The first term on the right-hand side of (2.3) goes to 0 as r increases to 1 by the first part of the proof. The left-hand side of (2.3) goes to 0 by hypothesis. Thus, $|S(0)| = |F(0)| = 1$. But $\|S\| = 1$, $\|F\| \leqslant 1$, and so both S and F are constants.

Theorem 2.2. *Let f be an inner function on Δ, f not constant or a finite Blaschke product. Then for $|w| < 1$, the function*

$$g_w(z) = \frac{f(z) - w}{1 - \bar{w} f(z)}$$

is an infinite Blaschke product except possibly for a set of w in Δ of logarithmic capacity zero.

Proof. Let $\mathbf{F}$ be those $w \in \Delta$ such that g_w is not an infinite Blaschke product. Suppose some compact subset $\mathbf{E}$ of $\mathbf{F}$ has positive logarithmic capacity. Then there is a positive measure ν of mass 1 supported on $\mathbf{E}$ such that

$$\int_{\mathbf{E}} \log|z - \zeta| \, d\nu(\zeta)$$

is bounded for all z and hence

$$u(z) = \int_{\mathbf{E}} \log \left| \frac{1 - \bar{\zeta} z}{z - \zeta} \right| d\nu(\zeta)$$

is bounded for all z. Note that u is continuous at $\mathbf{T}$ and $u = 0$ on $\mathbf{T}$. Further, $u(f(re^{i\theta}))$ converges to 0 a.e. $d\theta$ as $r \to 1$ since the nontangential limits of f lie in $\mathbf{T}$ a.e. $d\theta$. Hence, the dominated convergence theorem gives

$$0 = \lim_{r \uparrow 1} \int_{-\pi}^{\pi} u(f(re^{i\theta})) \frac{d\theta}{2\pi} = \lim_{r \uparrow 1} \int_{\mathbf{E}} \left\{ \frac{1}{2\pi} \int_{-\pi}^{\pi} \log \left| \frac{1 - \bar{\zeta} f(re^{i\theta})}{\zeta - f(re^{i\theta})} \right| d\theta \right\} d\nu(\zeta)$$

$$\geqslant \int_{\mathbf{E}} \liminf_{r \to 1} \frac{1}{2\pi} \int_{-\pi}^{\pi} \log \left| \frac{1 - \bar{\zeta} f(re^{i\theta})}{\zeta - f(re^{i\theta})} \right| d\theta \, d\nu(\zeta)$$

However, the inner integral is non-negative so that we must have

$$\lim_{r \to 1} \frac{1}{2\pi} \int_{-\pi}^{\pi} \log \left| \frac{1 - \bar{\zeta} f(re^{i\theta})}{\zeta - f(re^{i\theta})} \right| d\theta = 0$$

almost everywhere $d\nu$ on $\mathbf{E}$. This contradiction establishes the theorem.

Corollary 2.3. *If f is inner, then the range of f is all of Δ except possibly a set of logarithmic capacity zero.*

Corollary 2.4. *Each inner function may be uniformly approximated by a sequence of Blaschke products.*

Proof. Choose $\{w_n\}$ with $w_n \to 0$ and $(f(z) - w_n)(1 - \bar{w}_n f(z))^{-1}$ a Blaschke product.

EXAMPLE.

If $\mathbf{E}$ is a compact set of zero logarithmic capacity in Δ then there is an inner function whose range is precisely $\Delta \setminus \mathbf{E}$. For let $\Omega = \Delta \setminus \mathbf{E}$; then Ω is a domain since $\mathbf{E}$ must be totally disconnected. If $T: \Delta \to \Omega$ is the uniformizer of Ω then T is holomorphic and, of course, its range is precisely Ω so we will be finished after the next proposition.

Proposition 2.5. Let f be a nonconstant bounded holomorphic function on Δ. Let $\mathfrak{S}$ be a subset of $\mathbf{T}$ of positive measure on which f has radial limits and set $\mathbf{E} = \{f(e^{it}) : e^{it} \in \mathfrak{S}\}$. Then $\mathbf{E}$ has positive logarithmic capacity.

Proof. Suppose to the contrary that the capacity of $\mathbf{E}$ is zero. Then there is a probability measure σ on $\mathbf{E}$ whose potential

$$p(z) = -\int_{\mathbf{E}} \log|z - \zeta|\, d\sigma(\zeta)$$

satisfies

$$\lim\{p(z) : z \to \zeta\} = \infty, \qquad \text{all } \zeta \in \mathbf{E};$$

see Theorem 1.7.4.

Set $u(z) = p(f(z))$; then u is harmonic on Δ and u tends to ∞ as $z \to \zeta$ nontangentially at each point $\zeta \in \mathfrak{S}$. Let $^*u$ be the harmonic conjugate of u on Δ and put $F = \exp[-u - i\,^*u]$. Then $F \in H^\infty(\Delta)$ and $F(z) \to 0$ as z converges nontangentially to any point of $\mathfrak{S}$. Thus, $f \equiv 0$, a contradiction since u is not identically $+\infty$.

As an interesting application of the theorems of this section we conclude with a (version of a) theorem of E. L. Stout (extending a theorem of T. Radó) on removable singularities.

Theorem 2.6. *Let* $\mathbf{E}$ *be a set of capacity zero in* $\mathbb{C}$ *and let* K *be a relatively closed subset of* Δ. *Suppose* f *is holomorphic on* $\Delta \setminus K$, *bounded and continuous on* Δ, *and that* f *maps* K *into* $\mathbf{E}$. *Then* f *is holomorphic on* Δ *and* K *has capacity zero.*

Proof. Let Ω be a component of $\Delta \setminus K$ and let ϕ be the uniformizer of Ω. The boundary values of ϕ lie either in $\mathbf{T}$ or in K since ϕ is a covering map. Let $\mathfrak{S}$ be those points of $\mathbf{T}$ at which ϕ has a nontangential limit and this limit lies in K. Then $f \circ \phi$ is a bounded holomorphic function on Δ and $f \circ \phi$ maps $\mathfrak{S}$ into $\mathbf{E}$. Thus, Proposition 2.5 implies that $\mathfrak{S}$ has length zero (since $\mathbf{E}$ has capacity zero). Consequently, ϕ is inner and, as such, omits only a set of capacity zero; this is Corollary 2.3. Thus, Ω is all of Δ except a set of capacity zero and so K has capacity zero and thus is removable for bounded holomorphic functions.

Remark. Stout actually proved a sharper version of this result, one not requiring the á priori assumption that f is continuous on Δ. He also points out that *some* assumption is needed. For if $K = [-1, 1]$ and if $f \equiv 1$ in the upper half-plane and $f \equiv -1$ in the lower half-plane, then f "maps" K into the set $\mathbf{E} = \{-1, 1\}$ which surely has capacity zero but there is no way to extend f to be holomorphic on Δ.

5.3. APPROXIMATION BY INNER FUNCTIONS

We prove several theorems in this section each concerning approximation by inner functions or by quotients of inner functions. The setting is either the unit disc Δ and the unit circle $\mathbf{T}$ or, more generally, a domain Ω like that considered in Chapter 4, namely $\Gamma = \partial\Omega$ consists of $m + 1$ disjoint analytic simple closed curves. We shall drop the use of the asterisk to denote the boundary values of an element of $H^p(\Omega)$; that is, f will denote both the function in Ω and its values on Γ. We begin the section with a short discussion of some properties of the maximal ideal space of the Banach algebra $L^\infty(\Gamma, \omega) = L^\infty$ where, as usual, ω is harmonic measure on Γ for some particular point $z_0 \in \Omega$; we shall have need for only a few simple observations on this maximal ideal space in this section. In Chapter 6 there is another discussion of this topic concentrating on other aspects of the structure of this space.

Let $\mathbf{M}(L^\infty)$ be the maximal ideal space of L^∞; that is, $\mathbf{M}(L^\infty)$ consists of all continuous linear functionals m on L^∞ which are also multiplicative:

$$m(uv) = m(u)m(v), \qquad u, v \in L^\infty$$

Let χ stand for the identity function:

$$\chi(z) = z, \qquad z \in \Gamma$$

and, for $m \in \mathbf{M}(L^\infty)$, set $\lambda = m(\chi)$. Suppose that the complex number a is not in Γ. Then $(\chi - a)^{-1}$ is in L^∞ and so

$$1 = m\big((\chi - a)(\chi - a)^{-1}\big) = m(\chi - a)m\big((\chi - a)^{-1}\big)$$

$$= (\lambda - a)m\left(\frac{1}{\chi - a}\right)$$

Thus, $m(1/(\chi - a)) = 1/(\lambda - a)$ if $a \notin \Gamma$.

It follows immediately that λ itself must be a point of Γ. Further, if P is any polynomial then

$$m(P) = P(\lambda)$$

and if R is any rational function with no poles on Γ, then

$$m(R) = R(\lambda)$$

Since the rational functions with no poles on Γ are dense in the continuous functions on Γ we find

$$m(f) = f(\lambda), \qquad f \in C(\Gamma) \quad \text{and} \quad m(\chi) = \lambda$$

Each point $m \in \mathbf{M}(L^\infty)$ thus lies in exactly one fiber $\mathfrak{F}_\lambda$ defined by

$$\mathfrak{F}_\lambda = \{m \in \mathbf{M}(L^\infty) : m(\chi) = \lambda\}, \qquad \lambda \in \Gamma$$

Suppose now that $u \in L^\infty$ and $|u| = 1$ a.e. ω on Γ. Then

$$1 = m(1) = m(u\bar{u}) = m(u)m(\bar{u})$$

Hence, $|m(u)| = 1$ and $m(\bar{u}) = \overline{m(u)}$ if $|u| = 1$ a.e. ω. To conclude we show that the number $m(g)$ for $g \in L^\infty$, depends only on the values of g in an arbitrary neighborhood of λ when $m \in \mathfrak{F}_\lambda$. Suppose first that $v \in L^\infty$ and $v = 1$ a.e. ω in some interval $\mathcal{I}$ which contains λ in its interior. Let f be a continuous function on Γ with $f(\lambda) = 1$ and $f = 0$ off $\mathcal{I}$. Then $vf = f$ a.e. on Γ and $m(f) = f(\lambda) = 1$. Thus,

$$m(v) = m(vf) + m(v(1-f)) = m(f) + m(v)m(1-f)$$

$$= 1 + m(v) \cdot 0 = 1$$

Now let $u \in L^\infty$ and let $\mathcal{I}$ be as before. Define u_1 to be 0 on $\mathcal{I}$ and u off $\mathcal{I}$ and u_2 to be u on $\mathcal{I}$ and 0 off $\mathcal{I}$. Then

$$m(u) = m(u_1 + u_2) = m(u_1) + m(u_2) = m(u_2)$$

since $m(1 - u_1) = 1$ by the foregoing. This is what we wished to prove.

We summarize all this in the following theorem.

Theorem 3.1. *Each $m \in \mathbf{M}(L^\infty)$ lies in one fiber $\mathfrak{F}_\lambda$ over a point $\lambda \in \Gamma$. If $f \in \mathbf{C}(\Gamma)$, then $m(f) = f(\lambda)$. If $u \in L^\infty$ then the number $m(u)$ depends only on the values of u in an arbitrary neighborhood of λ. Finally, if $v \in L^\infty$ is unimodular, then $m(v) \in \mathbf{T}$ and $m(\bar{v}) = \overline{m(v)}$.*

With these preliminaries out of the way we begin our study of the approximation properties of inner functions with a theorem of Douglas and Rudin on the quotients of inner functions on the unit circle $\mathbf{T}$.

Theorem 3.2. *Let E_1 and E_2 be disjoint measurable sets in $\mathbf{T}$ with $E_1 \cup E_2 = \mathbf{T}$. Let λ_1 and λ_2 be distinct points of $\mathbf{T}$. If $\varepsilon > 0$ is given, then there are inner functions ψ_1, ψ_2 in $H^\infty(\Delta)$ with*

$$\left|\frac{\psi_1}{\psi_2} - \lambda_j\right| < \varepsilon \quad \text{a.e. on } E_j, \qquad j = 1, 2 \tag{3.1}$$

Proof. Let α_1 and α_2 be disjoint small closed arcs in $\mathbf{T}$ centered at λ_1 and λ_2, respectively, and set $\mathcal{U} = \mathbf{S}^2 - (\alpha_1 \cup \alpha_2)$. Then there is an annulus

$\mathcal{V}$, $\mathcal{V} = \{z : r_1 < |z| < r_2\}$ and a conformal mapping ψ from $\mathcal{V}$ onto $\mathcal{U}$; note that we may assume that ψ maps the circles $|z| = r_1$ and $|z| = r_2$ into α_1 and α_2, respectively. ψ has a simple pole at some point $z_0 \in \mathcal{V}$, but otherwise is holomorphic across $\partial\mathcal{V}$, by the reflection principle.

Let F be the Ahlfors function for $\mathcal{V}$ with zero at z_0 so that F has a simple zero at z_0 and is analytic over $\partial\mathcal{V}$ and has unit modulus on $\partial\mathcal{V}$; see Theorem 1.6. Now define $\phi = \psi F$ so that ϕ is analytic in $\mathcal{V}$ and across $\partial\mathcal{V}$, as well.

Let

$$u(x) = \begin{cases} \log r_1, & x \in E_1 \\ \log r_2, & x \in E_2 \end{cases}$$

so that u is in $L^1(\mathbf{T}, \sigma)$; let u denote, as well, the harmonic extension of u to Δ; let $^*u$ be the harmonic conjugate of u on Δ and set

$$h = \exp[u + i\,{}^*u]$$

Then $h \in H^\infty(\Delta)$ and h maps Δ into $\mathcal{V}$.

Define ψ_1 and ψ_2 by

$$\psi_1 = \phi \circ h, \qquad \psi_2 = F \circ h$$

Then ψ_1, ψ_2 are both inner functions in $H^\infty(\Delta)$ and for almost all $x \in E_1$ we have

$$\frac{\psi_1(x)}{\psi_2(x)} = \frac{\phi(h(x))}{F(h(x))} = \psi(h(x)) \in \alpha_1$$

Likewise, for almost all $x \in E_2$, $\psi_1(x)/\psi_2(x)$ lies in α_2. This proves the theorem.

Corollary 3.3. *The set of functions of the form $\bar{\phi}\psi$ where ϕ is an inner function in H^∞ and ψ is a finite linear combination of inner functions in $H^\infty(\Delta)$, is a norm dense linear subspace in $L^\infty(\mathbf{T}, \sigma)$.*

Proof. By Theorem 3.2 the closure of this set contains all unimodular functions with two, and hence finitely, many values. Such functions are dense in the set of unimodular functions. If E is a measurable set in $\mathbf{T}$, with characteristic function C_E, then $2C_E - 1$ is unimodular and so is a limit of functions of the required form. All that remains, then, is to show that the functions of the given type form a linear subspace. However, this is easy since

$$\bar{\phi}_1\psi_1 + \bar{\phi}_2\psi_2 = \overline{\phi_1\phi_2}\,(\phi_2\psi_1 + \phi_1\psi_2)$$

and the right-hand side has the desired form.

Corollary 3.4. *If m_1 and m_2 are distinct multiplicative linear functionals on $L^\infty(\mathbf{T}, \sigma)$, then there is an inner function $\phi \in H^\infty(\Delta)$ with $m_1(\phi) \neq m_2(\phi)$.*

Proof. Let u be an $L^\infty(\mathbf{T}, \sigma)$ function with $m_1(u) = 0$ and $m_2(u) = 1$. By Corollary 3.3 there is an inner function ϕ in $H^\infty(\Delta)$ and a finite linear combination of inner functions ψ with $\|u - \bar{\phi}\psi\|_\infty < \frac{1}{3}$. Thus,

$$|m_1(u) - \overline{m_1(\phi)}\, m_1(\psi)| < \tfrac{1}{3} \tag{3.2a}$$

$$|m_2(u) - \overline{m_2(\phi)}\, m_2(\psi)| < \tfrac{1}{3} \tag{3.2b}$$

By (3.2a) $|m_1(\psi)| < \frac{1}{3}$ and by (3.2b) $1 - |m_2(\psi)| < \frac{1}{3}$, so that $|m_2(\psi)| > \frac{2}{3}$. Thus, m_1 and m_2 must differ on some one of the inner functions whose linear combination makes up ψ.

Theorem 3.5. *Let $\Gamma = \partial\Omega$ consist of a finite number of disjoint analytic simple closed curves. Then the inner functions in $H^\infty(\Omega)$ separate the points of $\mathbf{M}(L^\infty(\Gamma, \omega))$, the maximal ideal space of $L^\infty(\Gamma, \omega)$.*

Proof. Suppose m_1, m_2 are distinct elements of $\mathbf{M}(L^\infty)$ and they lie in the same fiber $\mathfrak{F}_\lambda$ over some $\lambda \in \Gamma$, say $\lambda \in \Gamma_0$. Let F be the Ahlfors map for Ω and z_0 so that F is a homeomorphism of Γ_0 onto $\mathbf{T}$. Let F_0 be the restriction of F to Γ_0 and define a pair of linear functionals on $L^\infty(\mathbf{T}, \sigma)$ by

$$\tilde{m}_j(u) = m_j(u \circ F_0), \qquad j = 1, 2, \qquad u \in L^\infty(\mathbf{T}, \sigma)$$

Because F_0 is a homeomorphism and because the values of $m_j(v)$ depend only on the values of v near λ, we see that $\tilde{m}_1$ and $\tilde{m}_2$ are distinct elements of the maximal ideal space of $L^\infty(\mathbf{T}, \sigma)$. Thus, by Corollary 3.4 there is an inner function ϕ in $H^\infty(\Delta)$ with $\tilde{m}_1(\phi) \neq \tilde{m}_2(\phi)$. However, $\psi = \phi \circ F$ is an inner function in $H^\infty(\Omega)$ and we have $m_1(\psi) = \tilde{m}_1(\phi) \neq \tilde{m}_2(\phi) = m_2(\psi)$. This completes the proof in the case when m_1 and m_2 lie in the same fiber.

Suppose, then, that $m_j \in \mathfrak{F}_{\lambda_j}$, $j = 1, 2$, where $\lambda_1 \neq \lambda_2$. Theorem 4.6.5 asserts that there is an inner function $\phi \in \mathbf{A}(\Omega)$ with $\phi(\lambda_1) \neq \phi(\lambda_2)$ and so $m_1(\phi) = \phi(\lambda_1) \neq \phi(\lambda_2) = m_2(\phi)$ and the proof is complete.

The next two propositions are technical results needed in the proofs of the major results of the section, Theorems 3.9 and 3.10.

Proposition 3.6. Let $\Gamma = \partial\Omega$ consist of a finite number of disjoint analytic simple closed curves. Let $u \in L^\infty(\Gamma, \omega)$ and suppose that

$$\inf\{\|u + h\|_\infty : h \in H^\infty(\Omega)\} = d < 1$$

Then there is a function $v \in L^\infty(\Gamma, \omega)$ with

$$|v| = 1 \quad \text{a.e. } \omega \tag{3.3a}$$

$$u - v \in H^\infty(\Omega) \tag{3.3b}$$

Proof. Set

$$a = \sup\left\{\left|\int_\Gamma f\,d\omega\right| : f - u \in H^\infty(\Omega), \|f\|_\infty \leqslant 1\right\}$$

We know that $H^\infty(\Omega)$ is the dual space of L^1/A_0 where $A_0 = H_0^1(\Omega) + \mathbf{N}$ and that $H_0^\infty(\Omega)$ is the dual space of L^1/A where $A = H^1(\Omega) + \mathbf{N}$; the subscript naught is to indicate, as usual, that $\int_\Gamma f\,d\omega = 0$. These relationships are all discussed in Section 5 of Chapter 4. We can thus employ a weak-* compactness argument and conclude that there is at least one $v \in L^\infty$ with $v - u \in H^\infty(\Omega)$ and

$$a = \left|\int_\Gamma v\,d\omega\right|$$

Let us first note that

$$\inf\{\|v + g\| : g \in H_0^\infty(\Omega)\} = 1$$

For if there is a $g \in H_0^\infty(\Omega)$ with $\|v + g\| < 1$, then $\|v + g + c\| \leqslant 1$ for small constants c and

$$a < \left|\int_\Gamma (v + c)\,d\omega\right|,$$

a contradiction. Further, by hypothesis we have

$$\inf\{\|v + h\| : h \in H^\infty\} = d < 1$$

Equivalently, the dual versions are

$$\sup\left\{\left|\int_\Gamma vf\,d\omega\right| : f \in A, \|f\|_1 \leqslant 1\right\} = 1 \tag{3.4}$$

$$\sup\left\{\left|\int_\Gamma vh\,d\omega\right| : h \in A_0, \|h\|_1 \leqslant 1\right\} = d < 1 \tag{3.5}$$

Choose functions $h_n \in A$ with $\|h_n\|_1 \leqslant 1$ and

$$\lim\left|\int_\Gamma vh_n\,d\omega\right| = 1$$

Suppose E is a set of positive measure in Γ on which $|v| \leqslant 1 - \delta$ for some $\delta > 0$. Then

$$1 = \lim \left| \int_\Gamma v h_n \, d\omega \right| \leqslant \limsup \left\{ (1 - \delta) \int_E |h_n| \, d\omega + \int_{\Gamma \setminus E} |h_n| \, d\omega \right\}$$

$$< 1$$

a contradiction, unless

$$\int_E |h_n| \, d\omega \to 0 \tag{3.6}$$

Let us write $h_n = f_n + p_n$ where $f_n \in H^1$ and $p_n \in \mathbf{N}$. We shall show below that (3.6) implies that

$$\int_E |f_n| \, d\omega \to 0 \tag{3.7}$$

Assume (3.7) for the moment. Then

$$\int_E \log |f_n| \, d\omega \leqslant \log \int_E |f_n| \, d\omega \to -\infty$$

so that

$$\log \left| \int_\Gamma f_n \, d\omega \right| \leqslant \int_\Gamma \log |f_n| \, d\omega \to -\infty$$

Hence, if we put $g_n = h_n - \int_\Gamma f_n \, d\omega$, then $g_n \in A_0$ since $\int_\Gamma p \, d\omega = 0$ for all $p \in \mathbf{N}$. Further, the norm of g_n converges to 1 as $n \to \infty$ since $\int_\Gamma f_n \, d\omega \to 0$. Consequently,

$$1 = \lim \int_\Gamma v \frac{g_n}{\|g_n\|} \, d\omega$$

which contradicts (3.5). This establishes the proposition except for the implication that (3.6) implies (3.7). Since this implication has nothing to do with what's in the rest of the proposition we separate it as an independent entity.

Proposition 3.7. Suppose $\{f_n\}$ is a sequence of functions in $H^1(\Omega)$, $\{p_n\}$ is a sequence of functions in $\mathbf{N}$, and

$$\|f_n + p_n\|_1 \leqslant 1 \tag{3.8a}$$

$$\int_E |f_n + p_n| \, d\omega \to 0 \tag{3.8b}$$

where E is a (fixed) set of positive measure. Then

$$\int_E |f_n|\, d\omega \to 0 \tag{3.9}$$

Proof. The space $H^1(\Omega) + \mathbf{N}$ has an equivalent norm, namely

$$\|f + p\| = \|f\|_1 + \|p\|_1$$

since $\mathbf{N}$ is finite-dimensional. Hence, from (3.8a) we conclude that $\|f_n\| \leqslant M$, $\|p_n\| \leqslant M$ for some M. By passing to a subsequence if necessary we may assume that $p_n \to p \in \mathbf{N}$ and that the measures $f_n\, dz$ converge weak-* to a measure $d\mu$ on Γ. But then

$$0 = \int_\Gamma \frac{d\mu(\zeta)}{\zeta - z}, \qquad z \notin \Omega \cup \Gamma$$

and so $d\mu = f\,dz$ where $f \in H^1(\Omega)$ by Theorems 4.3.3 and 4.4.7. However, (3.8b) implies that $f + p = 0$ a.e. $d\omega$ on E. Let $t_1, \ldots, t_m$ be the critical points of the Green's function with pole at z_0 and let

$$P(z) = (z - t_1) \cdots (z - t_m)$$

We showed in Section 5 of Chapter 4 that

$$P(N + H^1) = H^1$$

Hence, $P(f + p) = h \in H^1$ and so $h = 0$ a.e. on E. Thus, h is zero a.e. on Γ and so $f + p = 0$ a.e. on E. However, we know again from Section 5 of Chapter 4 that $\mathbf{N} \cap H^1(\Omega)$ consists of just zero and so $f = p = 0$. Hence

$$\int_E |p_n|\, d\omega \leqslant \int_\Gamma |p_n|\, d\omega \to 0$$

and consequently

$$0 \leqslant \int_E |f_n|\, d\omega \leqslant \int_E |f_n + p_n|\, d\omega + \int_E |p_n|\, d\omega \to 0$$

Before we go on to derive the consequences of Proposition 3.6, it is worthwhile to show that a slightly modified version of it is true with $H^\infty(\Omega)$ replaced by $\mathbf{A}(\Omega)$. As usual we suppose that Ω is bounded by $m + 1$ disjoint analytic simple closed curves. Recall that $\mathbf{A}(\Omega)$ stands for those functions in $H^\infty(\Omega)$ which are continuous on $\Omega \cup \Gamma$.

Proposition 3.6′. Let ψ and $\phi_1, \ldots, \phi_N$ be inner functions in $\mathbf{A}(\Omega)$, let $c_1, \ldots, c_N$ be complex numbers, and suppose $\|\Sigma_1^N c_j\phi_j\| < 1$. Then there is a function G in $\mathbf{C}(\Gamma)$ such that

$$|G| \equiv 1 \quad \text{on } \Gamma \tag{3.10a}$$

$$\bar{\psi}\sum_1^N c_j\phi_j - G \quad \text{is in } \mathbf{A}(\Omega) \tag{3.10b}$$

Proof. Let us refer to Proposition 3.6 and its proof. There we produced a function $G \in L^\infty(\Gamma, \omega)$ with

$$|G| = 1 \quad \text{a.e. } \omega \text{ on } \Gamma, \qquad \bar{\psi}\sum_1^N c_j\phi_j - G = H \in H^\infty(\Omega)$$

Furthermore, we also obtained a sequence $\{h_n\}$ of elements from $A = H^1 + \mathbf{N}$ with $\|h_n\|_1 \leqslant 1$ and

$$\lim \int_\Gamma h_n G \, d\omega = 1 \tag{3.11}$$

We shall show that this G and this H are actually continuous on Γ.

To begin we may assume that the measures $\{h_n \, d\omega\}$ are weak-* convergent in the space of measures on Γ to a measure $d\nu$. We know that $h_n = f_n + p_n$ where $f_n \in H^1(\Omega)$ and $p_n \in \mathbf{N}$ and further

$$\|p_n\|_1 \leqslant M, \qquad \|f_n\|_1 \leqslant M$$

for some constant M. Again there is no loss in assuming that $p_n \to p \in \mathbf{N}$ in $L^1(\Gamma, \omega)$ and thus $\{f_n \, d\omega\}$ converges weak-* to the measure $d\nu - p\, d\omega$. However, Theorems 4.3.3 and 4.4.7 combine to show that $H^1(\Omega)$ is weak-* closed as a subspace of the space of measures on Γ. Hence, $d\nu - p\,d\omega = f\,d\omega, f \in H^1(\Omega)$. Thus, $\{h_n \, d\omega\}$ converges weak-* to $h\,d\omega$, $h = f + p$. In particular

$$\int h_n \bar{\psi}\sum_1^N c_j\phi_j \, d\omega \to \int h\bar{\psi}\sum_1^N c_j\phi_j \, d\omega \tag{3.12}$$

Furthermore,

$$\int h_n H \, d\omega = \int f_n H \, d\omega = H(z_0) f_n(z_0)$$

$$\to H(z_0) f(z_0) = \int fH \, d\omega = \int hH \, d\omega \tag{3.13}$$

Putting (3.11), (3.12), and (3.13) together we find

$$1 = \int_\Gamma \left(\bar{\psi} \sum_1^N c_j \phi_j - H \right) h \, d\omega \tag{3.14}$$

Equivalently,

$$1 = \int \left(\sum_1^N c_j \phi_j - \psi H \right) \left(\frac{h}{\psi} \right) d\omega$$

Now $F = \Sigma_1^N c_j \phi_j - \psi H$ is in $H^\infty(\Omega)$ and is unimodular on Γ a.e. ω since $F = \psi G$ on Γ. Further, h/ψ has $L^1(\Gamma, \omega)$ norm 1 and is locally H^1 on Γ; that is, each point $\zeta \in \Gamma$ has a neighborhood $\mathcal{U}$ such that $\mathcal{U} \cap \Omega$ is a simply connected domain and h/ψ is in $H^1(\mathcal{U} \cap \Omega)$; see Chapter 4. Thus, $Fh/\psi \geq 0$ on Γ and so by an argument familiar from Section 1, both F and h/ψ can be analytically continued over Γ. In particular, F is an inner function in $\mathbf{A}(\Omega)$ and so G lies in $\mathbf{C}(\Gamma)$ and hence so does H. This finishes the proof.

Proposition 3.8. Let Ω be bounded by $m + 1$ disjoint analytic simple closed curves and let $\mathbf{B}$ stand for either $H^\infty(\Omega)$ or $\mathbf{A}(\Omega)$. Let $\mathbf{W}$ consist of those functions f in $\mathbf{B}$ for which there is an inner function ϕ in $\mathbf{B}$ with $\phi\bar{f}$ again in $\mathbf{B}$. Then $\mathbf{W}$ is a subalgebra of $\mathbf{B}$ and each element of $\mathbf{W}$ of norm 1 or less is the uniform limit of a sequence of convex combinations of inner functions in $\mathbf{B}$.

Proof. If f_1, f_2 lie in $\mathbf{W}$, then there are inner functions ϕ_1, ϕ_2 in $\mathbf{B}$ with $h_j = \phi_j \bar{f}_j$ again in $\mathbf{B}$, $j = 1, 2$. Since

$$\overline{(f_1 + f_2)}\, \phi_1 \phi_2 = h_1 \phi_2 + h_2 \phi_1, \qquad \overline{f_1 f_2}\, \phi_1 \phi_2 = h_1 h_2$$

we see that $f_1 + f_2$ and $f_1 f_2$ are both in $\mathbf{W}$. Thus, $\mathbf{W}$ is an algebra.

If $f \in \mathbf{W}$ and $\|f\| < 1$, let ϕ be an inner function in $\mathbf{B}$ with $\phi\bar{f}$ in $\mathbf{B}$. For real t set

$$f_t = (f + \phi e^{it})(1 + \bar{f}\phi e^{it})^{-1}$$

Then f_t is an inner function in $\mathbf{B}$ and

$$f = \frac{1}{2\pi} \int_{-\pi}^{\pi} f_t \, dt$$

Since $\|f\| < 1$ the integral converges uniformly and f is the uniform limit of approximating Riemann sums, each of which is a convex combination of inner functions in $\mathbf{B}$.

Theorem 3.5 and Propositions 3.6 and 3.8 now combine to prove two major results of this section which are given in the next two theorems.

Theorem 3.9. *Let Ω be bounded by a finite number of disjoint analytic simple closed curves. Then the linear span of the inner functions in $H^\infty(\Omega)$ is dense in $H^\infty(\Omega)$. Indeed, the unit ball of $H^\infty(\Omega)$ is the closed convex hull of the inner functions in $H^\infty(\Omega)$.*

Proof. Let $\mathbf{M}(L^\infty)$ denote the maximal ideal space of $L^\infty(\Gamma, \omega)$. We know that the inner functions separate the points of $\mathbf{M}(L^\infty)$ by Theorem 3.5. Hence, the Stone-Weierstrass theorem implies that the self-adjoint algebra generated by the inner functions is dense in $L^\infty(\Gamma, \omega)$. If $\phi_1, \ldots, \phi_N$ and $\psi_1, \ldots, \psi_N$ are all inner functions, and $c_1, \ldots, c_N$ complex numbers, then

$$\sum_{j=1}^{N} c_j \phi_j \bar{\psi}_j = \bar{\psi} \sum_{1}^{N} c_j \tilde{\phi}_j$$

where ψ and $\tilde{\phi}_1, \ldots, \tilde{\phi}_N$ are again inner. Hence, if $f \in H^\infty(\Omega)$, then there are inner functions ψ and $\phi_1, \ldots, \phi_N$ and complex numbers $c_1, \ldots, c_N$ with

$$\left\| f - \bar{\psi} \sum_{1}^{N} c_j \phi_j \right\| < \varepsilon \quad \text{where} \quad 2\varepsilon < 1 - \|f\|$$

Let $g = \Sigma_1^N c_j \phi_j$. By Proposition 3.6 there is a unimodular function v with

$$g\bar{\psi} - \varepsilon v = h, \qquad h \in H^\infty(\Omega), \qquad \|h\| \leqslant 1$$

Thus, $g - \psi h = \varepsilon \psi v$ lies in $H^\infty(\Omega)$, so ψv is an inner function in $H^\infty(\Omega)$.

Now g and $v\psi$ are both in $\mathbf{W}$, defined in Proposition 3.8. Thus, ψh lies in $\mathbf{W}$ and so there is an inner function I in $H^\infty(\Omega)$ with

$$\overline{\psi h} I = H \in H^\infty(\Omega)$$

But

$$\bar{h} I = \bar{\psi} \bar{h} I \psi = H\psi \in H^\infty(\Omega)$$

and hence h itself lies in $\mathbf{W}$. Thus, h is the limit of linear combinations of inner functions. However,

$$\begin{aligned} \|f - h\| &\leqslant \|f - g\bar{\psi}\| + \|g\bar{\psi} - h\| \\ &< \varepsilon + \varepsilon = 2\varepsilon \end{aligned}$$

and we are finished.

Theorem 3.10. *Let Ω be bounded by $m + 1$ disjoint analytic simple closed curves. Then the unit ball of $\mathbf{A}(\Omega)$ is the closed convex hull of the inner functions in $\mathbf{A}(\Omega)$.*

Proof. The proof is nearly the same as that of Theorem 3.9. We know that the inner functions in $\mathbf{A}(\Omega)$ separate the points of Γ by Theorem 4.6.5. Hence, if $f \in \mathbf{A}(\Omega)$ and $\|f\| < 1$, then there are inner functions ψ and $\phi_1, \ldots, \phi_N$ in $\mathbf{A}(\Omega)$ and complex numbers $c_1, \ldots, c_N$ such that

$$\left\| f - \psi \sum_1^N c_j \phi_j \right\| < \varepsilon, \qquad \left\| \sum_1^N c_j \phi_j \right\| < 1$$

Let $g = \Sigma_1^N c_j \phi_j$. By Proposition 3.6′ there is a unimodular function G in $\mathbf{C}(\Gamma)$ and an $h \in \mathbf{A}(\Omega)$ related by

$$g\bar{\psi} - \varepsilon G = h, \qquad \|h\| \leqslant 1$$

Thus, $g - \psi h = \varepsilon G \psi$ lies in $\mathbf{A}(\Omega)$, so $G\psi$ is an inner function in $\mathbf{A}(\Omega)$. Now g and $G\psi$ are both in $\mathbf{W}$, defined in Proposition 3.8, so that ψh also is in $\mathbf{W}$ and thus there is an inner function I in $\mathbf{A}(\Omega)$ with $\overline{\psi h I} = H$ in $\mathbf{A}(\Omega)$. However, $\bar{h} I = \overline{\psi h I} \psi = H\psi$ which lies in $\mathbf{A}(\Omega)$. Hence, h itself lies in $\mathbf{W}$. Consequently, h is the limit of linear combinations of inner functions in $\mathbf{A}(\Omega)$. However,

$$\|f - h\| \leqslant \|f - g\bar{\psi}\| + \|g\bar{\psi} - h\| < \varepsilon + \varepsilon = 2\varepsilon$$

so f is the limit of linear combinations of inner functions. Thus, f lies in the closure of $\mathbf{W}$ and so f is itself the limit of convex combinations of inner functions in $\mathbf{A}(\Omega)$.

5.4. PICK-NEVANLINNA INTERPOLATION

Suppose Ω is a domain bounded by $m + 1$ disjoint analytic simple closed curves, $m \geqslant 0$; that is, Ω is either the unit disc Δ or the type of domain considered in Chapter 4. Let $z_1, \ldots, z_n$ be distinct points of Ω and set

$$\Lambda = \{(f(z_1), \ldots, f(z_n)) : f \in H^\infty(\Omega), \|f\| \leqslant 1\} \tag{4.1}$$

Λ is a closed convex subset of $\mathbf{C}^n$. The major results of this section characterize the boundary points of Λ and give a necessary and sufficient condition that an n-tuple $(w_1, \ldots, w_n)$ in $\mathbf{C}^n$ belongs to Λ.

Theorem 4.1. *A point* $\mathbf{P}$ *lies in the boundary of* Λ *if and only if there is precisely one* f *in the unit ball of* $H^\infty(\Omega)$ *with* $\mathbf{P} = (f(z_1), \ldots, f(z_n))$. *If this is the case, then* f *is a Blaschke product of degree at most* $n + m - 1$.

Proof. Let us suppose first that $\mathbf{P} = (F(z_1), \ldots, F(z_n))$ lies in the boundary of Λ. Then there are complex scalars $c_1, \ldots, c_n$, not all zero, such that

$$\operatorname{Re} \sum_1^n c_j F(z_j) \geqslant \operatorname{Re} \sum_1^n c_j f(z_j)$$

for all $f \in H^\infty(\Omega)$, $\|f\| \leqslant 1$. Set

$$\gamma = \operatorname{Re} \sum_1^n c_j F(z_j)$$

Replacing f by λf for all unimodular constants λ implies that

$$\gamma = \left| \sum_1^n c_j F(z_j) \right| \tag{4.2a}$$

$$\gamma \geqslant \left| \sum_1^n c_j f(z_j) \right|, \qquad f \in H^\infty(\Omega), \qquad \|f\| \leqslant 1 \tag{4.2b}$$

Let $\mathfrak{L}$ be the linear functional on $H^\infty(\Omega)$ given by

$$\mathfrak{L}(f) = \sum_1^n c_j f(z_j) \tag{4.3}$$

Then (4.2) says that $|\mathfrak{L}(F)| = \|L\| = \gamma$. Next, consider the same $\mathfrak{L}$ acting on $\mathbf{A}(\Omega)$, the space of functions continuous on $\Omega \cup \Gamma$ and holomorphic on Ω; as a linear functional on $\mathbf{A}(\Omega)$, $\mathfrak{L}$ has norm $\gamma' \leqslant \gamma$. Thus, there is a measure μ on Γ of norm γ' with

$$\int_\Gamma f \, d\mu = \sum_1^n c_j f(z_j), \qquad f \in \mathbf{A}(\Omega) \tag{4.4}$$

However, we also have

$$\frac{1}{2\pi i} \int_\Gamma f(z) \left\{ \sum_1^n c_j (z - z_j)^{-1} \right\} dz = \sum_1^n c_j f(z_j), \qquad f \in \mathbf{A}(\Omega) \tag{4.5}$$

and so the measure $d\mu - \sum_1^n c_j (z - z_j)^{-1} \, dz$ is orthogonal to $\mathbf{A}(\Omega)$. Thus, by Theorem 4.3.3, μ is absolutely continuous with respect to ω, where ω is harmonic measure for $z_0 \in \Omega$, say

$$d\mu = r(z) \, d\omega$$

But $d\omega = (i/2\pi) Q'(z) \, dz$ by Proposition 1.6.5 where Q' is analytic in a neighborhood of $\Omega \cup \Gamma$ except for a simple pole at z_0 and Q' has precisely m zeros on Ω and none on Γ. We thus see that

$$\left(r(z) \frac{1}{2\pi i} Q'(z) + \frac{1}{2\pi i} \sum_1^n c_k (z - z_k)^{-1} \right) dz$$

is orthogonal to $\mathbf{A}(\Omega)$. We now apply Theorem 4.4.7 and conclude that there is an $h \in H^1(\Omega)$ with

$$h(z) = r(z)Q'(z) + \sum_1^n c_k(z - z_k)^{-1}, \qquad z \in \Gamma \tag{4.6}$$

Solving for r we see that r has a meromorphic extension to Ω with at most $n + m$ poles and a zero at z_0. Further, we see that $\mathfrak{L}$ is given on $H^\infty(\Omega)$ by integration against $r\,d\omega$, since $\mathbf{A}(\Omega)$ is boundedly pointwise dense in $H^\infty(\Omega)$; see Proposition 4.4.2. We thus obtain

$$\gamma = |\mathfrak{L}(F)| = \left|\int_\Gamma Fr\,d\omega\right|$$

$$\leqslant \int_\Gamma |F||r|\,d\omega \leqslant \int_\Gamma |r|\,d\omega = \gamma' \leqslant \gamma$$

Hence, not only does $\gamma' = \gamma$ but as well we must have

$$Fr \geqslant 0 \quad \text{a.e. } d\omega \text{ on } \Gamma \tag{4.7a}$$

$$|F| = 1 \quad \text{a.e. } d\omega \tag{4.7b}$$

Just as in Section 1, we see that (4.7) implies that both F and r are analytic across Γ and $|F| = 1$ identically on Γ. Thus, the meromorphic function $F(z)r(z)$ has as many zeros as poles in Ω and so F has no more than $m + n - 1$ zeros and is a Blaschke product; see exercise 7 of Chapter 4.

We conclude that if $\mathbf{P} = (F(z_1), \dots, F(z_n))$ lies in $\partial\Lambda$ then F must be a (finite) Blaschke product. Suppose now that we also have $\mathbf{P} = (g(z_1), \dots, g(z_n))$ where $g \in H^\infty(\Omega)$ and $\|g\| \leqslant 1$. Then

$$\mathbf{P} = \left(\tfrac{1}{2}F(z_1) + \tfrac{1}{2}g(z_1), \dots, \tfrac{1}{2}F(z_n) + \tfrac{1}{2}g(z_n)\right)$$

and so $\frac{1}{2}F + \frac{1}{2}g$ is a finite Blaschke product; this implies that

$$1 = \left|\tfrac{1}{2}F(z) + \tfrac{1}{2}g(z)\right| \leqslant \tfrac{1}{2} + \tfrac{1}{2}|g(z)| \leqslant 1, \qquad z \in \Gamma$$

and so $F(z) = g(z)$ on Γ and thus on all of Ω.

Conversely, suppose $\mathbf{P}$ is not on the boundary of Λ. We shall show that $\mathbf{P}$ may be interpolated by infinitely many elements of the unit ball of $H^\infty(\Omega)$.

The linear mapping $\mathfrak{S}f = (f(z_1), \dots, f(z_n))$ maps $H^\infty(\Omega)$ onto $\mathbf{C}^n$ and Λ is the image of the unit ball of $H^\infty(\Omega)$ under $\mathfrak{S}$. Let $\mathfrak{K}$ be the kernel of $\mathfrak{S}$ so that $\tilde{\mathfrak{S}} : H^\infty/\mathfrak{K} \to \mathbf{C}^n$ is both one-to-one and onto. Further, Λ is the image under $\tilde{\mathfrak{S}}$ of those cosets, $f + \mathfrak{K}$, whose coset norm is 1 or less. If $\mathbf{P}$ is in the interior of Λ, then $\tilde{\mathfrak{S}}^{-1}(\mathbf{P})$ lies in the interior of the unit ball of $H^\infty/\mathfrak{K}$ so that there is some $f \in H^\infty(\Omega)$ with $\|f\| < 1$ and $\mathfrak{S}f = \mathbf{P}$. Let g be any function in $H^\infty(\Omega)$

with $g(z_j) = 0$ for $j = 1,\ldots, n$ but g not identically zero. Then $\mathbb{S}(f + \varepsilon g) = \mathbf{P}$ and for all small ε, $\|f + \varepsilon g\| \leqslant 1$.

Corollary 4.2. *Let $z_1,\ldots, z_n$ be distinct points of the open unit disc Δ. Suppose $w_1,\ldots, w_n$ are complex numbers for which the interpolation problem*

$$f(z_k) = w_k, \qquad k = 1,\ldots, n \tag{4.8}$$

is solvable for some $f \in H^\infty(\Delta)$, $\|f\| \leqslant 1$. Then the element F of $H^\infty(\Delta)$ of minimal norm which solves (4.8) *is unique and is a constant multiple of a Blaschke product of degree at most $n - 1$.*

We now turn to the question whether for given complex numbers $w_1,\ldots, w_n$ the interpolation problem

$$f(z_k) = w_k, \qquad k = 1,\ldots, n \tag{4.9}$$

has a solution f with $\|f\|_\infty \leqslant 1$. We begin by stating the result for the unit disc Δ.

Theorem 4.3. *The interpolation problem* (4.9) *has a solution f with $\|f\| \leqslant 1$ if and only if the matrix*

$$M = \left[\frac{1 - w_j\bar{w}_k}{1 - z_j\bar{z}_k}\right]_{j, k=1,\ldots,n}$$

is positive semidefinite; that is, $\langle x, Mx\rangle \geqslant 0$ for all $x \in \mathbb{C}^n$. The solution is unique if and only if $\det M = 0$.

We shall prove this result, indeed an extension of it, later in the section. First, however, we shall spend a minute examining it. The function $k(z; w) = (1 - z\bar{w})^{-1}$ is in $H^2(\Delta)$ and is the unique $H^2(\Delta)$ function such that

$$\langle f; k(\cdot\,; w)\rangle_{H^2(\Delta)} = f(w), \qquad f \in H^2(\Delta) \tag{4.10}$$

That is, $k(z, w)$ is the reproducing kernel for the point w in the Hilbert space $H^2(\Delta)$. Hence, Theorem 4.3 can be rephrased in this way: (4.9) is solvable if and only if the matrix

$$\left[(1 - w_j\bar{w}_k)k(z_j; z_k)\right]_{j, k=1,\ldots,n}$$

is positive semidefinite. With this reformulation in mind we can look for ways to extend the result (not yet proven for the disc) to other domains Ω.

We begin by considering again the class $MH^2(\Omega)$ discussed in Sections 6 and 7 of Chapter 4, but this time in a way that will make them single-valued.

For the remainder of this section Ω is a domain like that of Chapter 4, bounded by $m + 1$ disjoint analytic simple closed curves; as usual we write Γ for $\partial\Omega$. Let $\mathbf{C}_1, \ldots, \mathbf{C}_m$ be pairwise disjoint analytic cuts in Ω with union $\mathbf{C}$ such that the set $\Omega \setminus \mathbf{C}$ is simply connected. For $k = 1, \ldots, m$ let $\mathcal{U}_k$ and $\mathcal{V}_k$ be disjoint open sets in Ω with

$$\mathcal{U}_k \cap \mathcal{U}_j = \mathcal{V}_k \cap \mathcal{V}_j = \varnothing \quad \text{if } k \neq j \tag{4.11a}$$

$$\partial\mathcal{U}_k \cap \mathbf{C} = \partial\mathcal{V}_k \cap \mathbf{C} = \mathbf{C}_k \tag{4.11b}$$

see the accompanying figure.

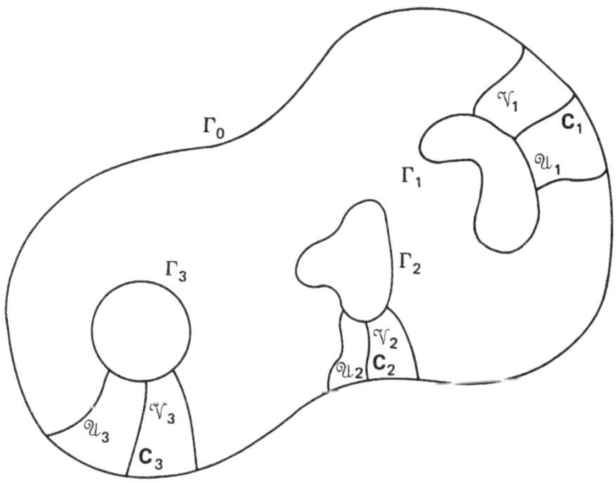

Let $\boldsymbol{\alpha} = (\alpha_1, \ldots, \alpha_m)$ be an m-tuple of unimodular complex numbers (that is, $\boldsymbol{\alpha} \in \mathbf{T}^m$) and let $H^p(\Omega, \boldsymbol{\alpha})$ be those complex functions on Ω such that

$$f \text{ is analytic on } \Omega \setminus \mathbf{C} \tag{4.12a}$$

$$\begin{aligned} &\text{If } 0 < p < \infty, \text{ then } |f|^p \text{ has a harmonic majorant on } \Omega; \\ &\text{if } p = \infty, \text{ then } |f| \text{ is bounded on } \Omega \end{aligned} \tag{4.12b}$$

$$\lim\{f(z) : z \to w,\, z \in \mathcal{V}_k\} = f(w), \qquad w \in \mathbf{C}_k \tag{4.12c}$$

$$\lim\{f(z) : z \to w,\, z \in \mathcal{U}_k\} = \alpha_k f(w), \qquad w \in \mathbf{C}_k \tag{4.12d}$$

A moment's thought shows that an element of $H^p(\Omega, \boldsymbol{\alpha})$ is just the restriction to $\Omega \setminus \mathbf{C}$ of a choice of a single branch of an element of $MH^p(\Omega)$ with periods $\alpha_1, \ldots, \alpha_m$ around $\Gamma_1, \ldots, \Gamma_m$, respectively. Conversely, if $f \in H^p(\Omega, \boldsymbol{\alpha})$, then f gives rise to an element of $MH^p(\Omega)$ by associating to f *all* those elements g of

$H^p(\Omega, \boldsymbol{\alpha})$ with $|g| = |f|$ on Ω. The advantage for us in considering $H^p(\Omega, \boldsymbol{\alpha})$ is that $H^p(\Omega, \boldsymbol{\alpha})$ consists of single-valued functions. In this section we shall work with $H^2(\Omega, \boldsymbol{\alpha})$; in Section 6 we will consider $H^\infty(\Omega, \boldsymbol{\alpha})$.

$H^2(\Omega, \boldsymbol{\alpha})$ is a Hilbert space and for each point $p \in \Omega$ the functional $f \mapsto f(p)$ is bounded. Hence, there is an element $k^{\boldsymbol{\alpha}}(z; p)$ in $H^2(\Omega, \boldsymbol{\alpha})$ such that

$$\langle f, k^{\boldsymbol{\alpha}}(\ ; p)\rangle = f(p), \qquad f \in H^2(\Omega, \boldsymbol{\alpha}) \tag{4.13}$$

With this background we can now give the Pick-Nevanlinna interpolation theorem for Ω.

Theorem 4.4. *Let $z_1, \ldots, z_n$ be distinct points of Ω, let $w_1, \ldots, w_n$ be complex numbers, and let A be a positive number. Then the interpolation problem*

$$f(z_j) = w_j, \qquad j = 1, \ldots, n \tag{4.14a}$$

$$f \in H^\infty(\Omega), \qquad \|f\|_\infty \leqslant A \tag{4.14b}$$

has a solution if and only if the matrix $M^{\boldsymbol{\alpha}}$ given by

$$\left[\left(A^2 - w_j \bar{w}_k\right) k^{\boldsymbol{\alpha}}(z_j, z_k)\right]_{j, k = 1, \ldots, n} \tag{4.15}$$

is positive semidefinite for all $\boldsymbol{\alpha} \in \mathbf{T}^m$. The solution is unique if and only if the determinant of some $M^{\boldsymbol{\alpha}}$ vanishes.

Proof. Each function $f \in H^2(\Omega, \boldsymbol{\alpha})$ or, for that matter, in $H^p(\Omega, \boldsymbol{\alpha})$, has boundary values a.e. on Γ. We shall employ the same letter f to stand both for the function in Ω and for its boundary values. In this way, $H^2(\Omega, \boldsymbol{\alpha})$ can be realized as closed subspace of $L^2(\Gamma, \omega)$ and, for this section only, we shall denote by $H^2(\Gamma, \boldsymbol{\alpha})$ the totality of boundary values of functions in $H^2(\Omega, \boldsymbol{\alpha})$. As usual we write ω for harmonic measure on Γ for the point $z_0 \in \Omega$. We shall also make use of the standard Hilbert space notation of $S^\perp$ to stand for the orthogonal complement of the subspace S of $L^2(\Gamma, \omega)$.

Let $k_t^{\boldsymbol{\alpha}}(\zeta) = k^{\boldsymbol{\alpha}}(\zeta; t)$ so that $k_t^{\boldsymbol{\alpha}} \in H^2(\Gamma, \boldsymbol{\alpha})$ for each $t \in \Omega$. Further, let $P_{\boldsymbol{\alpha}}$ be the orthogonal projection of $L^2(\Gamma, \omega)$ onto $H^2(\Gamma, \boldsymbol{\alpha})$. We begin by noting the relation

$$P_{\boldsymbol{\alpha}}\left(\bar{h} k_t^{\boldsymbol{\alpha}}\right) = \bar{h}(t) k_t^{\boldsymbol{\alpha}}, \qquad h \in H^\infty(\Omega) \tag{4.16}$$

which follows simply in this way. If $f \in H^2(\Gamma, \boldsymbol{\alpha})$, then

$$\begin{aligned} \langle f, P_{\boldsymbol{\alpha}}\left(\bar{h} k_t^{\boldsymbol{\alpha}}\right)\rangle &= \langle f, \bar{h} k_t^{\boldsymbol{\alpha}}\rangle = \langle fh, k_t^{\boldsymbol{\alpha}}\rangle \\ &= f(t) h(t) = \langle f, \overline{h(t)}\, k_t^{\boldsymbol{\alpha}}\rangle \end{aligned}$$

Now let $c_1, \ldots, c_n$ be complex numbers and set $K = \sum_{j=1}^{N} c_j k_{z_j}^{\boldsymbol{\alpha}}$; then

$$\|K\|_2^2 = \sum_{j,l=1}^{n} c_j \bar{c}_l k^{\boldsymbol{\alpha}}(z_j, z_l) \tag{4.17}$$

and, by (4.16) we find

$$\|P_{\boldsymbol{\alpha}}(\bar{h}k)\|_2^2 = \sum_{j,l=1}^{n} c_j \bar{c}_l h(z_j)\bar{h}(z_l) k^{\boldsymbol{\alpha}}(z_j, z_l) \tag{4.18}$$

Suppose $h(z_j) = w_j$, $1 \leqslant j \leqslant n$, and $\|h\|_\infty \leqslant A$. Then

$$\|P_{\boldsymbol{\alpha}}(\bar{h}k)\|_2^2 \leqslant A^2 \|k\|^2 \tag{4.19}$$

which is, by (4.17) and (4.18), the same as

$$\left[\left(A^2 - w_j \bar{w}_l\right) k^{\boldsymbol{\alpha}}(z_j, z_l)\right]_{j,l} \geqslant 0 \tag{4.20}$$

Conversely suppose (4.20) holds. We must find an $f \in H^\infty(\Omega)$ which satisfies (4.14). Note that (4.19) and (4.20) are equivalent. Let ϕ be any element of $H^\infty(\Omega)$ with $\phi(z_j) = w_j$ for $j = 1, \ldots, n$ and set $G(z) = (z - z_1) \cdots (z - z_n)$. Let $g \in H^2(\Gamma, \boldsymbol{\alpha})$ and let h be in $(GH^2(\Gamma, \boldsymbol{\alpha}))^\perp$. Define $\mathfrak{M}_{\boldsymbol{\alpha}}$ to be the span of the functions $\{k_{z_j}^{\boldsymbol{\alpha}}\}_{j=1}^{n}$. Then we surely have

$$H^2(\Gamma, \boldsymbol{\alpha}) = \mathfrak{M}_{\boldsymbol{\alpha}} \oplus GH^2(\Gamma, \boldsymbol{\alpha}) \tag{4.21}$$

so that $k = P_{\boldsymbol{\alpha}}(h)$ lies in $\mathfrak{M}_{\boldsymbol{\alpha}}$. Hence, using (4.19), we find

$$\begin{aligned}
\left|\int_\Gamma \bar{\phi}\bar{g}h \, d\omega\right| &= |\langle h, \phi g\rangle| = |\langle P_{\boldsymbol{\alpha}}(h), \phi g\rangle| \\
&= |\langle k, \phi g\rangle| = |\langle P_{\boldsymbol{\alpha}}(\bar{\phi}k), g\rangle| \\
&\leqslant \|P_{\boldsymbol{\alpha}}(\bar{\phi}k)\|_2 \|g\|_2 \\
&\leqslant A\|k\|_2 \|g\|_2 \\
&\leqslant A\|h\|_2 \|g\|_2
\end{aligned}$$

We shall presently show that each $f \in L^\infty(\omega) \cap (GH^2)^\perp$ has the form

$$f = \bar{g}h, \qquad |f| = |g|^2 = |h|^2 \quad \text{a.e.} \tag{4.22}$$

for some $\boldsymbol{\alpha} \in \mathbf{T}^m$, and some $g \in H^2(\Gamma, \boldsymbol{\alpha})$, $h \in (GH^2(\Gamma, \boldsymbol{\alpha}))^\perp$. Assuming this for the minute, we see that the foregoing computation implies that

$$\left|\int_\Gamma \bar{\phi} f\, d\omega\right| \leqslant A\|f\|_{L^1}, \qquad f \in L^\infty(\omega) \cap (GH^2)^\perp \tag{4.23}$$

Thus, the functional $f \mapsto \int_\Gamma \bar{\phi} f\, d\omega$ has norm no more than A on a certain subspace of $L^1(\Gamma, \omega)$ and so by the Hahn-Banach theorem it can be extended to all of $L^1(\Gamma, \omega)$ without increasing its norm. Thus, there is a $\psi \in L^\infty(\Gamma, \omega)$ with $\|\psi\|_\infty \leqslant A$, and

$$\int_\Gamma \bar{\psi} f\, d\omega = \int_\Gamma \bar{\phi} f\, d\omega, \qquad f \in L^\infty(\omega) \cap (GH^2)^\perp \tag{4.24}$$

Hence, $\phi - \psi$ is orthogonal in $L^2(\omega)$ to $(GH^2)^\perp$ and so lies in GH^2. This then implies that $\psi \in H^\infty(\Omega)$ and

$$\psi(z_j) = \phi(z_j) = w_j, \qquad 1 \leqslant j \leqslant n$$

$$\|\psi\|_\infty \leqslant A$$

The uniqueness assertion still remains but first we must prove (4.22). Let f be bounded and f orthogonal to GH^2. Thus $\bar{G}f$ is orthogonal to H^2 and hence $\log|\bar{G}f|$ is in $L^1(\Gamma, \omega)$; see Corollary 4.4.6. It follows that $\log|f|$ is in $L^1(\Gamma, \omega)$ and so there is an $\boldsymbol{\alpha} \in \mathbf{T}^m$ and an outer function $g \in H^2(\Gamma, \boldsymbol{\alpha})$ with $|g|^2 = f$. Set $h = f/\bar{g}$. Then $f = \bar{g}h$ and $|h|^2 = |f|^2|g|^{-2} = |f|$. We must show that h is orthogonal to $GH^2(\Gamma, \boldsymbol{\alpha})$. Since g is outer we may invoke Proposition 4.6 (which follows the theorem) to conclude that gH^∞ is a dense subspace of $H^2(\Gamma, \boldsymbol{\alpha})$. Thus, it suffices to prove that h is orthogonal to GgH^∞. But if $w \in H^\infty$, then

$$\langle h, wGg\rangle = \langle h\bar{g}, wG\rangle = \langle f, wG\rangle = 0$$

because f is orthogonal to GH^2.

The uniqueness assertion will depend on the fact that for fixed ζ, ξ in Ω, the function

$$\boldsymbol{\alpha} \mapsto k^{\boldsymbol{\alpha}}(\zeta, \xi) \tag{4.25}$$

is continuous from $\mathbf{T}^m$ to $\mathbf{C}$. We shall prove this later in a separate proposition but let us assume its validity for now. If $\det M^{\boldsymbol{\alpha}} \neq 0$ for all $\boldsymbol{\alpha}$, then it follows from the continuity of $\boldsymbol{\alpha} \mapsto k^{\boldsymbol{\alpha}}(\zeta, \xi)$ and the compactness of $\mathbf{T}^m$ that there is a positive ε such that $M^{\boldsymbol{\alpha}} - \varepsilon I \geqslant 0$ for all $\boldsymbol{\alpha} \in \mathbf{T}^m$, where I is the $n \times n$ identity matrix. Let $A^{\boldsymbol{\alpha}}$ be the matrix $[k^{\boldsymbol{\alpha}}(z_i, z_j)]$; each $A^{\boldsymbol{\alpha}}$ is positive definite (since the values 0 can be interpolated by a function bounded by 1) and so the continuity of $\boldsymbol{\alpha} \mapsto A^{\boldsymbol{\alpha}}$ and the compactness of $\mathbf{T}^m$ again can be invoked to get a positive δ such that $\varepsilon I - \delta A^{\boldsymbol{\alpha}} \geqslant 0$ for all $\boldsymbol{\alpha}$. Hence,

$$M^{\boldsymbol{\alpha}} - \delta A^{\boldsymbol{\alpha}} \geqslant 0, \qquad \text{all } \boldsymbol{\alpha} \in \mathbf{T}^m \tag{4.26}$$

which implies that the matrix

$$\left[\left(A^2 - \delta - w_i \bar{w}_j\right) k^{\boldsymbol{\alpha}}(z_i, z_j)\right]_{i,j=1}^{n}$$

is positive definite. Consequently, by the first part of the theorem there is an $f \in H^\infty(\Omega)$ with $f(z_j) = w_j$ $1 \leqslant j \leqslant n$, and $\|f\|_\infty^2 \leqslant A^2 - \delta$. If $G(z) = (z - z_1) \cdots (z - z_n)$ and if η lies between $(A^2 - \delta)^{1/2}$ and A, then $f + \eta G/\|G\|_\infty$ satisfies (4.14); that is, the interpolating function is not unique.

Conversely, suppose $\|f\| \leqslant A$, $f(z_j) = w_j$, $1 \leqslant j \leqslant n$, and $\det M^{\boldsymbol{\alpha}} = 0$ for some $\boldsymbol{\alpha}$. It follows that there are complex numbers $c_1, \ldots, c_n$, not all zero, with

$$\sum_{j,l=1}^{n} c_j \bar{c}_l \left(A^2 - f(z_j)\overline{f(z_l)}\right) k^{\boldsymbol{\alpha}}(z_j, z_l) = 0$$

Equivalently, if $K = \sum_{j=1}^{n} c_j k_{z_j}^{\boldsymbol{\alpha}}$, then

$$\|P_{\boldsymbol{\alpha}}(\bar{f}k)\|_2 = A\|k\|_2$$

However, $\|f\|_\infty \leqslant A$ and so we have

$$A\|k\|_2 = \|P_{\boldsymbol{\alpha}}(\bar{f}k)\|_2 \leqslant \|\bar{f}k\|_2 \leqslant A\|k\|_2$$

Consequently, $\bar{f}k = g$, for some $g \in H^2(\Gamma, \boldsymbol{\alpha})$ and $|f| = A$ a.e. ω. If f_1 is another interpolating function, then so is $f_2 = \frac{1}{2}(f + f_1)$ and so $|f_2| = A$ a.e. ω. Thus, $f_1 = f = f_2$ a.e. ω and f is uniquely determined.

Corollary 4.5. *Let $\mathfrak{E}$ be a subset of Ω and let f be a function on $\mathfrak{E}$. Then there is a function h in $H^\infty(\Omega)$, $\|h\|_\infty \leqslant A$, with $h = f$ on $\mathfrak{E}$ if and only if the function*

$$\left[\left(A^2 - f(z)\overline{f(w)}\right) k^{\boldsymbol{\alpha}}(z, w)\right] \tag{4.27}$$

is positive semidefinite on $\mathfrak{E} \times \mathfrak{E}$, for all $\boldsymbol{\alpha} \in \mathbf{T}^m$.

Proof. If such an h exists, then the conclusion follows directly from Theorem 4.4.

Conversely, suppose the function of z and w given in (4.27) is positive semidefinite. If $\mathfrak{E}$ is finite then once again the conclusion is direct from Theorem 4.4. If $\mathfrak{E}$ is infinite let $\{z_j\}_{j=1}^{\infty}$ be a countable dense set in $\mathfrak{E}$. For each

n, there is an $h_n \in H^\infty(\Omega)$ with

$$\|h_n\|_\infty \leqslant A \tag{4.28a}$$

$$h_n(z_j) = f(z_j), \qquad j = 1,\dots, n \tag{4.28b}$$

Let n go to infinity and apply a normal families argument to produce an $h \in H^\infty(\Omega)$ with

$$\|h\| \leqslant A \tag{4.29a}$$

$$h(z_j) = f(z_j), \qquad j = 1, 2,\dots \tag{4.29b}$$

Now this h actually interpolates f at all points of $\mathfrak{E}$. For if $p \in \mathfrak{E}$ but $p \neq z_j$ for all j and if $\tilde{h}$ is an element of $H^\infty(\Omega)$ constructed as above to have

$$\|\tilde{h}\| \leqslant A \tag{4.30a}$$

$$\tilde{h}(z_j) = f(z_j), \qquad j = 1, 2,\dots \tag{4.30b}$$

$$\tilde{h}(p) = f(p) \tag{4.30c}$$

then $h - \tilde{h}$ vanishes on the points $\{z_j\}$ and so $h - \tilde{h}$ is also zero at the point p since $\{z_j\}$ is dense in $\mathfrak{E}$.

To finish the main result, Theorem 4.4, we need to prove two rather technical propositions; neither is difficult but because of the fact that we are working in $H^2(\Omega, \boldsymbol{\alpha})$ we need to use some results from Section 6 of Chapter 4 to "correct" errant periods.

Proposition 4.6. Suppose Ω is bounded by $m + 1$ disjoint analytic simple closed curves. Suppose $F \in H^2(\Omega)$ and F is outer. Then $FH^\infty(\Omega)$ is a dense subspace of $H^2(\Omega)$.

Proof. It suffices to prove that each $h \in H^\infty(\Omega)$ is the $H^2(\Omega)$ limit of a sequence of the form $\{Fh_n\}$, $h_n \in H^\infty(\Omega)$. Let $F = \exp[\tilde{u} + i\,{*\tilde{u}}]$ where $\tilde{u}$ is a real-valued harmonic function on Ω and u is in $L^1(\Gamma, \omega)$. Set $u_n = \max\{u, -n\}$ on Γ; then $u_1 \geqslant u_2 \geqslant \cdots \geqslant u$ and $u_n \to u$ a.e. on Γ. Thus, $u_n \to u$ in $L^1(\Gamma, \omega)$ and so if $\tilde{u}_n$ is the harmonic extension to Ω of u_n, then $\tilde{u}_n \to \tilde{u}$ uniformly on compact sets in Ω. In particular, the periods of ${*\tilde{u}_n}$ about $\Gamma_1,\dots, \Gamma_m$ converge to those of ${*\tilde{u}}$. Let $\{\nu_n\}$ be a sequence of positive measures on Γ, all supported at the $m + 1$ points $x_0,\dots, x_m$ in Γ, with these properties: (a) $\|\nu_n\| \to 0$, (b) the period of ${*\tilde{\nu}_n}$ about Γ_k is exactly that of ${*\tilde{u}_n} - {*\tilde{u}}$ for $k = 1,\dots, m$ and $m = 1, 2,\dots$; such measures exist by Theorem 4.6.3. Let $S_n = \exp[-\tilde{\nu}_n - i\,{*\tilde{\nu}_n}]$

for $n = 1, 2\ldots$ and, finally, put

$$h_n = S_n h \exp[-\tilde{u}_n - i{}^*\tilde{u}_n]$$

Then h_n is single-valued and $h_n \in H^\infty(\Omega)$. Furthermore, we certainly know that $h_n \to h/F$ uniformly on compact sets in Ω. Set $F_n = \exp[\tilde{u}_n + i{}^*\tilde{u}_n]$ so that $h_n = S_n h/F_n$. Then,

$$\int_\Gamma |h - Fh_n|^2\, d\omega = \int_\Gamma |h|^2 \left|1 - \frac{S_n F}{F_n}\right|^2 d\omega$$

$$\leqslant \|h\|_\infty^2 \int_\Gamma \left|1 - \frac{S_n F}{F_n}\right|^2 d\omega$$

However,

$$\int_\Gamma \left|1 - \frac{S_n F}{F_n}\right|^2 d\omega = 1 - 2\,\mathrm{Re}\left(\frac{S_n F}{F_n}\right)(z_0) + \int_\Gamma \left|\frac{F}{F_n}\right|^2 d\omega$$

and

$$\left|\frac{F}{F_n}\right| = \exp[u - u_n] \leqslant 1 \quad \text{and} \quad \left|\frac{F}{F_n}\right| \to 1 \quad \text{a.e. on } \Gamma$$

as $n \to \infty$. Hence,

$$\int_\Gamma \left|1 - \frac{S_n F}{F_n}\right|^2 d\omega \to 0 \quad \text{as} \quad n \to \infty$$

and the proposition is proved.

Proposition 4.7. Let ζ, ξ be points of Ω. Then the map $\alpha \mapsto k^\alpha(\zeta, \xi)$ is continuous from $\mathbf{T}^m$ into $\mathbb{C}$.

Proof. Suppose $\alpha_n \to \alpha$ in $\mathbf{T}^m$. Let $f_n(z) = k^{\alpha_n}(z, \xi)$ for $z \in \Omega \cup \Gamma$. We first show that the $L^2(\Gamma, \omega)$ norms of $\{f_n\}$ remain uniformly bounded and that $\{f_n\}$ is uniformly bounded on compact subsets of Ω. We certainly know from Sections 6 and 7 of Chapter 4 that if $\mathbf{E}$ is a compact set in Ω, then there is a constant C depending only on $\mathbf{E}$ and Ω such that

$$\sup_{\xi \in \mathbf{E}} |f(\xi)| \leqslant C\|f\|, \qquad f \in MH^2(\Omega) \tag{4.31}$$

Hence, $\|f_n\|_{L^2(\Gamma, \omega)} \leqslant C$, as well. We may suppose, therefore, that $\{f_n\}$ converges weakly in $L^2(\Gamma, \omega)$ to some function h. However, we also see from (4.31) that $\{f_n\}$ is uniformly bounded on compact sets in Ω and so we may suppose that $f_n \to f$ uniformly on compact sets in Ω. Clearly, $f \in H^2(\Omega, \boldsymbol{\alpha})$. Further, the boundary values of f must be h a.e. ω, by the following argument. By using Proposition 4.6.1 we can find a $\delta > 0$ and a sequence $\{g_n\}$ of functions satisfying

$$g_n \in H^\infty(\Omega, \boldsymbol{\alpha} - \boldsymbol{\alpha}_n), \qquad \delta \leqslant |g_n| \leqslant \frac{1}{\delta} \quad \text{on } \Omega \cup \Gamma$$

and $g_n \to 1$ pointwise in Ω and in $L^2(\Gamma, \omega)$. Thus, for $z \in \Omega$, we have

$$\begin{aligned}\int_\Gamma h\bar{k}_z^{\boldsymbol{\alpha}}\, d\omega &= \lim \int_\Gamma f_n \bar{k}_z^{\boldsymbol{\alpha}}\, d\omega \\ &= \lim \int_\Gamma f_n g_n \bar{k}_z^{\boldsymbol{\alpha}}\, d\omega \\ &= \lim f_n(z) g_n(z) = f(z)\end{aligned}$$

Here we make use of the fact that $k_z^{\boldsymbol{\alpha}}$ is actually bounded. Virtually the same argument shows that h is actually in $H^2(\Omega, \boldsymbol{\alpha})$ and thus, $f = h$ on Γ. Finally, we must show that f is $k_\xi^{\boldsymbol{\alpha}}$. Let $g \in H^\infty(\Omega, \boldsymbol{\alpha})$; then

$$\begin{aligned}\int_\Gamma g\bar{f}\, d\omega &= \lim \int_\Gamma g\bar{f}_n\, d\omega \\ &= \lim \int_\Gamma g \frac{1}{g_n} \bar{f}_n\, d\omega \\ &= \lim \frac{g(\xi)}{g_n(\xi)} = g(\xi)\end{aligned}$$

Since $H^\infty(\Omega, \boldsymbol{\alpha})$ is dense in $H^2(\Omega, \boldsymbol{\alpha})$ we are finished.

5.5. INTERPOLATION SEQUENCES

Suppose that $\{z_k\}$ is a sequence of distinct points in Δ whose only accumulation points lie in $\mathbf{T}$. If $f \in H^\infty(\Delta)$, then $\{f(z_k)\}$ is an element of l^∞, the space of bounded sequences of complex numbers. Conversely, given an element $\{w_k\}$ of l^∞, we can ask if there is a function $f \in H^\infty(\Delta)$ with

$$f(z_k) = w_k, \qquad k = 1, 2, \ldots \tag{5.1}$$

This section contains the derivation of a condition on the sequence $\{z_k\}$ that is both necessary and sufficient to assure that (5.1) is solvable for *all* elements $\{w_k\}$ of l^∞; note the distinction between this and the interpolation problem of Pick-Nevanlinna discussed in Section 4. If (5.1) is solvable for each bounded sequence $\{w_k\}$ then $\{z_k\}$ is termed an *interpolation sequence*.

Suppose (5.1) has a solution f for each $\{w_k\}$ in l^∞. Then the linear operator $f \mapsto \{f(z_k)\}$ from $H^\infty(\Delta)$ to l^∞ is both bounded and onto and hence there is a bounded linear operator L from l^∞ into $H^\infty(\Delta)$ with $L(\{w_k\})$ equal to some f in $H^\infty(\Delta)$ satisfying (5.1). Let δ^{-1} be the norm of L, $\delta > 0$. If k is fixed then there is an element f_k in $H^\infty(\Delta)$ with

$$\|f_k\|_\infty \leqslant \frac{1}{\delta} \tag{5.2a}$$

$$f_k(z_j) = \delta_{jk}, \qquad j = 1, 2, \ldots \tag{5.2b}$$

Let B_k be the Blaschke product

$$B_k(z) = \prod_{\substack{j=1 \\ j \neq k}}^{\infty} \left(\frac{z - z_j}{1 - \bar{z}_j z} \right) \left(\frac{-\bar{z}_j}{|z_j|} \right)$$

Thus, $g_k = f_k / B_k$ has sup norm no more than $1/\delta$ on Δ and so $|g_k(z_k)| \leqslant 1/\delta$; equivalently,

$$\delta \leqslant \prod_{\substack{j=1 \\ j \neq k}}^{\infty} \left| \frac{z_k - z_j}{1 - \bar{z}_j z_k} \right|, \qquad k = 1, 2, \ldots \tag{5.3}$$

Hence, condition (5.3) is necessary for interpolation. We say that the sequence $\{z_k\}$ is *uniformly separated* if (5.3) is satisfied. We shall now show that (5.3) is also sufficient for interpolation. That is, $\{z_k\}$ is an interpolation sequence for $H^\infty(\Delta)$ if and only if (5.3) holds. We begin, however, with an auxiliary result.

Theorem 5.1. *If $\{z_k\}$ satisfies (5.3), then*

$$\sum_{k=1}^{\infty} (1 - |z_k|)|g(z_k)| \leqslant C\|g\|_1 \tag{5.4}$$

for all $g \in H^1(\Delta)$ where C is some constant.

Proof. The quantity on the left side of the inequality sign in (5.4) is obviously increased, while the right side is unchanged, if g is zero-free. Thus, (5.4) will be proved if we can show

$$\sum_{k=1}^{\infty} (1 - |z_k|)|h(z_k)|^2 \leqslant C\|h\|_2^2, \qquad h \in H^2(\Delta) \tag{5.5}$$

since if g is a zero-free H^1 function then $g = h^2$ where $h \in H^2(\Delta)$.

Suppose first that whenever $\{w_k\}$ is an n-tuple of complex numbers there is an $f \in H^2(\Delta)$ with

$$\left(1 - |z_k|^2\right)^{1/2} f(z_k) = w_k, \qquad 1 \leqslant k \leqslant n \tag{5.6a}$$

$$\|f\|_2^2 \leqslant C' \sum_1^n |w_k|^2 \tag{5.6b}$$

where C' is a constant independent of n and of $w_1, \ldots, w_n$. We will prove (5.6) presently but let's assume it for the moment and show how (5.5) follows. Let h be an arbitrary element of $H^2(\Delta)$ and let f be an element of $H^2(\Delta)$ satisfying (5.6) for an n-tuple $w_1, \ldots, w_n$ with $\Sigma_1^n |w_k|^2 \leqslant 1$. Put $B(z) = \Pi_1^n[(z - z_k)/(1 - z\bar{z}_k)]$; then we have

$$\left| \sum_{k=1}^n w_k \left(1 - |z_k|^2\right)^{-1/2} \left(B'(z_k)\right)^{-1} h(z_k) \right| = \left| \frac{1}{2\pi i} \int_{\mathbf{T}} h(z) f(z) B^{-1}(z)\, dz \right|$$

$$\leqslant \|h\|_2 \|f\|_2 \leqslant \sqrt{C'}\, \|h\|_2$$

Taking the maximum over all $\{w_k\}_1^n$ with $\Sigma_1^n |w_k|^2 \leqslant 1$ we find

$$\sum_{k=1}^n |h(z_k)|^2 \left(1 - |z_k|^2\right)^{-1} |B'(z_k)|^{-2} \leqslant C' \|h\|_2^2$$

However,

$$\left(1 - |z_k|^2\right) |B'(z_k)| = \prod_{j \neq k} \left| \frac{z_j - z_k}{1 - z_j \bar{z}_k} \right| \leqslant 1$$

so we have the desired conclusion (5.5). Thus, it remains only to prove (5.6). We begin with a lemma.

Lemma. (a) If $\{z_k\}$ satisfies (5.3), then there is a constant A such that

$$\sum_{j=1}^{\infty} \frac{\left(1 - |z_j|^2\right)\left(1 - |z_k|^2\right)}{|1 - z_j \bar{z}_k|^2} \leqslant A \quad \text{for } k = 1, 2, \ldots$$

(b) If a_{jk} are complex numbers with $a_{jk} = \bar{a}_{kj}$, and

$$\sum_{j=1}^n |a_{jk}| \leqslant M, \qquad k = 1, \ldots, n$$

then

$$\left| \sum_{j,k=1}^n a_{jk} x_j \bar{x}_k \right| \leqslant M \sum_{j=1}^n |x_j|^2$$

Proof. (a) We have

$$1-\left|\frac{z_k-z_j}{1-z_k\bar{z}_j}\right|^2=\frac{\left(1-|z_j|^2\right)\left(1-|z_k|^2\right)}{|1-z_j\bar{z}_k|^2}$$

Thus, the desired sum is less than

$$\sum_{j=1}^{\infty}\left(1-\left|\frac{z_k-z_j}{1-z_k\bar{z}_j}\right|^2\right)\leqslant -\log\left(\prod_{j\neq k}\left|\frac{z_k-z_j}{1-z_k\bar{z}_j}\right|^2\right)$$

$$\leqslant -2\log\delta$$

The proof of (b) is just the following computation.

$$\sum_{j,k=1}^{n}|a_{jk}x_j\bar{x}_k|\leqslant\sum_{k=1}^{n}\left(\sum_{j=1}^{n}|a_{jk}||x_j|^2\right)^{1/2}\left(\sum_{j=1}^{n}|a_{jk}||x_k|^2\right)^{1/2}$$

$$\leqslant M^{1/2}\sum_{k=1}^{n}\left(\sum_{j=1}^{n}|a_{jk}||x_j|^2\right)^{1/2}|x_k|$$

$$\leqslant M^{1/2}\left(\sum_{k=1}^{n}|x_k|^2\right)^{1/2}\left(\sum_{k=1}^{n}\sum_{j=1}^{n}|a_{jk}||x_j|^2\right)^{1/2}$$

$$\leqslant M\sum_{k=1}^{n}|x_k|^2$$

With the lemma in hand, the proof of (5.6) is direct. Set

$$B(z)=\prod_{j=1}^{n}\left(\frac{z-z_j}{1-\bar{z}_jz}\right)$$

and

$$b_k=\left(1-|z_k|^2\right)B'(z_k),\qquad k=1,\ldots,n$$

Then define

$$g_k(z)=\left(1-|z_k|^2\right)^{3/2}B^2(z)(z-z_k)^{-2}$$

and

$$f(z)=\sum_{1}^{n}w_kb_k^{-2}g_k(z)$$

It follows that

$$\left(1-|z_k|^2\right)^{1/2} f(z_k) = w_k, \qquad k = 1,\ldots,n$$

and

$$\|f\|_2^2 = \sum_{j,k=1}^{n} w_j \bar{w}_k b_j^{-2} \bar{b}_k^{-2} \langle g_j, g_k \rangle$$

However,

$$\langle g_j, g_k\rangle = \frac{1}{2\pi}\int_{-\pi}^{\pi} g_j(e^{i\theta})\bar{g}_k(e^{i\theta})\,d\theta$$

$$= \left(1-|z_j|^2\right)^{3/2}\left(1-|z_k|^2\right)^{3/2}\left(1+z_j\bar{z}_k\right)\left(1-z_j\bar{z}_k\right)^{-3}$$

so

$$|\langle g_j, g_k\rangle| \leqslant 2\left(1-|z_j|^2\right)\left(1-|z_k|^2\right)|1-z_j\bar{z}_k|^{-2}$$

We now apply (a) of the lemma to conclude that

$$\sum_{j=1}^{n} |\langle g_j, g_k\rangle| \leqslant A$$

where A is independent of n and then (b) of the lemma to learn

$$\|f\|_2^2 \leqslant \delta^{-4} A \sum_{1}^{n} |w_k|^2$$

which is exactly (5.6) with $C' = \delta^{-4}A$, a constant independent of n.

Now that Theorem 5.1 is proved we can begin the investigation of the interpolation problem (5.1). We shall do this by looking very carefully at the finite interpolation problem

$$f(z_k) = w_k, \qquad k = 1,\ldots,n \tag{5.7}$$

Proposition 5.2. Let $B(z) = \prod_1^n[(z-z_k)/(1-\bar{z}_k z)]$. Then

$$\inf\{\|f\|_\infty : f \text{ satisfies } (5.7)\} = \sup\left\{\left|\sum_1^n \frac{G(z_k)}{B'(z_k)} w_k\right| : G \in H^1, \|G\|_1 = 1\right\} \tag{5.8}$$

Proof. Note first that if f satisfies (5.7) and $G \in H^1(\Delta)$, then

$$\sum_1^n \frac{G(z_k)}{B'(z_k)} w_k = \frac{1}{2\pi i} \int_{\mathbf{T}} \frac{G(\zeta)}{B(\zeta)} f(\zeta)\, d\zeta$$

so that the right-hand side of (5.8) is no more than the left-hand side.

Now fix $q < \infty$ and let f_q be the unique element of $H^q(\Delta)$ with $f_q(z_k) = w_k$ for $k = 1, \ldots, n$ and $\|f_q\|_q$ a minimum. Any other function f in $H^q(\Delta)$ which also satisfies $f(z_k) = w_k$ for $k = 1, \ldots, n$ has the form $f = f_q + Bg$ where $g \in H^q(\Delta)$ and thus for small real ε we find

$$\int_{\mathbf{T}} |f_q|^q\, d\sigma \leqslant \int_{\mathbf{T}} |f_q + \varepsilon B g|^q\, d\sigma$$

$$= \int_{\mathbf{T}} |f_q|^q + q\varepsilon \operatorname{Re} \int_{\mathbf{T}} |f_q|^{q-2} \bar{f}_q B g\, d\sigma + o(\varepsilon)$$

and so

$$0 = \int_{\mathbf{T}} |f_q|^{q-2} \bar{f}_q B g\, d\sigma, \qquad \text{all } g \in H^q(\Delta)$$

Select for g in turn the functions z^m, $m = 0, 1, \ldots$; we see that there is a G_q in H_0^1 with

$$G_q = |f_q|^{q-2} \bar{f}_q B$$

Let $m_q = \|f_q\|$ and set $F_q(e^{it}) = e^{-it} G_q(e^{it}) m_q^{1-q}$. Then

$$\int_{\mathbf{T}} |F_q|\, d\sigma = \left(\int_{\mathbf{T}} |f_q|^{q-1}\, d\sigma \right) m_q^{1-q}$$

$$\leqslant 1$$

Further,

$$m_q = m_q^{1-q} \int_{\mathbf{T}} |f_q|^q\, d\sigma$$

$$= \int_{\mathbf{T}} f_q \frac{F_q}{B} e^{it}\, d\sigma$$

$$= \sum_1^n \frac{F_q(z_k)}{B'(z_k)} w_k$$

Let $q \to \infty$; the functions $\{F_q\}$ form a normal family by Proposition 3.2.4 and so at least a subsequence converges uniformly on compact subsets to a function F which must be in $H^1(\Delta)$, $\|F\|_1 \leqslant 1$. Clearly, m_q increases to m_∞, the left-hand side of (5.8). Thus,

$$m_\infty = \sum_1^n \frac{F(z_k)}{B'(z_k)} w_k$$

and so equality holds in (5.8) and the proof is complete.

Now consider the variational problem

$$\rho = \sup\left\{ \sum_1^n \lambda_k |g(z_k)| : g \in H^1(\Delta), \|g\|_1 \leqslant 1 \right\} \tag{5.9}$$

where the $\{\lambda_k\}$ are some given positive numbers. Suppose G attains the maximum in (5.9); such a G can be shown to exist by using problem 2 of Chapter 4. G must be outer since the presence of an inner factor of G only reduces $|G(z_k)|$ while not changing the $H^1(\Delta)$ norm of G. Let $G = \exp[u + i\,{}^*u]$. Let v be any real bounded function on $\mathbf{T}$ with

$$0 < \sum_1^n v(z_k) \lambda_k e^{u(z_k)} = \int_{\mathbf{T}} v \left(\sum_1^n \lambda_k e^{u(z_k)} P_k \right) d\sigma \tag{5.10}$$

where P_k is the Poisson kernel for $z_k = r_k e^{it_k}$:

$$P_k(e^{i\theta}) = P(r_k, \theta - t_k) = \frac{1 - r_k^2}{|e^{i\theta} - z_k|^2}$$

Then $G_\varepsilon = G\exp[\varepsilon(v + i\,{}^*v)]$ is in $H^1(\Delta)$ and, for small positive ε,

$$\sum_1^n \lambda_k |G_\varepsilon(z_k)| = \rho + \varepsilon \sum_1^n v(z_k) \lambda_k e^{u(z_k)} + o(\varepsilon)$$

$$> \rho$$

Thus, the $H^1(\Delta)$ norm of G_ε must exceed 1. Equivalently,

$$0 \leqslant \int e^u v \, d\sigma$$

Combining this with (5.10) we see that e^u must be a constant multiple of $\sum_1^n \lambda_k e^{u(z_k)} P_k$; the multiple must be $1/\rho$ to make the norms work out correctly. Hence,

$$\rho |G(e^{i\theta})| = \sum_1^n \lambda_k e^{u(z_k)} \frac{1 - r_k^2}{|e^{i\theta} - z_k|^2} \tag{5.11}$$

We now make the special choice of

$$\lambda_k = |B'(z_k)|^{-1}, \qquad k = 1,\dots,n$$

where $B(z) = \prod_1^n[(z - z_j)/(1 - \bar{z}_j z)]$. Then from (5.8) we have

$$M_n = \sup\left\{\sum_1^n \left|\frac{g(z_k)}{B'(z_k)}\right| : \|g\|_1 = 1\right\}$$

$$= \sup\left\{\sup\left\{\sum_1^n w_k \frac{g(z_k)}{B'(z_k)} : |w_k| \leqslant 1\right\} : \|g\|_1 \leqslant 1\right\}$$

$$= \sup\{\inf\{\|f\|_\infty : f(z_k) = w_k\} : |w_k| \leqslant 1\}$$

Thus, M_n is the minimum that we can make the norm of $H^\infty(\Delta)$ functions and still be assured that all elements of the unit ball of l_n^∞ can be interpolated at $z_1,\dots,z_n$.

Consider next

$$F_m(z) = \frac{B(z)}{B'(z_m)(z - z_m)} \frac{G(z_m)}{G(z)} \frac{1 - r_m^2}{1 - z\bar{z}_m}$$

where B and G are as the preceding. We have

$$F_m(z_j) = \delta_{jm}, \qquad j, m - 1,\dots,n$$

Further, on $\mathbf{T}$ we find

$$\sum_1^m |F_m(e^{i\theta})| = \frac{1}{|G(e^{i\theta})|} \sum_{m=1}^n \frac{|G(z_m)|}{|B'(z_m)||e^{i\theta} - z_m|} \frac{1 - r_m^2}{|1 - e^{i\theta} z_m|}$$

$$= M_n \quad \text{for all } \theta$$

from (5.11). Now set

$$f(z) = \sum_1^n F_m(z) w_m \tag{5.12}$$

Then

$$|f(z)| \leqslant \sum_1^n |F_m(z)| \leqslant M_n, \qquad |z| < 1 \tag{5.13a}$$

$$f(z_k) = w_k, \qquad k = 1,\dots,n \tag{5.13b}$$

Thus, f defined by (5.12) gives an interpolant whose norm, maximized over all $\{w_k\}$ in the unit ball of l_n^∞, is as small as possible.

The final step is to pass to the limit as $n \to \infty$. Suppose that $\{z_k\}$ satisfies condition (5.3) and let

$$C(z) = \prod_{1}^{\infty}\left(\frac{-\bar{z}_k}{|z_k|}\right)\left(\frac{z - z_k}{1 - \bar{z}_k z}\right)$$

Then

$$(1 - |z_k|^2)|C'(z_k)| = \prod_{\substack{j=1 \\ j \neq k}}^{\infty}\left|\frac{z_k - z_j}{1 - \bar{z}_j z_k}\right| \tag{5.14}$$

and so $|C'(z_k)|^{-1} \leqslant A(1 - |z_k|)$ where A is some constant, independent of k. Hence, the numbers M_n are no larger than

$$M_n \leqslant A \sup\left\{\sum_{1}^{n}(1 - |z_k|)|g(z_k)| : g \in H^1(\Delta), \|g\| \leqslant 1\right\}$$

However, we know from Theorem 5.1 that whenever $\{z_k\}$ satisfies (5.3), then

$$\sum_{1}^{\infty}(1 - |z_k|)|g(z_k)| \leqslant C', \qquad g \in H^1(\Delta), \qquad \|g\| \leqslant 1$$

and so $M_n \leqslant C''$ for all n. We can thus pass to the limit as $n \to \infty$ by a normal families argument.

We have, in fact, established a considerably more powerful result.

Theorem 5.3. *Let $\{z_k\}$ satisfy* (5.3). *Then there are functions $F_1, F_2, \ldots$ in $H^\infty(\Delta)$ with*

$$F_j(z_k) = \delta_{jk}, \qquad j, k = 1, 2, \ldots \tag{5.15a}$$

$$\sum_{1}^{\infty}|F_j(z)| \leqslant M, \qquad |z| \leqslant 1 \tag{5.15b}$$

where M is a constant dependent only on $\{z_k\}$.

Corollary 5.4. *Let $\{z_j\}$ satisfy condition* (5.3) *and let B be the Blaschke product for $\{z_j\}$. Then there is a closed subspace S of $H^\infty(\Delta)$ with*

$$H^\infty(\Delta) = S \oplus BH^\infty(\Delta)$$

Proof. Let S consist of those functions g in $H^\infty(\Delta)$ of the form

$$g(z) = \sum_{k=1}^{\infty} w_k F_k(z)$$

where $F_1, F_2, \ldots$ are the functions from Theorem 5.3 and $\{w_1, w_2, \ldots\}$ is a bounded sequence of complex numbers. If f is any $H^\infty(\Delta)$ function, then

$$f(z) = \sum_{k=1}^{\infty} f(z_k) F_k(z) + B(z)h(z)$$

for some $h \in H^\infty(\Delta)$ and, furthermore, this representation is clearly unique. Finally, S is a closed subspace of $H^\infty(\Delta)$ since if

$$\sum_{k=1}^{\infty} w_k^{(m)} F_k(z) \to H(z) \quad \text{as } m \to \infty$$

uniformly in Δ, then $\{w_k^{(m)}\}_{k=1}^{\infty} \to \{H(z_k)\}_{k=1}^{\infty}$ uniformly and hence

$$\left\| \sum_{k=1}^{\infty} w_k^{(m)} F_k - \sum_{k=1}^{\infty} H(z_k) F_k \right\| \leqslant M \sup_k |w_k^{(m)} - H(z_k)| \to 0$$

as $m \to \infty$ which certainly implies that $H(z) = \sum_{k=1}^{\infty} H(z_k) F_k$.

We may also apply Proposition 3.6 to the interpolation problem (5.1). Suppose (5.3) holds so that any sequence $\{w_k\} \in l^\infty$ can be interpolated by some $f \in H^\infty$. Let B be the Blaschke product for the points $\{z_k\}$.

Suppose $\{w_k\} \in l^\infty$ and there is some $f \in H^\infty(\Delta)$ with

$$f(z_k) = w_k, \qquad k = 1, 2, \ldots \tag{5.16a}$$

$$\|f\|_\infty < 1 \tag{5.16b}$$

Then $\inf\{\|\bar{B}f + h\|_\infty : h \in H^\infty\} < 1$ and so by Proposition 3.6 there is a unimodular function $v \in L^\infty$ with

$$v - \bar{B}f = g \in H^\infty$$

Equivalently,

$$Bv = f + Bg$$

But Bv is unimodular and $f + Bg$ is not only in H^∞, but also $(f + Bg)(z_k) = f(z_k) = w_k$, $k = 1, 2, \ldots$. Thus, Bv is an inner function in H^∞ which also accomplishes the interpolation. This proves the following proposition.

Proposition 5.5. If $\{w_k\} \in l^\infty$ and if there is some $f \in H^\infty(\Delta)$ satisfying (5.16), then there is an inner function G in $H^\infty(\Delta)$ with $G(z_j) = w_j$, $1 \leqslant j < \infty$.

Remark. Earle (1970) has shown something even better than the conclusion of Proposition 5.5: if $\{z_k\}$ is uniformly separated then each $\{w_k\} \in l^\infty$ can be interpolated by a constant multiple of a Blaschke product. His proof shows directly that condition (5.3) implies interpolation is possible for all $\{w_k\} \in l^\infty$ but it does not produce the sequence $\{F_k\}$ given in Theorem 5.3.

5.6. THE MAXIMUM PRINCIPLE FOR MULTIPLE-VALUED BOUNDED ANALYTIC FUNCTIONS

In this section we shall formulate and then solve an extremal problem much like that in (1.1) which led to the Ahlfors function. However, instead of working in $H^\infty(\Omega)$ we shall work in the space $H^\infty(\Omega, \boldsymbol{\alpha})$ [for the definition, see (4.12) when Ω is finitely connected]; this will lead to a criterion on the size of $|f(p)|$ if $f \in H^\infty(\Omega, \boldsymbol{\alpha})$ and $|f|$ is bounded by one. Hence, the title of the section.

We begin by reviewing the relation between the fundamental group, $\pi_1(\Omega)$, of Ω and $MH^p(\Omega)$. Let $f \in MH^p(\Omega)$; that is, $|f|$ is single-valued in Ω, $|f|^p$ has a harmonic majorant if $0 < p < \infty$ or is bounded if $p = \infty$, and each point $z_0 \in \Omega$ has a neighborhood $\mathcal{U}$ and a single-valued holomorphic function $g = g_{\mathcal{U}}$ on $\mathcal{U}$ with $|f| = |g|$ on $\mathcal{U}$. (There may be many such g for each $\mathcal{U}$.) If γ is an element of $\pi_1(\Omega)$, then analytic continuation of $g_{\mathcal{U}}$ along γ results in a function element h defined on some domain $\mathcal{V}$ in $\mathcal{U}$ for which $|h| = |g|$ on $\mathcal{V}$. Thus, $h = e^{ia}g$ on $\mathcal{V}$. That is, with γ we can associate the unimodular number e^{ia}. If γ' is another element of $\pi_1(\Omega)$ with the same base point as γ, then it is easy to see that the unimodular complex number associated to $\gamma'\gamma$ is just $e^{i(a'+a)}$. Hence, the map $\gamma \mapsto e^{ia}$ is a homomorphism of $\pi_1(\Omega)$ into $\mathbf{T}$; that is, each $f \in MH^\infty(\Omega)$ produces a character on $\pi_1(\Omega)$, an element of $\pi_1(\Omega)^*$, the character group of $\pi_1(\Omega)$. Thus, if $\boldsymbol{\alpha} \in \pi_1(\Omega)^*$ we write $H^p(\Omega, \boldsymbol{\alpha})$ for those elements of $MH^p(\Omega)$ whose associated character is $\boldsymbol{\alpha}$. It is entirely possible that $H^p(\Omega, \boldsymbol{\alpha})$ may be empty for some $\boldsymbol{\alpha} \in \pi_1(\Omega)^*$ and not empty for other $\boldsymbol{\alpha}$. Indeed, a major point of the section is to decide whether, for a given domain Ω, the classes $H^\infty(\Omega, \boldsymbol{\alpha})$, $\boldsymbol{\alpha} \in \pi_1(\Omega)$, are all nonempty. It is apparent from the discussions in Section 6 of Chapter 4 that $H^p(\Omega, \boldsymbol{\alpha})$ is nonempty for all $\boldsymbol{\alpha} \subset \pi_1(\Omega)^*$ if $\partial\Omega$ consists of a finite number of disjoint analytic simple closed curves. We begin the discussion by analyzing this finitely connected case in detail and finding the best possible constant for the associated extremal problem. We conclude with some extensions of the results to certain infinitely connected domains Ω.

In the material that follows when Ω is a domain bounded by a finite number of disjoint analytic simple closed curves, we use the notation $H^p(\Omega, \boldsymbol{\alpha})$ both for the functions on Ω and for their boundary values on $\Gamma = \partial\Omega$.

Suppose then that Ω is a domain whose boundary, Γ, consists of $m + 1$ disjoint analytic simple closed curves. For each point p in Ω and each $\boldsymbol{\alpha} \in \mathbf{T}^m$ define

$$m(p, \boldsymbol{\alpha}, \Omega) = \sup\{|f(p)| : f \in H^\infty(\Omega, \boldsymbol{\alpha}), \|f\|_\infty \leqslant 1\} \tag{6.1}$$

Note first that if $\boldsymbol{\alpha}_0 = (1, \dots, 1)$, then $m(p, \boldsymbol{\alpha}_0, \Omega) = 1$ for all $p \in \Omega$. Also take note that the assumption on the nature of Γ implies that there are many f in $H^\infty(\Omega, \boldsymbol{\alpha})$ with $f(p) \neq 0$ and hence $m(p, \boldsymbol{\alpha}, \Omega) > 0$ for all $\boldsymbol{\alpha} \in \mathbf{T}^m$ and all $p \in \Omega$. The number $m(p, \boldsymbol{\alpha}, \Omega)$ has the property that

$$|f(p)| \leqslant m(p, \boldsymbol{\alpha}, \Omega)\|f\|_\infty, \qquad f \in H^\infty(\Omega, \boldsymbol{\alpha}), \qquad p \in \Omega \tag{6.2}$$

We shall be interested in determining how small $m(p, \boldsymbol{\alpha}, \Omega)$ becomes as $\boldsymbol{\alpha}$ varies over $\mathbf{T}^m$; the precise answer is given in Theorem 6.2.

Next, for $p \in \Omega$ and $\boldsymbol{\beta} \in \mathbf{T}^m$ set

$$m'(p, \boldsymbol{\beta}, \Omega) = \inf\left\{\frac{1}{2\pi}\int_\Gamma |h(\zeta)||\zeta - p|^{-2}|d\zeta|\right\} \tag{6.3}$$

where the infimum is taken over all h satisfying

$$h \in H^1(\Omega, \boldsymbol{\beta}) \tag{6.4a}$$

$$h(p) = 0 \tag{6.4b}$$

$$h'(p) = 1 \tag{6.4c}$$

Once again, the assumption about Γ implies that there are many functions h satisfying (6.4). (Here we may assume that p is not in one of the cuts that define the classes $H(\Omega, \boldsymbol{\alpha})$; see Section 4 of this chapter.) A normal families argument guarantees that there is at least one $F \in H^\infty(\Omega, \boldsymbol{\alpha})$ with

$$\|F\|_\infty = 1 \tag{6.5a}$$

$$F(p) = m(p, \boldsymbol{\alpha}, \Omega) \tag{6.5b}$$

and at least one $H \in H^1(\Omega, \boldsymbol{\beta})$ with

$$H(p) = 0 \tag{6.6a}$$

$$H'(p) = 1 \tag{6.6b}$$

$$\frac{1}{2\pi}\int_\Gamma |H(\zeta)||\zeta - p|^{-2}|d\zeta| = m'(p, \boldsymbol{\beta}, \Omega) \tag{6.6c}$$

Theorem 6.1. *Suppose $\partial\Omega$ consists of $m + 1$ disjoint analytic simple closed curves. For each $\boldsymbol{\alpha} \in \mathbf{T}^m$ and each $p \in \Omega$ we have*

$$m(p, \boldsymbol{\alpha}, \Omega) = m'(p, -\boldsymbol{\alpha}, \Omega) \tag{6.7}$$

Further, the F satisfying (6.5) is uniquely determined. If $\boldsymbol{\alpha} \neq (1, \ldots, 1)$, then F is a finite Blaschke product with m or fewer zeros in Ω, say at $z_1, \ldots, z_r$, and

$$m(p, \boldsymbol{\alpha}, \Omega) = \exp\left[-\sum_1^r g(z_j; p)\right] \tag{6.8}$$

where $g(\cdot; p)$ is the Green's function for Ω with singularity at p.

Proof. It is easy to prove that the left-hand side of (6.7) is smaller than the right. For suppose $f \in H^\infty(\Omega, \boldsymbol{\alpha})$ with $\|f\|_\infty \leqslant 1$ and that h satisfies (6.4) with

$\beta = -\alpha$. Then $fh \in H^1(\Omega)$ and

$$f(p) = (fh)'(p) = \frac{1}{2\pi i}\int_\Gamma (fh)(\zeta)(\zeta - p)^{-2}\, d\zeta$$

so that

$$|f(p)| \leqslant \|f\|_\infty \frac{1}{2\pi}\int_\Gamma |h(\zeta)||\zeta - p|^{-2}|d\zeta|$$

and this immediately implies that $m(p, \alpha, \Omega) \leqslant m'(p, -\alpha, \Omega)$.

To obtain the reverse inequality let $d\mu$ be the measure $(\zeta - p)^{-2}\, d\zeta$ on Γ. We know that $H^1(\Omega, -\alpha)$ is a subspace of $L^1(\Gamma, |\mu|)$ and that the norm of the linear functional $h \mapsto h'(p)$ acting on those $h \in H^1(\Omega, -\alpha)$ with $h(p) = 0$ is exactly $[m'(p, -\alpha, \Omega)]^{-1}$. The Hahn-Banach theorem implies that there is a norm preserving extension of this functional to all of $L^1(\Gamma, |\mu|)$. Thus, there is a function $g \in L^\infty(\Gamma, |\mu|)$ with

$$\|g\|_\infty = [m'(p, -\alpha, \Omega)]^{-1} \tag{6.9a}$$

$$h'(p) = \frac{1}{2\pi i}\int_\Gamma h(\zeta)g(\zeta)\, d\mu(\zeta), \quad h \in H^1(\Omega, -\alpha), \quad h(p) = 0 \tag{6.9b}$$

In particular,

$$0 = \int_\Gamma h(\zeta)g(\zeta)\, d\zeta, \qquad h \in H^1(\Omega, -\alpha) \tag{6.10}$$

Fix a function u_0 in $H^\infty(\Omega, -\alpha)$ with $|u_0| \geqslant \delta > 0$ on Ω for some $\delta > 0$; this is possible by Proposition 4.6.1. Then (6.10) implies

$$0 = \int_\Gamma u_0(\zeta)g(\zeta)\frac{1}{\zeta - z}\, d\zeta, \qquad \text{all } z \notin \Omega \cup \Gamma$$

and so by Theorem 4.4.7 there is a $v \in H^\infty(\Omega)$ with $v = u_0 g$ a.e. ds on Γ; equivalently there is an $F \in H^\infty(\Omega, \alpha)$ with $F = (m'(p, -\alpha, \Omega))g$ a.e. ds on Γ. Hence, $\|F\|_\infty = 1$. Now the function FH is in $H^1(\Omega)$ and we have

$$\begin{aligned} F(p) = (FH)'(p) &= \frac{1}{2\pi i}\int_\Gamma (FH)(\zeta)(\zeta - p)^{-2}\, d\zeta \\ &= m'(p, -\alpha, \Omega)\frac{1}{2\pi i}\int_\Gamma g(\zeta)H(\zeta)\, d\mu(\zeta) \\ &= m'(p, -\alpha, \Omega)H'(p) = m'(p, -\alpha, \Omega) \end{aligned}$$

Thus, $m'(p, -\alpha, \Omega) \leqslant m(p, \alpha, \Omega)$ and we're done with (6.7).

We further have

$$\begin{aligned} m(p,\alpha,\Omega) &= F(p) = (FH)'(p) \\ &= \frac{1}{2\pi i}\int_\Gamma F(\zeta)H(\zeta)(\zeta-p)^{-2}\,d\zeta \\ &\leqslant \frac{1}{2\pi}\int_\Gamma |F(\zeta)||H(\zeta)||\zeta-p|^{-2}|d\zeta| \\ &\leqslant \|F\|_\infty \frac{1}{2\pi}\int_\Gamma |H(\zeta)||\zeta-p|^{-2}|d\zeta| \\ &= m'(p,-\alpha,\Omega) = m(p,\alpha,\Omega) \end{aligned}$$

From this we learn that

$$F(\zeta)H(\zeta)(\zeta-p)^{-2}\,d\zeta \geqslant 0 \quad \text{as a measure on } \Gamma \tag{6.11a}$$

$$|F(\zeta)| = 1 \quad \text{a.e. } ds \text{ on } \Gamma \text{ where } H \neq 0 \tag{6.11b}$$

However, H is nonzero a.e. ds on Γ so that $|F| = 1$ a.e. ds on Γ. Moreover, we know from Proposition 1.6.5 that

$$\frac{i}{2\pi}Q'(\zeta)\,d\zeta = d\omega_p(\zeta)$$

where $Q(z) = g(z;p) + ih(z;p)$ and ω_p is harmonic measure for p. Thus,

$$\frac{F(\zeta)H(\zeta)}{i(\zeta-p)^2Q'(\zeta)} \geqslant 0 \quad \text{on } \Gamma \tag{6.12}$$

Of course, (6.12) implies both F and H can be locally analytically continued over Γ, just as we saw earlier in Sections 1 and 4 of this chapter. Hence, F is a finite Blaschke product with zeros at $z_1,\dots,z_r$. Now Q' has a pole of order 1 at p with residue 1 and H vanishes at p. Further, Q' has m zeros on Ω. Thus, the function

$$\frac{(FH)(z)}{i(z-p)^2Q'(z)}$$

is meromorphic (and single-valued) on Ω and non-negative on Γ; it has m or fewer poles in Ω and therefore also m or fewer zeros in Ω. Thus, F has no more than m zeros in Ω; that is, $r \leqslant m$. Let

$$|F(z)| = \exp\left[-\sum_1^r g(z;z_j)\right]$$

Then $m(p,\alpha,\Omega) = |F(p)| = \exp[-\sum_1^r g(z_j;p)]$.

To prove the uniqueness of F is easy. We've shown that each function satisfying (6.5) has unit modulus on Γ. If two functions satisfied (6.5) then so would one-half their sum. But one-half the sum of unimodular functions can not be unimodular on Γ unless the two functions are identical.

Theorem 6.2. *Let Ω be bounded by $m+1$ disjoint analytic simple closed curves. As $\boldsymbol{\alpha}$ varies over $\mathbf{T}^m$, the numbers $m(p, \boldsymbol{\alpha}, \Omega)$ vary over the entire closed interval*

$$\left\{t\colon \exp\left\{-\sum_1^m g(z_j^*;p)\right\} \leqslant t \leqslant 1\right\} \tag{6.13}$$

where $z_1^*, \ldots, z_m^*$ are the critical points of $g(\cdot\,;p)$.

Proof. Clearly $m(p, \boldsymbol{\alpha}, \Omega) = \exp[-\Sigma_1^r g(z_j; p)] \leqslant 1$ for all choices of $\boldsymbol{\alpha}$ and the upper limit of 1 is attained when $\boldsymbol{\alpha} = \boldsymbol{\alpha}_0 = (1, \ldots, 1)$. Let us show that $m(p, \boldsymbol{\alpha}, \Omega)$ always exceeds the left-hand endpoint of the interval in (6.13).

Let B be the element of $MH^\infty(\Omega)$ defined by

$$|B(z)| = \exp\left[-\sum_1^m g(z; z_j^*)\right]$$

and let $Q(z) = g(z;p) + ih(z;p)$ as before. Take any $\boldsymbol{\alpha} \in \mathbf{T}^m$ and let H be the function satisfying (6.6). Then

$$m(p, \boldsymbol{\alpha}, \Omega) = \frac{1}{2\pi}\int_\Gamma |H(\zeta)||\zeta - p|^{-2}|d\zeta|$$

$$= \int_\Gamma |H(\zeta)||\zeta - p|^{-2}|B(\zeta)||Q'(\zeta)|^{-1}\,d\omega_p(\zeta)$$

where we again used Proposition 1.6.5 and the fact that $|B(\zeta)| = 1$ identically on Γ. However, the function

$$v(z) = |H(z)||z-p|^{-2}|B(z)||Q'(z)|^{-1}, \qquad z \in \Omega$$

is subharmonic in Ω and continuous on $\Omega \cup \Gamma$. Thus, the value of v at p is no more than $\int_\Gamma v\,d\omega_p = m(p, \boldsymbol{\alpha}, \Omega)$. Hence,

$$|B(p)| = v(p) \leqslant \int_\Gamma v\,d\omega_p = m(p, \boldsymbol{\alpha}, \Omega)$$

which shows that $\exp[-\Sigma_1^m g(z_j^*; p)]$ is a lower bound for all $m(p, \boldsymbol{\alpha}, \Omega)$. The lower bound can actually be attained. Let

$$h(z) = Q'(z)\frac{B(p)}{B(z)}(z-p)^2, \qquad z \in \Omega$$

Then h lies in $H^\infty(\Omega, \boldsymbol{\beta})$ for some choice of $\boldsymbol{\beta} \in \mathbf{T}^m$ and $h(p) = 0$, $h'(p) = 1$. Hence,

$$m(p, \boldsymbol{\beta}, \Omega) \leqslant \frac{1}{2\pi} \int_\Gamma |h(\zeta)||\zeta - p|^{-2} |d\zeta|$$

$$= |B(p)| \frac{1}{2\pi} \int_\Gamma |Q'(\zeta)||d\zeta|$$

$$= |B(p)| \int_\Gamma d\omega_p = |B(p)|$$

To finish the proof, we need only prove that $m(p, \boldsymbol{\alpha}, \Omega)$ depends continuously on $\boldsymbol{\alpha}$, $\boldsymbol{\alpha} \in \mathbf{T}^m$. Suppose, then, that $\boldsymbol{\alpha}_n \to \boldsymbol{\alpha}$ in $\mathbf{T}^m$; we must show $m(p, \boldsymbol{\alpha}_n, \Omega) \to m(p, \boldsymbol{\alpha}, \Omega)$. A simple normal families argument shows that

$$m(p, \boldsymbol{\alpha}, \Omega) \geqslant \limsup\{m(p, \boldsymbol{\alpha}_n, \Omega) : n \to \infty\}$$

On the other hand, by Proposition 4.6.1 there is a sequence $\{h_n\}$ of functions with these properties

$$h_n \in H^\infty(\Omega, \boldsymbol{\alpha}_n - \boldsymbol{\alpha}) \tag{6.14a}$$

$$\|h_n\|_\infty \leqslant 1 \tag{6.14b}$$

$$h_n(z) \to 1 \text{ uniformly on compact sets in } \Omega \tag{6.14c}$$

Let F satisfy (6.5) and consider $g_n = Fh_n$. Then g_n lies in $H^\infty(\Omega, \boldsymbol{\alpha}_n)$, $\|g_n\| \leqslant 1$, and so

$$m(p, \boldsymbol{\alpha}_n, \Omega) \geqslant |g_n(p)| = |F(p)||h_n(p)|$$

$$= m(p, \boldsymbol{\alpha}, \Omega)|h_n(p)|$$

Hence, (6.14c) implies $\liminf\{m(p, \boldsymbol{\alpha}_n, \Omega) : n \to \infty\} \geqslant m(p, \boldsymbol{\alpha}, \Omega)$. This completes the proof of Theorem 6.2.

Let us define

$$m(p, \Omega) = \inf\{m(p, \boldsymbol{\alpha}, \Omega) : \boldsymbol{\alpha} \in \mathbf{T}^m\}$$

so that Theorem 6.2 gives the result

$$m(p, \Omega) = \exp\left[-\sum_1^m g(z_j^*, p)\right] \tag{6.15}$$

provided that $\Gamma = \partial\Omega$ consists of $m + 1$ disjoint analytic simple closed curves. The equality in (6.15) holds as well, by conformal invariance, if the complement of Ω is composed of $m + 1$ disjoint nontrivial continua. We shall now briefly look into extensions of (6.15) to (certain) infinitely connected domains.

Let us begin by defining

$$m(p, \Omega) = \inf\{m(p, \alpha, \Omega) : \alpha \in \pi_1(\Omega)^*\} \tag{6.16}$$

Note that if $m(p, \Omega)$ is positive then each of sets $H^\infty(\Omega, \alpha)$ is surely nonempty and hence any criteria which give us a formula for the size of $m(p, \Omega)$ can help us decide whether there are $H^\infty(\Omega, \alpha)$ functions for every $\alpha \in \pi_1(\Omega)^*$.

The next two results concern the dependence of $m(p, \Omega)$ on Ω. We shall have to make use of the fact that if H is a subgroup of the topological group G then each character on H may be extended to a character on G; see Rudin (1962).

Proposition 6.3. Suppose Ω and Λ are domains with $\Lambda \subset \Omega$ and each component of Λ^C contains a component of Ω^C. If $p \in \Lambda$, then

$$m(p, \Lambda) \geqslant m(p, \Omega)$$

Proof. Let $\alpha \in \pi_1(\Lambda)^*$ be chosen so that

$$m(p, \Lambda) + \varepsilon \geqslant m(p, \alpha, \Lambda)$$

Then note that

$$m(p, \alpha, \Lambda) \geqslant m(p, \tilde{\alpha}, \Omega)$$

where $\tilde{\alpha}$ is any extension to $\pi_1(\Omega)^*$ of α. But $m(p, \tilde{\alpha}, \Omega) \geqslant m(p, \Omega)$. This completes the proof.

Definition. Let Ω be a domain. A *completely regular exhaustion* of Ω is a sequence $\{\Omega_n\}$ of subdomains of Ω such that

(a) $\Omega_n \subset \mathrm{CL}\,\Omega_n \subset \Omega_{n+1}$, $n = 1, 2, \ldots$ and $\cup\Omega_n = \Omega$
(b) $\partial\Omega_n$ is a finite number of disjoint analytic simple closed curves for $n = 1, 2, \ldots$
(c) each component of Ω_n^C contains a component of Ω^C for $n = 1, 2, \ldots$

We leave it as an exercise to prove that each domain Ω has a completely regular exhaustion.

Proposition 6.4. Suppose $\{\Omega_n\}$ is a completely regular exhaustion of Ω. Then for each $p \in \Omega$,

$$\lim\{m(p, \Omega_n) : n \to \infty\} = m(p, \Omega) \tag{6.17}$$

Proof. We know from Proposition 6.3 that $m(p, \Omega_n) \geqslant m(p, \Omega)$ for all n so that

$$\liminf\{m(p, \Omega_n): n \to \infty\} \geqslant m(p, \Omega)$$

Now fix $\boldsymbol{\alpha} \in \pi_1(\Omega)^*$ and let $\boldsymbol{\alpha}_n$ be any character on $\pi_1(\Omega_n)$ which extends to $\boldsymbol{\alpha}$. Let f_n be an element of $H^\infty(\Omega_n, \boldsymbol{\alpha}_n)$ with

$$\|f_n\| \leqslant 1, \qquad f_n(p) = m(p, \boldsymbol{\alpha}_n, \Omega_n)$$

A normal families argument produces an f with $\|f\| \leqslant 1$, $f(p) = \limsup\{m(p, \boldsymbol{\alpha}_n, \Omega_n): n \to \infty\}$. Since each cycle in Ω is a compact set in Ω, we also must have $f \in H^\infty(\Omega, \boldsymbol{\alpha})$. Hence,

$$m(p, \boldsymbol{\alpha}, \Omega) \geqslant |f(p)| \geqslant \limsup\{m(p, \boldsymbol{\alpha}_n, \Omega_n): n \to \infty\}$$

$$\geqslant \limsup\{m(p, \Omega_n): n \to \infty\}$$

Thus, $m(p, \Omega) \geqslant \limsup\{m(p, \Omega_n): n \to \infty\}$ and the proof is finished.

Suppose Ω is a domain for which the Dirichlet problem is solvable; let $g(z; p)$ be the Green's function for Ω with pole at p and for $t > 0$ set

$$\Omega_t = \{z \in \Omega : g(z; p) > t\} \tag{6.18}$$

Then the closure of Ω_t is exactly those points z at which $g(z; p) \geqslant t$ and this is a compact set in Ω. Further, Ω_t is a domain, for if $\mathcal{U}$ is a component of Ω_t which does not contain p then $g(z; p)$ is harmonic in $\mathcal{U}$, continuous on CL $\mathcal{U}$, and $g(z; p) = t$ identically on $\partial\mathcal{U}$, which implies $g(z; p) = t$ in $\mathcal{U}$, a contradiction. Further, the domain Ω_t has only a finite number of components in its complement. This follows from the readily established fact that the boundary of Ω_t remains at a positive distance from $\partial\Omega$, because $g(z; p)$ is continuous at $\partial\Omega$ and vanishes there.

These remarks prepare the way for the next theorem.

Theorem 6.5. *Let Ω be a domain for which the Dirichlet problem is solvable. For each positive number t let*

$$\Omega_t = \{z \in \Omega : g(z, p) > t\}$$

and let $B(t)$ be one less than the number of components of the complement of Ω_t. Then

$$m(p, \Omega) = \exp\left[-\int_0^\infty B(t)\, dt\right] \tag{6.19}$$

Proof. Assume first that Ω is bounded by $m + 1$ disjoint nontrivial continua. Then Theorem 6.2 gives

$$m(p, \Omega) = \exp\left[-\sum_1^m g(z_j^*; p)\right]$$

where g is the Green's function for Ω and $z_1^*, \ldots, z_m^*$ are the critical points of $g(z; p)$. Now Ω_t has Green's function $g(z; p) - t$ so the number of points among $\{z_1^*, \ldots, z_m^*\}$ at which $g(z_j^*; p)$ exceeds t is exactly $B(t)$. Thus,

$$\sum_1^m g(z_j^*; p) = \int_0^\infty B(t)\, dt$$

and the formula (6.19) follows. For a general Ω, let $\{\Omega_n\}$ be a completely regular exhaustion of Ω and let $B_n(t)$ be the counting function for Ω_n, $n = 1, 2, \ldots$. Then

$$B_n(t) \leqslant B_{n+1}(t) \quad \text{and} \quad \lim_{n \to \infty} B_n(t) = B(t)$$

Thus, the monotone convergence theorem and Proposition 6.4 combine to give

$$\begin{aligned} \exp\left[-\int_0^\infty B(t)\, dt\right] &= \lim_{n \to \infty} \exp\left[-\int_0^\infty B_n(t)\, dt\right] \\ &= \lim_{n \to \infty} m(p, \Omega_n) \\ &= m(p, \Omega) \end{aligned}$$

Remark. Theorem 6.5 is true without the assumption that Ω is regular for the Dirichlet problem but it is necessary to know that Ω_t is a domain. This is true but more difficult to prove; see Widom (1971).

Theorem 6.6. *Let Λ be a domain on the sphere whose complement, $\mathbf{K}$, is a nontrivial continuum. Let S be a finite or countable set in Λ accumulating only in $\mathbf{K}$ (if at all) and set $\Omega = \Lambda \setminus S$. Then*

$$m(p, \Omega) = \exp\left[-\sum_{s \in S} g(s; p)\right] \tag{6.20}$$

where $g(z; p)$ is the Green's function for Λ.

Proof. The Green's function for Ω is also $g(z; p)$ since S is at most countable. Further.

$$\Omega_t = \{z \in \Omega : g(z, p) > t\}$$

is obtained by deleting from the simply connected set $\{z \in \Delta : g(z, p) > t\}$ those points $s \in S$, finite in number, at which $g(s, p) > t$. Hence, Ω_t is a domain with finitely many complementary components and (6.19) holds for Ω, even though Ω is not regular for the Dirichlet problem. Further, $B(t)$ is exactly equal to the number of points $s \in S$ at which $g(s; p) > t$; thus, (6.20) follows immediately.

Definition. A *Blaschke domain* is a domain Ω obtained from the open unit disc by deleting a countable collection $\{\mathbf{K}_j\}$ of disjoint continua with these two properties:

if $z_j \in \mathbf{K}_j$ for $j = 1, 2, \ldots$, then all limit points of $\{z_j\}$ lie in $\mathbf{T}$ (6.21a)

there is some choice of $s_j \in \mathbf{K}_j$, for $j = 1, 2, \ldots$ for which $\sum_{j=1}^{\infty} (1 - |s_j|) < \infty$ (6.21b)

Corollary 6.7. *If Ω is a Blaschke domain then $m(p, \Omega) > 0$ for each $p \in \Omega$.*

Proof. Consider the domain $\tilde{\Omega} = \Delta - \{s_j\}_1^{\infty}$; we know from Theorem 6.6 that

$$m(p, \tilde{\Omega}) = \exp\left[-\sum_1^{\infty} \log\left|\frac{1 - \bar{s}_j p}{s_j - p}\right|\right]$$

$$= \prod_1^{\infty} \left|\frac{s_j - p}{1 - \bar{s}_j p}\right| > 0$$

However, since $\Omega \subset \tilde{\Omega}$ and each component of Ω^C contains a component of $\tilde{\Omega}^C$ we have

$$m(p, \Omega) \geqslant m(p, \tilde{\Omega})$$

by Proposition 6.3.

ADDITIONAL READINGS AND NOTES

The Ahlfors function was first investigated, but not named, by Ahlfors (1947, 1950). The proof of uniqueness given in Theorem 1.1 is from Fisher (1969b); earlier proofs of uniqueness for the general case can be found in Havinson (1964) and Carleson (1967). More on dominating sets can be found in Rubel-Shields (1966). Theorem 1.3 is from Fisher (1972), while Theorem 1.4 is from Fisher (1969b), as is Theorem 1.5. Ahlfors proved Theorem 1.6 in his seminal work (1947) on the subject. The Garabedian function is extensively studied in Garnett's lecture notes (Garnett, 1972); also see the references therein. Theorem 2.1 is due to Akutowitz (1956). Theorems 2.2 and its corollaries are due to Frostman (1935). Theorem 2.6 is in Stout (1968). Theorem 3.2 was proved by Douglas and Rudin (1969); the remainder of the results of Section 3 are from papers of Marshall (1976), Garnett (1977), and Bernard, Garnett, Marshall (1977). Theorem 3.10 was originally proved in Fisher (1969a). The material on Pick-Nevanlinna interpolation on the disc is classical; see Goluzin (1969) and, for an elementary proof, Marshall (1974). The version given here for a finitely connected domain Ω is from Abrahamse (1979). The characterization of interpolation sequences in Δ is due to Carleson (1958); he also proved Proposition 5.2. The construction of the functions described in Theorem 5.3 is to be found in Carleson's paper (1962a); Carleson attributes it to Pehr Beurling. The discussion of interpolation sequences given here by no means exhausts the subject; in particular, it was carried out without recourse to *Carleson measures*. This important topic is given a full treatment in the recent book of Garnett (1981).

EXERCISES

1. Let Ω be bounded by $m + 1$ disjoint analytic simple closed curves. Let $f \in MH^\infty(\Omega)$ with $|f| \leqslant 1$ on Ω. Show that f is a constant multiple of a Blaschke product if and only if

$$\lim_{n\to\infty} \int_{\partial\Omega_n} \log|f(\zeta)|\, d\omega_n(\zeta) = 0$$

for each regular exhaustion $\{\Omega_n\}$ of Ω; here ω_n is harmonic measure on $\partial\Omega_n$ for $z_0 \in \Omega$.

2. Let Ω be as in exercise 1 and let $f \in H^\infty(\Omega)$, $|f| \leqslant 1$. Suppose T is the uniformizer of Ω, T: $\Delta \to \Omega$. Show that f is inner if and only if $f \circ T$ is inner on Δ. Does T preserve Blaschke products? singular inner functions?

3. Extend Theorem 2.2 and Corollaries 2.3 and 2.4 to the case when Ω is bounded by a finite number of disjoint analytic simple closed curves.

4. Let Ω be bounded by $m + 1$ disjoint analytic simple closed curves and let $\{z_j\}_1^\infty$ be a sequence in Ω. Prove the following assertions to be equivalent.

a. $\{z_j\}$ is an interpolation sequence for $H^\infty(\Omega)$.

b. there is an $M < \infty$ such that $\sum_{j=1, j\neq l}^{\infty} g(z_j, z_l) \leqslant M$, $l = 1,2,\ldots$ where $g(z, \zeta)$ is the Green's function for Ω with singularity at ζ.

c. If $\{z_j\}_1^\infty = S_0 \cup \cdots \cup S_m$ where $S_j \cap S_k = \varnothing$ for $j \neq k$ and all limit points of S_j lie on Γ_j, $0 \leqslant j \leqslant m$, then S_j is an interpolation sequence for $H^\infty(\mathcal{U}_j)$, $j = 0,\ldots, m$ (notation that of Chapter 4).

5. Let $z_0,\ldots, z_n$ be distinct points of Ω, Ω as in exercise 4. For $w_0,\ldots, w_n$ complex numbers with $\Sigma_0^n |w_j|^2 = 1$ let

$$\sigma(\mathbf{w}) = \inf\{\|f\|_\infty : f \in H^\infty(\Omega), f(z_j) = w_j, 0 \leqslant j \leqslant n\}$$

Show:

a. $\sigma(\mathbf{w})$ varies continuously as $\mathbf{w}$ varies over $\mathbf{S}^{2n+1}$.

b. There is an $F_{\mathbf{w}} \in H^\infty(\Omega)$ with $\|F_{\mathbf{w}}\| = \sigma(\mathbf{w})$ and $F_{\mathbf{w}}(z_j) = w_j$, $0 \leqslant j \leqslant n$.

c. $F_{\mathbf{w}}$ is unique and $[\sigma(\mathbf{w})]^{-1}F_{\mathbf{w}}$ is a Blaschke product of degree at most $n + m$.

d. The map $\mathbf{w} \mapsto [\sigma(\mathbf{w})]^{-1}F_{\mathbf{w}}$ is continuous from $\mathbf{S}^{2n+1}$ into the set of Blaschke products of degree at most $n + m$ where the latter is given the topology of uniform convergence on compact subsets on Ω.

6. Let $\Omega = \{z : 0 < |z| < 1\}$; then $\pi_1(\Omega) = \mathbb{Z}$ and $\pi_1(\Omega)^* = \mathbf{T}$. Give an explicit function f in $H^\infty(\Omega, \lambda)$, for each $\lambda \in \mathbf{T}$. Do the same if Ω is obtained from Δ by deleting a finite number of points.

In exercises 7–13, Ω is a domain bounded by $m + 1$ disjoint analytic simple closed curves.

7. Let $G(z) = r(z)Q'(z)$ where $r(z)$ is given in (1.8). Show that G has these properties:

a. G is holomorphic on a neighborhood of $\Omega \cup \Gamma$ except at p.

b. $\lim\{(z - p)^2 G(z); z \to p\} = 1$.

c. $iG(z)F(z)\,dz \geqslant 0$ as a measure on Γ; F is the Ahlfors function for p and Ω.

d. $(1/2\pi)\int_\Gamma |G(z)|\,ds = \gamma$, γ given by (1.1).

e. G is not zero on $\Omega \setminus \{p\}$.

f. The change in $\arg[(z - p)^2 G(z)]$ along each component Γ_j of Γ, $0 \leqslant j \leqslant m$, is zero.

8. Show that there is a holomorphic function K on Ω with $(K(z))^2 = (z - p)^2 G(z)$, $K(p) = 1$. Use exercise 7(f).

9. If $h \in H^2(\Omega)$ and $h(p) = 1$, then

$$\gamma \leqslant \frac{1}{2\pi}\int \frac{|h(\zeta)|^2}{|\zeta - p|^2}\,d|\zeta|$$

with equality only for $h = K$, K from exercise 8.

10. Use exercise 9 to show that K is uniquely determined as the element of the convex set $\Lambda = \{h \in H^2(\Omega) : h(p) = 1\}$ for which

$$\int_\Gamma \frac{|h(z)|^2}{|\zeta - p|^2}\,ds$$

is a minimum. Conclude as well that G is unique.

11. Let $\mathcal{W}$ consist of all holomorphic functions on Ω whose real part is positive. Fix distinct points $z_1, \ldots, z_n$ in Ω and set

$$\Lambda = \{(f(z_1), \ldots, f(z_n)) : f \in \mathcal{W}\}$$

Describe the boundary points of Λ. If $(w_1, \ldots, w_n) \in \Lambda$ describe the element F of $\mathcal{W}$ which satisfies

a. $F(z_j) = w_j, \qquad 1 \leqslant j \leqslant n$

b. $F(z_0) = \min\{f(z_0) : f \text{ satisfies (a)}, f \in \mathcal{W}\}$

12. What happens in Theorem 4.3 (or 4.4) if the points $z_1, \ldots, z_n$ coalesce? That is, is it possible to give a necessary and sufficient condition for there to be a solution of

$$f^{(j)}(z_0) = w_j, \qquad 0 \leqslant j \leqslant n - 1$$

$$\|f\|_\infty \leqslant 1?$$

13. Let $z_1, \ldots, z_n$ be distinct points of Ω and set

$$\Lambda_1 = \{(f(z_1), \ldots, f(z_n)) : f \in H^1(\Omega), \|f\|_1 \leqslant 1\}$$

Describe the boundary of Λ_1 in the manner of Theorem 4.1; see Goluzin (1969, chapter XI, section 7) for the case of the unit disc.

14. Suppose $z_1, \ldots, z_n$ are distinct points on $\mathbf{T}$ and $w_1, \ldots, w_n$ are points in $\mathbf{T}$. Does Theorem 4.3 give any help in deciding if there is an $f \in \mathbf{A}(\Delta)$ with $f(z_j) = w_j$, $1 \leqslant j \leqslant n$? Prove directly that there is an inner function f in $\mathbf{A}(\Delta)$ with $f(z_j) = w_j$ for $j = 1, \ldots, n$; see Abrahamse and Fisher (1980).

15. If B is a finite Blaschke product on the unit disc Δ of degree 2 or more, then B' has a zero somewhere in Δ; see Carathéodory (1960).

16. Show that a rational function G maps the upper half-plane $U = \{z = x + iy : y > 0\}$ into itself and is real on $\mathbb{R}$ if and only if it has the form

$$G(z) = c + \int \frac{1 + tz}{t - z}\, d\mu(t)$$

where $c \in \mathbb{R}$ and μ is a non-negative measure on $\mathbb{R}$ whose support is a finite number of points. These functions are called Cayley inner functions and are the upper half-plane equivalent of a finite Blaschke product.

17. Prove each domain Ω has a completely regular exhaustion. HINT: let $\{\Omega_n'\}$ be an exhaustion of Ω, let $p \in \Omega_0$, and let $g_n(z)$ be the Green's function for Ω_n' with pole at p. Set $\Omega_n = \{z \in \Omega_n' : g_n(z) > \varepsilon_n\}$ where ε_n goes to zero rapidly with n.

18. If Ω is regular for the Dirichlet problem, prove

$$m(p, \Omega) = \exp\left[-\sum g(z_j^*; p)\right]$$

where the sum is extended over all those points $z_1^*, z_2^*, \ldots$ which are critical points of $g(z; p)$ (that is, the complex derivative of $g(z; p) + ih(z; p)$ vanishes at z_j^*).

19. Let $\{z_j\}_1^\infty$ be distinct points in Δ and let $\{w_1\}_{j=1}^\infty$ be points of Δ. Show that there is an $f \in H^\infty(\Delta)$ with

$$\begin{cases} \|f\|_\infty \leqslant 1 \\ f(z_j) = w_j, \qquad 1 \leqslant j < \infty \end{cases}$$

if and only if

$$\sum_{n,m=1}^{N} \alpha_n \bar{\alpha}_m \frac{1 - w_n \bar{w}_m}{1 - z_n \bar{z}_m} \geqslant 0 \quad \text{for all } \alpha_1, \ldots, \alpha_n \in \mathbf{C}, \text{ all } N$$

20. Let Ω be a Blaschke domain and let T be its uniformizer. Prove that a function $f \in H^\infty(\Omega)$, $\|f\|_\infty = 1$, is an extreme point of the unit ball of $H^\infty(\Omega)$ if and only if the function $g = f \circ T$ has the property that $u = \log(1 - |g|)$ is *not* in $L^1(\mathbf{T}, \sigma)$.

6

THE MAXIMAL IDEAL SPACE OF $H^\infty(\Omega)$

6.1. INTRODUCTION

In this chapter we study the structure of the space $\mathbf{M}(\Omega)$ consisting of those continuous linear functionals ϕ on $H^\infty(\Omega)$ which are multiplicative:

$$\phi(fg) = \phi(f)\phi(g), \qquad f, g \in H^\infty(\Omega) \tag{1.1}$$

The obvious example of such a ϕ is "evaluation at p"

$$\phi_p(f) = f(p), \qquad f \in H^\infty(\Omega), \qquad p \in \Omega \tag{1.2}$$

There are, however, a multitude more ϕ which are not given by evaluation at a point of the domain (or even a point of Ω^*). Indeed, here we will see for the first time an example of a phenomenon that is present in certain infinitely connected domains Ω that is not ever present in a finitely connected domain: the "distinguished" homomorphisms. The plan of the chapter is to begin with some general considerations and constructions and then to proceed with the analysis of $\mathbf{M}(\Omega)$. A further section is devoted to describing some properties of $\mathbf{S}(\Omega)$, the Shilov boundary of $H^\infty(\Omega)$. The final section discusses the corona theorem and presents a particularly simple proof of this result for the unit disc.

6.2. PEAK POINTS AND PARTS

Throughout this section, $\mathbf{X}$ will be a compact Hausdorff space and $\mathbf{A}$ an algebra of continuous functions on $\mathbf{X}$ which contains the constants and which is closed in the uniform norm on $\mathbf{X}$. The results of this section, while different in character to most of what has gone before, will find application in the study of the structure of the maximal ideals of $H^\infty(\Omega)$.

Definition. A closed set $\mathbf{E} \subset \mathbf{X}$ is a *peak set* for $\mathbf{A}$ if there is a function $f \in \mathbf{A}$ with $f = 1$ on $\mathbf{E}$ and $|f(y)| < 1$ is $y \in \mathbf{X}$, $y \notin \mathbf{E}$. In the case when $\mathbf{E}$ is a single point x, then we say x is a *peak point* of $\mathbf{A}$.

Theorem 2.1. *Suppose the point x is a G_δ. Then x is a peak point for $\mathbf{A}$ if and only if, given any open set $\mathcal{U}$ containing x, there is a function g in $\mathbf{A}$ with*

$$\|g\| \leqslant 1 \tag{2.1a}$$

$$|g(x)| \geqslant \tfrac{3}{4} \tag{2.1b}$$

$$|g(y)| \leqslant \tfrac{1}{4}, \qquad y \notin \mathcal{U} \tag{2.1c}$$

Proof. Suppose x is a peak point; let $f \in \mathbf{A}$ satisfy $f(x) = 1$ and $|f(y)| < 1$ for all $y \in \mathbf{X}$, $y \neq x$. Given a neighborhood $\mathcal{U}$ of x we have $|f(y)| \leqslant 1 - \delta$ for all $y \notin \mathcal{U}$ for some $\delta > 0$. Hence, a sufficiently high power, $g = f^N$, of f will have the properties given in (2.1).

Conversely, suppose (2.1) holds. Let $\{\mathcal{V}_j\}_0^\infty$ be a nested sequence of open sets with $\cap_0^\infty \mathcal{V}_j = \{x\}$ and put $\mathcal{F}_n = \mathbf{X} \setminus \mathcal{V}_n$ for $n = 0, 1, 2, \ldots$. Let $\{\varepsilon_n\}$ be a sequence of positive numbers decreasing to zero chosen so that $6\varepsilon_n(2^n - 1) < 1$. We now define a sequence of functions $\{h_j\}$ and closed sets $\{\mathcal{W}_j\}$ by induction. Let h_1 be an element of $\mathbf{A}$ with $h_1(x) = 1$, $\|h_1\| \leqslant \frac{4}{3}$, and $|h_1(y)| \leqslant \frac{1}{3}$ if $y \in \mathcal{F}_0$. Set

$$\mathcal{W}_1 = \{y \in X : |h_1(y)| \geqslant 1 + \varepsilon_1\}$$

Now choose $h_2 \in A$ with $h_2(x) = 1$, $\|h_2\| \leqslant \frac{4}{3}$, and $|h_2(y)| \leqslant \frac{1}{3}$ if $y \in \mathcal{W}_1 \cup \mathcal{F}_1$. In general, if $h_1, \ldots, h_n$ and $\mathcal{W}_1, \ldots, \mathcal{W}_{n-1}$ have been chosen, set

$$\mathcal{W}_n = \left\{ y \in \mathbf{X} : \max_{1 \leqslant j \leqslant n} |h_j(y)| \geqslant 1 + \varepsilon_n \right\}$$

and choose $h_{n+1} \in \mathbf{A}$ with $h_{n+1}(x) = 1$, $\|h_{n+1}\| \leqslant \frac{4}{3}$, and $|h_{n+1}(y)| \leqslant \frac{1}{3}$ if $y \in \mathcal{W}_n \cup \mathcal{F}_n$. Put

$$h = \sum_1^\infty 2^{-j} h_j$$

Then $h \in \mathbf{A}$ and $h(x) = 1$. We must show that $|h(y)| < 1$ for all $y \in X$, $y \neq x$. To begin, suppose $y \notin \cup_1^\infty \mathcal{W}_j$. Then $|h_j(y)| \leqslant 1$ for all j and since $y \in \mathcal{F}_m$ for all sufficiently large m we have $|h_m(y)| < \frac{1}{3}$ if $m \geqslant m_0$; thus, $|h(y)| < 1$. To finish suppose that $y \in \cup_{j=1}^\infty \mathcal{W}_j$. Since $\mathcal{W}_1 \subset \mathcal{W}_2 \subset \cdots$ there is a first index N with $y \notin \mathcal{W}_N$, $y \in \mathcal{W}_{N+1}$. Then

$$|h_j(y)| \leqslant 1 + \varepsilon_N, \qquad 1 \leqslant j \leqslant N$$

$$|h_{N+1}(y)| \leqslant \tfrac{4}{3}$$

$$|h_j(y)| \leqslant \tfrac{1}{3}, \qquad N + 2 \leqslant j < \infty$$

Thus,

$$|h(y)| \leqslant (1+\varepsilon_N)\sum_1^N 2^{-j} + 2^{-N-1}(\tfrac{4}{3}) + (\tfrac{1}{3})\sum_{N+2}^{\infty} 2^{-j}$$

$$= \varepsilon_N(1-2^{-N}) - (\tfrac{1}{6})2^{-N} + 1 < 1$$

Remark. It should be apparent that the choice $\frac{3}{4}$ and $\frac{1}{4}$ is not essential; any two numbers $c_1, c_2, 0 < c_1 < c_2 < 1$, in place of $\frac{1}{4}$ and $\frac{3}{4}$, respectively, will do.

Corollary 2.2. *Suppose* **X** *is also a metric space. Then the set of peak points of* **A** *is a* G_δ *set.*

Proof. Let $\mathfrak{U}_n$ be those points $x \in X$ for which there is a function $f \in A$ with $\|f\| = 1$, $|f(x)| > \frac{3}{4}$, and $|f(y)| < \frac{1}{4}$ if the distance from x to y is $1/n$ or more. Clearly $\mathfrak{U}_n$ is open and just as clearly all the peak points of **A** lie in each $\mathfrak{U}_n$. But Theorem 2.1 shows that each point in $\cap_{n=1}^{\infty} \mathfrak{U}_n$ is a peak point.

Definition. Let l be a linear functional on **A**; a positive measure μ on **X** for which

$$l(f) = \int f\,d\mu, \qquad f \in \mathbf{A} \tag{2.2}$$

will be called a *representing measure* for l. If l is the particular linear functional

$$l(f) = f(x), \qquad f \in \mathbf{A}$$

for some $x \in \mathbf{X}$, then a positive measure μ satisfying

$$f(x) = \int f\,d\mu, \qquad f \in \mathbf{A} \tag{2.3}$$

will be called a *representing measure* for x.

Theorem 2.3. *Let x be a point of* **X** *which is also a* G_δ. *Then x is a peak point for* **A** *if and only if the only representing measure for x is the point mass at x.*

Proof. Suppose first that $f \in$ **A** peaks at x. If μ is any representing measure for x, then $\{f^n\}$ converges boundedly and pointwise to the characteristic function of $\{x\}$ and so

$$1 = \int f^n\,d\mu \to \mu(\{x\})$$

which is the desired conclusion since $\|\mu\| = 1$.

For the converse we must use the following lemma. In the lemma, **A** is an algebra of the type considered in this section and ϕ is any continuous multiplicative linear functional on **A**.

Lemma. Let v be a real-valued continuous function on **X**. Let Σ_ϕ be the set of all representing measures for ϕ. Then

$$\sup\{\operatorname{Re}\phi(f) : \operatorname{Re} f \leqslant v\} = \inf\left\{\int v\, d\mu : \mu \in \Sigma_\phi\right\} \tag{2.4}$$

Proof. If $f \in \mathbf{A}$ and $u = \operatorname{Re} f \leqslant v$ and $\mu \in \Sigma_\phi$, then

$$\operatorname{Re}\phi(f) = \operatorname{Re}\int f\, d\mu = \int u\, d\mu \leqslant \int v\, d\mu$$

so that the right-hand side of (2.4) exceeds the left-hand side.

On the other hand, let **B** be the subspace of $\mathbf{C}_r(\mathbf{X})$ spanned by v and the real parts of functions in **A**. Let c be the left-hand side of (2.4). Then set $\tilde{\phi}(v) = c$ so that $\tilde{\phi}$ is a positive linear functional on **B**. Extend $\tilde{\phi}$ from **B** to a positive linear functional on all of $\mathbf{C}_r(\mathbf{X})$; this extension is represented by a positive measure μ_0 which must lie in Σ_ϕ. Thus, there is a $\mu_0 \in \Sigma_\phi$ with $\int v\, d\mu_0 = c$, and so the proof of the lemma is complete.

With the lemma in hand, the theorem follows quickly. Suppose that the point mass at x is the only representing measure for x. If $\mathcal{U}$ is any neighborhood of x, let v be a continuous real-valued function on **X** with $v(x) > \log(\frac{3}{4})$, $v(y) < \log(\frac{1}{4})$ for all $y \notin \mathcal{U}$ and $v < 0$ on all of **X**. Then the lemma and the hypothesis imply that there is an element $g \in \mathbf{A}$ with $\operatorname{Re} g \leqslant v$ on **X** but $\operatorname{Re} g(x) > \log(\frac{3}{4})$. Set $f = \exp(g)$; then $f \in A$, $\|f\| \leqslant 1$, $|f(x)| > \frac{3}{4}$, while $|f(y)| < \frac{1}{4}$ if $y \notin U$. By Theorem 2.1, x is a peak point for **A**.

The issue certainly remains as to whether there are any peak points at all. The following argument shows that there are. Let Λ consist of those linear functionals l on **A** with

$$\|l\| = l(1) = 1$$

The set Λ is nonempty since Λ contains all linear functionals of the form

$$l(f) = \int f\, d\rho$$

where ρ is a non-negative measure of mass 1. Further, Λ is convex and weak-* compact in the dual space of **A**. Let l_0 be an extreme point of Λ and let μ be any positive measure of X which represents l_0. We shall show that μ is the point mass at some single point x_0 of X and hence x_0 is a peak point of **A**. Suppose the closed support of μ can be written as the union of two disjoint sets

S_1 and S_2 each with positive μ measure, say $\mu(S_1) = a$ and $\mu(S_2) = 1 - a$, $1 > a > 0$. Then $l_0 = al_1 + (1-a)l_2$ where

$$l_1(f) = \frac{1}{a}\int_{S_1} f\,d\mu, \qquad f \in \mathbf{A}$$

$$l_2(f) = \frac{1}{1-a}\int_{S_2} f\,d\mu, \qquad f \in \mathbf{A}$$

and each of l_2, l_2 lie in Λ. This contradiction shows that the support of μ is a single point x_0 and hence x_0 is a peak point by Theorem 2.3, provided it is a G_δ. This last condition is assured if $\mathbf{X}$ is metric so we can state the following.

Proposition 2.4. Let $\mathbf{X}$ be a compact metric space and let $\mathbf{A}$ be a closed subalgebra of $\mathbf{C(X)}$ with $1 \in \mathbf{A}$. Then $\mathbf{A}$ has at least one peak point.

EXAMPLE.

Let $\mathbf{K}$ be a compact set in $\mathbb{C}$ of positive area and let $\mathbf{A}(\mathbf{K}, \mathbf{S}^2)$ denote those functions f which are continuous on the sphere $\mathbf{S}^2$ and holomorphic off $\mathbf{K}$. As we noted earlier (exercise 6, Chapter 3), the function

$$f(z) = \int_{\mathbf{K}} \frac{1}{z - \zeta}\,d\sigma\,d\tau, \qquad \zeta = \sigma + i\tau$$

lies in $\mathbf{A}(\mathbf{K}, \mathbf{S}^2)$ and is not constant, so that this algebra is nontrivial. It's true that almost all points of $\mathbf{K}$ (with respect to area measure) are peak points for $\mathbf{A}(\mathbf{K}, \mathbf{S}^2)$. We certainly know that the set of peak points is a G_δ; see Corollary 2.2. Let $\mathfrak{U}$ be any open set containing all the peak points; if $\mathbf{K}_1 = \mathbf{K} \setminus \mathfrak{U}$ has positive area, the algebra $\mathbf{A}(\mathbf{K}_1, \mathbf{S}^2)$ is nontrivial and is, of course, a subalgebra of $\mathbf{A}(\mathbf{K}, \mathbf{S}^2)$. Hence, there is a peak point for $\mathbf{A}(\mathbf{K}_1, \mathbf{S}^2)$ (and hence also for $\mathbf{A}(\mathbf{K}, \mathbf{S}^2)$) in $\mathbf{K}_1$, a contradiction. Thus, $\mathfrak{U}$ has measure equal to that of $\mathbf{K}$ and the result follows.

We now turn to a technique for dividing up the maximal ideal space $\mathbf{M(A)}$ of an algebra $\mathbf{A}$ of the type discussed in this section. The original motivation was to find subsets of $\mathbf{M(A)}$ on which the functions in $\mathbf{A}$ are "analytic" in some sense; the relationship with Schwarz's lemma for analytic functions on Δ will be apparent in the proof of Theorem 2.5 and in the example following Corollary 1.6.

Definition. Let $\mathbf{M(A)}$ denote the maximal ideal space of $\mathbf{A}$ and let ϕ, ψ be two elements of $\mathbf{M(A)}$. We say that ϕ and ψ lie in the same *part* of $\mathbf{M(A)}$ if

$$\|\phi - \psi\| < 2 \tag{2.5}$$

where the norm is that of $\mathbf{A}^*$.

The following theorem gives two other equivalent conditions for ϕ and ψ to lie in the same part of **M(A)**.

Theorem 2.5. *Let* $\phi, \psi \in \mathbf{M(A)}$. *The following are equivalent*:

(*a*) ϕ, ψ *lie in the same part*

(*b*) *there is a constant* $\rho \in (0, 1)$ *with*

$$\sup\{|\psi(f)| : \|f\| \leqslant 1, \phi(f) = 0\} \leqslant \rho \tag{2.6}$$

(*c*) *if* $\{f_n\}$ *is a sequence of elements in the unit ball of* **A** *with* $|\phi(f_n)| \to 1$, *then* $|\psi(f_n)| \to 1$ *as well*

Proof. Suppose (a) holds but (b) does not. Thus, there is a sequence $\{f_n\}$ in the unit ball of **A** with $\psi(f_n) = 0$ but $\phi(f_n) \to 1$. Let $b_n = \phi(f_n)$; there is no loss in supposing that b_n is positive for each n. Define $c_n = ((1 - b_n)^{1/2} - 1)((1 - b_n)^{1/2} + 1)^{-1}$ and

$$g_n = (f_n + c_n)(1 + c_n f_n)^{-1}$$

Then g_n is in the unit ball of **A** and we also have $\psi(g_n) = c_n \to -1$ as $n \to \infty$ while

$$\begin{aligned}\phi(g_n) &= (b_n + c_n)(1 + c_n b_n)^{-1}\\ &= \left((b_n + 1) - (1 - b_n)^{1/2}\right)\left(b_n + 1 + (1 - b_n)^{1/2}\right)^{-1}\\ &\to 1 \quad \text{as } n \to \infty\end{aligned}$$

which contradicts (a). Hence, (b) holds.

Suppose next that (b) holds and let $\phi(f_n) = b_n$ so that $|b_n| \to 1$. Set

$$g_n = \frac{f_n - b_n}{1 - \bar{b}_n f_n}$$

Then g_n is again in the unit ball of **A** and $\phi(g_n) = 0$. Thus,

$$\left|\frac{\psi(f_n) - b_n}{1 - \bar{b}_n \psi(b_n)}\right| \leqslant \rho < 1, \qquad \text{all } n$$

and hence $|\psi(f_n)|$ must tend to 1.

Finally, suppose (c) holds. If (a) fails, there are elements $\{f_n\}$ in the unit ball of **A** with

$$|\phi(f_n) - \psi(f_n)| \to 2$$

Thus, by passing to a subsequence if necessary and adjusting by a unimodular constant we may assume that $\phi(f_n) \to 1$ while $\psi(f_n) \to -1$. Let

$$g_n = \exp[f_n - 1]$$

Then g_n lies in the unit ball of **A**, $\phi(g_n) = \exp[\phi(f_n) - 1]$ and these numbers tend to 1 whereas $\phi(\psi_n) = \exp[\psi(f_n) - 1]$ and these numbers converge to $\exp[-2]$, contradicting (c).

Corollary 2.6. *"Being in the same part" is an equivalence relation on* **M(A)**.

Proof. If ϕ and ψ are in the same part and ψ and ω are also in the same part, we must show ϕ and ω are in the same part. To do this we use (c) of Theorem 2.5. Suppose $\{f_n\}$ is a sequence in the unit ball of **A** with $|\phi(f_n)| \to 1$. Then $|\psi(f_n)| \to 1$ and so $|\omega(f_n)| \to 1$ as well, and we're done.

Needless to say, the equivalence classes in **M(A)** determined by $\|\phi - \psi\| < 2$ are called the *parts* of **M(A)**.

EXAMPLE.

Let Ω be a domain in $\mathbb{C}$ for which $H^\infty(\Omega)$ is nontrivial. $H^\infty(\Omega)$ is a commutative Banach algebra with unit satisfying $\|f^2\| = \|f\|^2$ and so $H^\infty(\Omega)$ is isometrically isomorphic to its Gelfand transform considered as an algebra of continuous functions on its maximal ideal space $\mathbf{M}(\Omega)$. The parts of $\mathbf{M}(\Omega)$ are difficult to visualize even when $\Omega = \Delta$, the open unit disc. However, it is easy to show that all of Ω lies in a single part. Suppose p and q are distinct points of Ω and $f \in H^\infty(\Omega)$, $\|f\| \leqslant 1$, and $f(p) = 0$. Then clearly $|f(q)| \leqslant \rho < 1$ where ρ does not depend on f. For otherwise, there are elements f_n in the unit ball of $H^\infty(\Omega)$ with $f_n(p) = 0$, $f_n(q) \to 1$. A normal families argument yields $f_n \to 1$ uniformly on compact subsets of Ω, in particular at p, a contradiction.

6.3. THE FIBERS OF M(Ω)

We shall suppose throughout the chapter that Ω is a domain on the sphere which supports nonconstant bounded analytic functions. We shall assume as well that each point in $\partial\Omega$ is essential; this obviously involves no loss of generality. We begin this section with two results on approximation of bounded analytic functions and then show how these results may be used to great advantage in analyzing some of the structure of the maximal ideal space of $H^\infty(\Omega)$.

Let u be a $\mathbf{C}^1$ function on the plane with

$$u(z) \equiv 1 \quad \text{if} \quad |z| \leqslant 1 \tag{3.1a}$$

$$u(z) \equiv 0 \quad \text{if} \quad |z| \geqslant 2 \tag{3.1b}$$

$$\left|\frac{\partial u}{\partial \bar{z}}\right| \leqslant \tfrac{1}{2} \quad \text{for all points} \tag{3.1c}$$

One way to obtain such a function is to set $u(re^{i\theta}) = \frac{1}{2}w(r/2)$ where

$$w(t) = \begin{cases} 2, & 0 \leqslant t \leqslant 2 \\ 2 - (2-t)^2, & 2 \leqslant t \leqslant 3 \\ (t-4)^2, & 3 \leqslant t \leqslant 4 \\ 0, & t \geqslant 4 \end{cases}$$

Then

$$2\frac{\partial u}{\partial \bar{z}} = \frac{\partial u}{\partial x} + i\frac{\partial u}{\partial y} = \frac{\partial u}{\partial r}\left(\frac{\partial r}{\partial x} + i\frac{\partial r}{\partial y}\right)$$

so that

$$2\left|\frac{\partial u}{\partial \bar{z}}\right| \leqslant 2\left|\frac{\partial u}{\partial r}\right| = \frac{1}{2}|w'(r/2)| \leqslant 1$$

Set $u_\delta(z) = u(z/\delta)$; we shall use u_δ to define a family of linear operators on $L^\infty(\mathbf{C}, dx\,dy)$. Suppose $h \in L^\infty(\mathbf{C}, dx\,dy)$; set

$$(T_\delta h)(\zeta) = \frac{1}{\pi}\iint \frac{h(\zeta) - h(z)}{\zeta - z}\frac{\partial u_\delta}{\partial \bar{z}}\,dx\,dy, \qquad z = x + iy \tag{3.2}$$

By Green's theorem

$$(T_\delta h)(\zeta) = h(\zeta)u(\zeta) + \frac{1}{\pi}\iint \frac{h(z)}{z - \zeta}\frac{\partial u_\delta}{\partial \bar{z}}\,dx\,dy \tag{3.3}$$

The integral term on the right-hand side of (3.3) is a continuous function of ζ being the convolution of the locally integrable function $1/z$ with a bounded function. Thus

$$T_\delta h \text{ is bounded} \tag{3.4a}$$

$$T_\delta h \text{ is analytic off the disc } |z| \leqslant 2\delta \tag{3.4b}$$

$$T_\delta h \text{ vanishes at } \infty \tag{3.4c}$$

$$T_\delta h - h \text{ is analytic on the disc } |z| < \delta \tag{3.4d}$$

Further,

$$T_\delta h \text{ is analytic wherever } h \text{ is analytic} \tag{3.5}$$

and

$$\|T_\delta h\|_\infty \leqslant 2\sup\{|h(z) - h(w)| : |z|, |w| < 2\delta\} \tag{3.6}$$

The estimate in (3.6) follows from the definition of $T_\delta h$ given in (3.2) since

$$|T_\delta h(\zeta)| \leqslant \sup\{|h(z) - h(\zeta)| : |z| \leqslant 2\delta\}\left(\frac{1}{2\delta}\right)\left(\frac{1}{\pi}\iint_{|z|\leqslant 2\delta} \frac{1}{|\zeta - z|}dx\,dy\right)$$

But the integral term can be estimated by

$$\iint_{|z|\leqslant 2\delta} \frac{1}{|\zeta - z|}dx\,dy \leqslant \iint_{|z|\leqslant 2\delta} \frac{1}{|z|}dx\,dy = 4\pi\delta$$

Thus, (3.6) holds. Note that (3.6) implies

$$\|T_\delta h\|_\infty \leqslant 4\|h\|_\infty \tag{3.7}$$

Let $L_\delta h$ be the constant

$$L_\delta h = \frac{1}{\pi}\iint \frac{h(z)}{z}\frac{\partial u_\delta}{\partial \bar{z}}dx\,dy \tag{3.8}$$

Then we have the final estimate

$$|T_\delta h(\zeta) - h(\zeta) - L_\delta h| \leqslant 4|\zeta|\|h\|_\infty \delta^{-1}, \qquad |\zeta| < \delta \tag{3.9}$$

This last inequality just reflects the fact that

$$H(\zeta) = \frac{1}{\pi}\iint_{|z|\leqslant 2\delta} \frac{h(z)}{z - \zeta}\frac{\partial u_\delta}{\partial \bar{z}}dx\,dy$$

is analytic in $|\zeta| < \delta$ and $Lh = H(0)$; whence

$$|H(\zeta) - H(0)| \leqslant 2\|H\|_\infty|\zeta|\delta^{-1} \leqslant 4\|h\|_\infty|\zeta|\delta^{-1}$$

We now use (3.4)–(3.9) to prove the desired approximation results.

Proposition 3.1. Let Ω be a domain, let $\lambda \in \partial\Omega$, and let $\mathcal{U}$ be a neighborhood of λ. Suppose f is bounded and holomorphic in $\Omega \cap \mathcal{U}$. Then there is an $F \in H^\infty(\Omega)$ with

$$\|F\|_\Omega \leqslant 8\|f\|_{\Omega\cap\mathcal{U}} \tag{3.10}$$

$h = F - f$ is holomorphic in a disc centered at λ and $h(\lambda) = 0$ (3.11)

Proof. Let $v(z) = u((z - \lambda)/\delta)$ where u has the properties described in (3.1) and δ is chosen so small that the disc of radius 2δ centered at λ lies in $\mathcal{U}$.

Set $f = 0$ off $\mathcal{U}$ and define

$$F(z) = T_v f - L_v f$$

where T_v and L_v are just as in (3.2) and (3.8), respectively, with v replacing u_δ. The properties (3.10) and (3.11) now follow immediately from (3.7) and (3.9).

Proposition 3.2. Suppose $f \in H^\infty(\Omega)$ and $\lambda \in \partial\Omega$. Then there is a family $\{f_\delta\}$ of elements of $H^\infty(\Omega)$ with these properties:

$$f_\delta \text{ is analytic in a neighborhood of } \lambda \text{ for each } \delta \tag{3.12a}$$

$$\|f_\delta\| \leqslant 5\|f\| \tag{3.12b}$$

$$\text{if } \mathcal{V} \text{ is any neighborhood of } \lambda, \text{ then } f_\delta \to f \text{ uniformly on } \Omega \setminus \mathcal{V} \text{ as } \delta \to 0 \tag{3.12c}$$

$$\text{if } f \text{ is continuous at } \lambda, \text{ then } f_\delta \to f \text{ uniformly on } \Omega \tag{3.12d}$$

Proof. Let $v_\delta(z) = u((z - \lambda)/\delta)$ where u is again the function from (3.1). Denote by T_δ the operator given by (3.2) but with v_δ in place of u_δ, and set $f_\delta = T_\delta f - f$. Then (3.12a) follows from (3.4d) and (3.12b) follows from (3.7). Further, if $|\zeta - \lambda| \geqslant \eta > 0$, then for δ small enough

$$|T_\delta f(\zeta)| \leqslant \frac{1}{\pi}\|h\|_\infty(\eta - \delta)^{-1}\left(\frac{2}{\delta}\right)\pi\delta^2 = 0(\delta)$$

and this proves (3.12c). Finally, (3.12d) follows from (3.6).

We now outline a delineation of the elements of $\mathbf{M}(\Omega)$. By a simple inversion, there is no loss in assuming that ∞ is an interior point of Ω. Let ϕ be a point in $\mathbf{M}(\Omega)$, ϕ not equal to "evaluation at ∞," and let h be any element of $H^\infty(\Omega)$ with $\phi(h) = 1$ and $h(\infty) = 0$. Thus, $zh(z)$ is also in $H^\infty(\Omega)$. Define the complex number λ by $\lambda = \phi(zh(z))$. We first shall show that λ is independent of h and then that λ lies in $\Omega \cup \partial\Omega$. As for the first, suppose $g \in H^\infty(\Omega)$, $g(\infty) = 0$, and $\phi(g) = 1$. Then

$$\begin{aligned}\phi(zg(z)) &= \phi(zg(z)h(z))\\ &= \phi(zh(z)g(z))\\ &= \phi(zh(z))\phi(g)\\ &= \lambda\end{aligned}$$

As for the second, suppose that $\lambda \notin \Omega \cup \partial\Omega$. Then $p(z) = 1/(z - \lambda)$ lies in

$H^\infty(\Omega)$ so that if $h \in H^\infty(\Omega)$ vanishes at ∞ and has $\phi(h) = 1$, then

$$1 = \phi\big((z-\lambda)h(z)(z-\lambda)^{-1}\big)$$
$$= \phi((z-\lambda)h(z))\phi(p) = 0 \cdot \phi(p)$$

an obviously ridiculous situation. On the other hand, if $\lambda \in \Omega$, then for any $f \in H^\infty(\Omega)$ and h as before we have

$$\phi(f - f(\lambda)) = \phi(h(f - f(\lambda))) = \phi((z-\lambda)h)\phi\left(\frac{f(z) - f(\lambda)}{z - \lambda}\right)$$
$$= 0$$

Thus, $\phi(f) = f(\lambda)$ for all $f \in H^\infty(\Omega)$.

Suppose next that $\lambda \in \partial\Omega$. If f is holomorphic in some neighborhood of λ, then the preceding argument shows that $\phi(f) = f(\lambda)$. Hence, the same holds if only f is continuous at λ by (3.12d).

The last step is to show that if λ' is point of $\Omega \cup \partial\Omega$ for which $\phi(f) = f(\lambda')$ whenever f is continuous at λ' (and in $H^\infty(\Omega)$) then $\lambda' = \lambda$. Suppose $\lambda \neq \lambda'$. Certainly the construction in Proposition 3.1 can be used to produce a function f in $H^\infty(\Omega)$ which is holomorphic in a neighborhood of both λ and λ' and $f(\infty) = 0$. If $f(\lambda) = f(\lambda') \neq 0$, then $zf(z)$ distinguishes λ and λ'. On the other hand if $f(\lambda) = f(\lambda') = 0$, then f has a zero of order $m \geqslant 1$ at λ so that $f(z)/(z-\lambda)^m$ then distinguishes λ and λ'. This contradiction shows that $\lambda = \lambda'$.

This construction leads us to the following definitions.

Definition. Let ρ be the mapping of $\mathbf{M}(\Omega)$ into $\Omega \cup \partial\Omega$ given by

$$\rho(\phi) = \lambda$$

where λ is the unique point of $\Omega \cup \partial\Omega$ such that

$$\phi(f) = f(\lambda), \qquad f \in H^\infty(\Omega), \qquad f \text{ analytic near } \lambda.$$

Definition. The set $\mathbf{M}_\lambda = \rho^{-1}(\lambda)$ is called the *fiber* of λ in $\mathbf{M}(\Omega)$.

Proposition 3.3. The mapping ρ is continuous from $\mathbf{M}(\Omega)$ with the weak-* topology into the extended complex plane.

Proof. Suppose $\{\phi_\alpha\}$ is a net in $\mathbf{M}(\Omega)$ which converges to $\phi \in \mathbf{M}(\Omega)$. Then by definition of the weak-* topology

$$\phi_\alpha(f) \to \phi(f), \qquad f \in H^\infty(\Omega)$$

Let λ be any limit point of $\{\lambda_\alpha\}$. If f is analytic in a neighborhood of λ then we have

$$f(\lambda) = \lim_j f(\lambda_{\alpha_j}) = \lim_j \phi_{\alpha_j}(f) = \phi(f)$$

so that $\rho(\phi) = \lambda$, as desired.

Definition. Let $f \in H^\infty(\Omega)$ and let $\lambda \in \partial\Omega$. The *cluster set* of f at λ is the set of all complex numbers w for which there is a sequence $\{z_n\}$ with $z_n \to \lambda$ and $f(z_n) \to w$. Clearly, the cluster set of f at λ is closed.

Theorem 3.4. *Let $f \in H^\infty(\Omega)$ and $\lambda \in \partial\Omega$. Then the cluster set of f at λ coincides with the range of $\hat{f}$, the Gelfand transform of f, on the fiber $\mathbf{M}_\lambda$.*

Proof. Suppose $z_n \to \lambda$ and $f_n(z) \to w$; then $f - w$ approaches 0 along $\{z_n\}$. Let $\mathfrak{I}$ be the ideal in $H^\infty(\Omega)$ of functions which have limit zero along $\{z_n\}$. $\mathfrak{I}$ lies in a maximal ideal $\mathfrak{J}$ and the multiplicative linear functional ϕ whose kernel is $\mathfrak{J}$ clearly lies in $\mathbf{M}_\lambda$. Thus, $\phi(f - w) = 0$ or $\hat{f}(\phi) = w$.

Conversely, suppose $\phi \in \mathbf{M}_\lambda$ and $w = \phi(f)$. If w is not in the cluster set of f at λ there must be a positive ε and a positive δ such that $|f(z) - w| \geqslant \varepsilon$ if $|z - \lambda| < \delta$, $z \in \Omega$. Let $g(z) = 1/(f(z) - w)$ for $|z - \lambda| < \delta$, $z \in \Omega$. By Proposition 3.1 there is an $F \in H^\infty(\Omega)$ such that $F - g = h$ extends to be holomorphic at λ and $h(\lambda) = 0$. Thus,

$$F(z)(f(z) - w) - 1 = (f(z) - w)h(z), \qquad |z - \lambda| < \delta, \qquad z \in \Omega$$

However, $F(z)(f(z) - w) - 1$ is in $H^\infty(\Omega)$ and since this equals $(f(z) - w)h(z)$ near λ, we see that $F(z)(f(z) - w) - 1$ tends to 0 at λ. Hence, this function is in the kernel of ϕ. So

$$0 = \phi(F(f - w) - 1) = \phi(F)(\phi(f) - w) - 1 = -1$$

This contradiction shows that w must be in the cluster set of f at λ.

Let $f \in H^\infty(\Omega)$ and let $\{f_\delta\}$ be the family of functions from Proposition 3.2. We shall need the following strengthening of (3.12c).

Proposition 3.5. Let $\mathcal{V}$ be any neighborhood of $\mathbf{M}_\lambda$ in $\mathbf{M}(\Omega)$. Then $f_\delta \to f$ uniformly on $\mathbf{M}(\Omega) \setminus \mathcal{V}$.

Proof. Suppose $\phi \in \mathbf{M}(\Omega) \setminus \mathcal{V}$; there is a positive number η independent of ϕ such that $|\rho(\phi) - \lambda| \geqslant \eta$. Let $g \in H^\infty(\Omega)$ satisfy

$$g(\infty) = 0$$

$$\phi(g) = 1$$

$$\|g\| \leqslant 2$$

Then $(z-\lambda)g(z)$ lies in $H^\infty(\Omega)$, has norm at most $4M$, where $M = \max\{|z| : z \in \partial\Omega\}$, and is very small for z near λ. Thus, if δ is small enough we may conclude from (3.12c) that

$$\|(z-\lambda)g(z)(f_\delta(z) - f(z))\|_\Omega < \varepsilon, \qquad \delta \leqslant \delta_0$$

Hence,

$$\begin{aligned} \varepsilon &> |\phi((z-\lambda)g(z))(f_\delta(z) - f(z))| \\ &= |\rho(\phi) - \lambda||\phi(f_\delta) - \phi(f)| \\ &\geqslant \eta|\phi(f_\delta) - \phi(f)| \end{aligned}$$

which proves the proposition.

We shall have need of just one more aspect of the topology of $H^\infty(\Omega)$ and that is the weak-* topology of $H^\infty(\Omega)$ which $H^\infty(\Omega)$ inherits as a subspace of $L^\infty(\Omega, dx\,dy)$. The Krein-Smulian theorem [see Hoffman and Rossi (1967) for an appropriate version] allows us to conclude that a convex subset of $L^\infty(\Omega, dx\,dy)$ is weak-* closed if and only if it is weak-* sequentially closed. We thus obtain the following proposition.

Proposition 3.6. A bounded sequence $\{f_n\}$ of elements of $H^\infty(\Omega)$ converges weak-* to $f \in H^\infty(\Omega)$ if and only if

$$f_n(z) \to f(z) \text{ uniformly on compact sets in } \Omega \tag{3.13}$$

Consequently, $H^\infty(\Omega)$ is a weak-* closed subalgebra of $L^\infty(\Omega, dx\,dy)$.

Proof. Suppose $f_n \to f$ weak-*, $\|f_n\| \leqslant M$. To see (3.13) it suffices to prove $f_n(z) \to f(z)$ pointwise and then apply a normal families argument. Let $z \in \Omega$ and let $\mathcal{V}$ be a disc of radius δ in Ω with closure still in Ω. The linear functional $l(u) = \iint_{\mathcal{V}} u\,dx\,dy$ is weak-* continuous on $L^\infty(\Omega, dx\,dy)$ and on $H^\infty(\Omega)$, we have $l(f) = \pi\delta^2 f(z)$. Hence, $l(f_n) \to l(f)$ implies $f_n(z) \to f(z)$.

Conversely, if (3.13) holds, then the dominated convergence theorem implies that $\int f_n g\,dx\,dy \to \int fg\,dx\,dy$ for each $g \in L^1(\Omega, dx\,dy)$ and so $f_n \to f$ weak-*.

6.4. DISTINGUISHED HOMOMORPHISMS

To understand what will be the theme of this section it is worthwhile to begin with an example which displays the rather odd phenomenon which we shall examine.

EXAMPLE.

Let $\{r_j\}$ and $\{c_j\}$ be sequences of positive numbers satisfying

$$1 > c_1 > c_2 > \cdots > 0, \qquad \lim c_j = 0 \tag{4.1a}$$

$$r_1 > r_2 > \cdots > 0, \qquad \lim r_j = 0 \tag{4.1b}$$

$$r_{j+1} + c_{j+1} < c_j - r_j, \qquad j = 1, 2, \ldots; r_1 + c_1 < 1 \tag{4.1c}$$

$$\sum_1^\infty \frac{r_j}{c_j} < \infty \tag{4.1d}$$

Let Ω be the domain obtained from the set $\{z : 0 < |z| < 1\}$ by deleting the closed discs Δ_j centered at c_j and of radius r_j for $j = 1, 2, \ldots$. Let $\Delta_0 = \Delta = \{z : |z| < 1\}$.

Suppose f is in $H^\infty(\Omega)$. Let γ_j be the circle centered at the origin of radius ρ_j,

$$\rho_j = \tfrac{1}{2}[c_{j+1} + r_{j+1} + c_j - r_j]$$

For any $z \in \Omega$ and N large enough we have by Cauchy's formula

$$f(z) = \sum_{k=0}^N \frac{1}{2\pi i} \int_{\partial\Delta_k} \frac{f(w)}{w - z} dw + \frac{1}{2\pi i} \int_{\gamma_N} \frac{f(w)}{w - z} dw$$

The integral over γ_N is bounded above by $C\|f\|\rho_N$ where C is a constant depending only on the distance from z to γ_N. Thus, the integral over γ_N goes to zero as $N \to \infty$ and we find that

$$f(z) = \frac{1}{2\pi i} \int_{\partial\Omega} \frac{f(w)}{w - z} dw, \qquad z \in \Omega$$

Note that the assumption (4.1d) assures us that $|w|^{-1}$ and also $|w - z|^{-1}$ are in $L^1(\partial\Omega, ds)$. Suppose now that z lies in the negative real axis $-\frac{1}{2} < z < 0$; then $|w - z|^{-1} \leqslant |w|^{-1}$ for all w with $\operatorname{Re} w \geqslant 0$ and $|w - z|^{-1} \leqslant 2$ if $|w| = 1$. Thus we can apply the dominated convergence theorem and conclude that

$$\lim\{f(x) : x \to 0, x < 0\} = \frac{1}{2\pi i} \int_{\partial\Omega} \frac{f(w)}{w} dw \tag{4.2}$$

for each $f \in H^\infty(\Omega)$; here we again are making use of (4.1d). Let us denote by ϕ_0 the limit in (4.2):

$$\phi_0(f) = \lim\{f(x) : x \to 0, x < 0\}$$

Clearly, ϕ_0 is a multiplicative linear functional on $H^\infty(\Omega)$ and $\phi_0(1) = 1$ so that ϕ_0 is not identically zero. Furthermore, ϕ_0 is not evaluation at any point of Ω since $\phi_0(I) = 0$ where $I(z) = z$. Finally, if x is a point in $(-1, 0)$, then

$$|f(x) - \phi_0(f)| \leqslant \|f\|\varepsilon(x)$$

where $\varepsilon(x)$ is the $L^1(\partial\Omega, ds)$ distance from $(w - z)^{-1}$ to w^{-1}. Since we know that $\varepsilon(x) \to 0$ as $x \to 0$ we conclude that ϕ_0 belongs to the same part of $H^\infty(\Omega)$ as Ω. To repeat this astounding fact again: if (4.1a)–(4.1d) hold, then there is an element ϕ_0 of $\mathbf{M}(\Omega)$ which belongs to the same part for $H^\infty(\Omega)$ as does Ω yet ϕ_0 is not evaluation at any point of Ω. Further,

$$\phi_0(f) = \lim\{f(x): x \to 0, x < 0\}, \qquad f \in H^\infty(\Omega)$$

This ϕ_0 is an example of a *distinguished* homomorphism of $H^\infty(\Omega)$; it is the purpose of this section to explore this idea for a general domain Ω in $\mathbb{C}$.

The first theorem gives a necessary and sufficient condition that there is a distinguished homomorphism in the fiber $\mathbf{M}_\lambda$ over some $\lambda \in \partial\Omega$.

Theorem 4.1. *Let λ be a point of $\partial\Omega$. Then there is a $\phi_\lambda \in \mathbf{M}_\lambda$ which lies in the same part for $H^\infty(\Omega)$ as Ω if and only if $\mathbf{M}_\lambda$ is not a peak set for $H^\infty(\Omega)$. Further, if $\mathbf{M}_\lambda$ is not a peak set, then there is exactly one such ϕ_λ in $\mathbf{M}_\lambda$. Finally, if $\phi_\lambda \in \mathbf{M}_\lambda$ does lie in the same part for $H^\infty(\Omega)$ as Ω and if*

$$A_{\varepsilon,\delta} = \{z \in \Omega : |z - \lambda| < \delta \quad \text{and} \quad \|\phi_z - \phi_\lambda\| < \varepsilon\}$$

then $A_{\varepsilon,\delta}$ has positive area for each positive ε and δ. Indeed,

$$\frac{\operatorname{area}(A_{\varepsilon,\delta})}{\pi\delta^2} \to 1 \quad \text{as } \delta \to 0$$

for each $\varepsilon > 0$.

Proof. One direction is quite easy. If $\phi_\lambda \in \mathbf{M}_\lambda$ does lie in the same part as Ω then there is a constant ρ in the interval (0, 1) such that

$$|\phi_\lambda(f)| \leqslant \rho, \qquad f \in H^\infty(\Omega), \qquad \|f\| \leqslant 1, \qquad f(z_0) = 0 \tag{4.3}$$

Hence, no $f \in H^\infty(\Omega)$ can peak on $\mathbf{M}_\lambda$.

Conversely, assume that $\mathbf{M}_\lambda$ is not a peak set for $H^\infty(\Omega)$. Let $\mathbf{B}(\lambda)$ be the closed subalgebra of $H^\infty(\Omega)$ consisting of those functions which are continuous at λ. The maximal ideal space of $\mathbf{B}(\lambda)$ is just that of $H^\infty(\Omega)$ with $\mathbf{M}_\lambda$ identified to a point. The assumption that $\mathbf{M}_\lambda$ is not a peak set for $H^\infty(\Omega)$ is just the assumption that λ is not a peak point for $\mathbf{B}(\lambda)$. Hence, there is a positive measure ν of mass 1 on $\mathbf{M}(\Omega)$ which has no mass on $\mathbf{M}_\lambda$ and for which

$$\int f\,d\nu = f(\lambda), \qquad f \in \mathbf{B}(\lambda) \tag{4.4}$$

Let ϕ_λ be the linear functional on $H^\infty(\Omega)$ given by

$$\phi_\lambda(h) = \int h\,d\nu, \qquad h \in H^\infty(\Omega) \tag{4.5}$$

The functional ϕ_λ is not identically zero since $\phi(1) = 1$. Further, ϕ_λ is actually multiplicative. For if g and h are in $H^\infty(\Omega)$, then there are sequences $\{g_n\}$ and $\{h_n\}$ of elements of $\mathbf{B}(\lambda)$ with $g_n \to g$ and $h_n \to h$ boundedly and pointwise on $\mathbf{M}(\Omega)\setminus\mathbf{M}_\lambda$; see Proposition 3.5. Hence,

$$\begin{aligned}
\phi_\lambda(gh) &= \int gh\,d\nu = \lim_n \int g_n h_n\,d\nu \\
&= \lim_n g_n(\lambda) h_n(\lambda) \\
&= \lim_n \left(\int g_n\,d\nu\right)\left(\int h_n\,d\nu\right) \\
&= \phi_\lambda(g)\phi_\lambda(h)
\end{aligned}$$

Here we have made use of the fact that the support of ν lies in $\mathbf{M}(\Omega)\setminus\mathbf{M}_\lambda$. Thus, $\phi_\lambda \in \mathbf{M}_\lambda$. Further, ϕ_λ is the only element of $\mathbf{M}_\lambda$ with a representing measure with no mass on $\mathbf{M}_\lambda$. For if ψ is also such an element of $\mathbf{M}_\lambda$ and if $f \in H^\infty(\Omega)$, then there is a sequence $\{f_n\}$ of elements of $\mathbf{B}(\lambda)$ with $f_n \to f$ boundedly and pointwise on $\mathbf{M}(\Omega)\setminus\mathbf{M}_\lambda$. Hence,

$$\phi_\lambda(f) = \lim \phi_\lambda(f_n) = \lim f_n(\lambda) = \lim \psi(f_n) = \psi(f)$$

The difficult part of the proof remains before us—we must show that ϕ_λ lies in the same part of $\mathbf{M}(\Omega)$ as Ω. Indeed, we shall show considerably more. Let Δ_δ be the disc of radius δ centered at λ and let

$$A_{\varepsilon,\delta} = \{z \in \Omega \cap \Delta_\delta : \|\phi_\lambda - \phi_z\| < \varepsilon\} \tag{4.6}$$

We shall show that

$$\lim_{\delta\to 0} \frac{\operatorname{area}(A_{\varepsilon,\delta})}{\operatorname{area}(\Delta_\delta)} = 1$$

for each $\varepsilon > 0$.

Let η be the measure on $\Omega \cup \partial\Omega$ defined by

$$\int u\,d\eta = \int u \circ \rho\,d\nu, \qquad u \in C(\Omega \cup \partial\Omega) \tag{4.7}$$

where ρ is the projection of $\mathbf{M}(\Omega)$ onto $\Omega \cup \partial\Omega$. Then η is a positive measure of mass 1 and η has no mass at the point λ since ν has no mass on $\mathbf{M}_\lambda$. Set

$$\hat{\eta}(z) = \int (\zeta - z)^{-1}\, d\nu(\zeta), \qquad z \in \mathbf{C}$$

$$\tilde{\eta}(z) = \int |\zeta - z|^{-1}\, d\nu(\zeta), \qquad z \in \mathbf{C}$$

It is known that $\tilde{\eta}$ is finite a.e. $dx\,dy$ in $\mathbf{C}$ (this is just an application of Green's theorem; see exercise 3). Let dm be Lebesgue area measure, set $\Delta_n = \{z : |z - \lambda| \leqslant 1/n\}$, and

$$F_n(w) = \frac{1}{m(\Delta_n)} \iint_{\Delta_n} \left| \frac{z - \lambda}{z - w} \right| dm(z) \tag{4.8}$$

Then $F_n(\lambda) = 1$ and

$$F_n(w) \leqslant (n|w - \lambda| - 1)^{-1} \quad \text{if} \quad |w - \lambda| \geqslant \frac{1}{n}$$

so that $F_n(w) \to 0$ as $n \to \infty$ if $w \neq \lambda$. Further, for all w,

$$|F_n(w)| \leqslant \frac{1}{m(\Delta_n)} \frac{1}{n} \iint_{\Delta_n} \frac{dm(z)}{|z - w|}$$

$$\leqslant \frac{n}{\pi} \iint_{\Delta_n} \frac{dm(z)}{|z - \lambda|} = 2$$

Thus, the dominated convergence theorem implies

$$0 = \lim \int F_n(w)\, d\eta(w)$$

However,

$$\int F_n(w)\, d\eta(w) = \int \frac{1}{m(\Delta_n)} \iint_{\Delta_n} \left| \frac{z - \lambda}{z - w} \right| dm(z)\, d\eta(w)$$

$$= \frac{1}{m(\Delta_n)} \iint_{\Delta_n} |z - \lambda| \tilde{\eta}(z)\, dm(z)$$

so we learn the important relation

$$0 = \lim_{n \to \infty} \frac{1}{m(\Delta_n)} \iint_{\Delta_n} |z - \lambda| \tilde{\eta}(z)\, dm(z)$$

Equivalently,

$$0 = \lim_{\delta \to 0} \frac{1}{m(\Delta_\delta)} \iint_{\Delta_\delta} |z - \lambda| \tilde{\eta}(z)\, dm(z) \tag{4.9}$$

Suppose now that z is a point of Ω for which

$$\tilde{\eta}(z)|z - \lambda| < \frac{\varepsilon}{2 + \varepsilon} \tag{4.10}$$

We shall show that $\|\phi_\lambda - \phi_z\| < \varepsilon$. Set

$$c = \int \frac{w - \lambda}{w - z}\, d\eta(w) = 1 + (z - \lambda)\hat{\eta}(z)$$

so

$$|c| \geqslant 1 - |z - \lambda||\hat{\eta}(z)| \geqslant 1 - |z - \lambda|\tilde{\eta}(z) > 0$$

If $h \in H^\infty(\Omega)$, then

$$g(\zeta) = \frac{h(\zeta) - h(z)}{\zeta - z}(\zeta - \lambda), \qquad \zeta \in \Omega \tag{4.11}$$

is also in $H^\infty(\Omega)$ and, indeed, lies in $\mathbf{B}(\lambda)$ and vanishes at λ. Consequently, with ρ the projection of $\mathbf{M}(\Omega)$ into $\Omega \cup \Gamma$ given in Section 3 we have

$$0 = \int g\, d\nu = \int \frac{h - h(z)}{\rho - z}(\rho - \lambda)\, d\nu$$

Equivalently,

$$h(z) = \int h \frac{1}{c} \frac{\rho - \lambda}{\rho - z}\, d\nu, \qquad h \in H^\infty(\Omega)$$

Hence, the distance in the norm of $H^\infty(\Omega)^*$ from ϕ_λ to ϕ_z does not exceed the total variation of the measure $(1 - c^{-1}(\rho - \lambda)(\rho - z)^{-1})\, d\nu$. However, this is equal to

$$|c|^{-1} \int \left| \frac{w - \lambda}{w - z} - c \right| d\eta(w)$$

$$= |z - \lambda||c|^{-1} \int |w - z|^{-1} |w - \lambda - (1 + (z - \lambda)\hat{\eta}(z))(w - z)|\, d\eta(w)$$

$$\leqslant |c|^{-1}|z - \lambda|[\tilde{\eta}(z) + |\hat{\eta}(z)|]$$

$$\leqslant 2|z - \lambda|(1 - |z - \lambda||\hat{\eta}(z)|)^{-1}\tilde{\eta}(z)$$

$$\leqslant \left(\frac{2\varepsilon}{2 + \varepsilon}\right)\left(1 - \frac{\varepsilon}{2 + \varepsilon}\right)^{-1} = \varepsilon$$

Now let us review the foregoing computation in two slightly different contexts. Suppose first that (4.10) holds for some $z \notin \Omega \cup \partial\Omega$. If $h \in H^\infty(\Omega)$, then

$$g(\zeta) = \frac{h(\zeta)}{(\zeta - z)}$$

also lies in $H^\infty(\Omega)$ and the entire argument above can be repeated, leading to the absurd conclusion that

$$|\phi_\lambda(h)| \leqslant \varepsilon, \qquad \text{all } h \in H^\infty, \qquad \|h\| \leqslant 1$$

Hence, (4.10) does not hold if $z \notin \Omega \cup \partial\Omega$. Suppose next that (4.10) holds for some $z \in \partial\Omega$, $z \neq \lambda$. If h is in $H^\infty(\Omega)$ and is holomorphic in a neighborhood of z, then $g(\zeta)$, given by (4.11), is also in $H^\infty(\Omega)$ and is holomorphic at z. The very same argument then yields

$$|h(z) - \phi_\lambda(h)| \leqslant \varepsilon\|h\|$$

Thus,

$$|f(z) - \phi_\lambda(f)| \leqslant 5\varepsilon\|f\|, \qquad f \in \mathbf{B}(z)$$

In particular, if $5\varepsilon < \frac{1}{2}$ (which is safe to assume) we see that

$$|f(z) - \phi_\lambda(f)| \leqslant \tfrac{1}{2}, \qquad f \in \mathbf{B}(z), \qquad \|f\| \leqslant 1$$

However, we have shown in the example following Proposition 2.4 that almost all points in $\partial\Omega$ (with respect to area measure) are peak points for the algebra $\mathbf{A}(\partial\Omega, S^2)$ which is a subalgebra of $H^\infty(\Omega)$. Hence, the set of those points in $\partial\Omega$ at which (4.10) holds must have area measure zero. Thus, we must conclude that almost all points at which (4.10) holds must be points of Ω. Hence,

$$m(A_{\varepsilon,\delta}) \geqslant m\left(\left\{z \in \Omega \cap \Delta_\delta : |z - \lambda|\tilde{\eta}(z) < \frac{\varepsilon}{2+\varepsilon}\right\}\right)$$

so that

$$1 \geqslant \frac{m(A_{\varepsilon,\delta})}{m(\Delta_\delta)} \geqslant 1 - \frac{2+\varepsilon}{\varepsilon}\,\frac{1}{m(\Delta_\delta)} \iint_{\Delta_\delta} |z - \lambda|\tilde{\eta}(z)\, dm(z)$$

However, if we apply (4.9) we see that

$$\lim_{\delta \to 0} \frac{m(A_{\varepsilon,\delta})}{m(\Delta_\delta)} = 1 \tag{4.12}$$

as desired. Actually, we've only proved (4.12) when ε is less than $\frac{1}{10}$ but since $A_{\varepsilon,\delta}$ increases as ε increases (4.12) holds for all $\varepsilon \leqslant 2$.

We next connect the distinguished homomorphisms to the weak-* topology of $H^\infty(\Omega)$.

Theorem 4.2. *Let $\phi \in \mathbf{M}_\lambda$. The following three statements are equivalent:*

(*a*) *ϕ lies in the same part of $\mathbf{M}(\Omega)$ as Ω*

(*b*) *ϕ is weak-* continuous on $H^\infty(\Omega)$ (in the weak-* topology of $L^\infty(\Omega, dx\,dy)$)*

(*c*) *ϕ has a complex representing measure supported on Ω*

Proof. If (a) holds, then we know from Theorem 4.1 that (4.12) holds as well, and thus there are points $z_n \in \Omega$ with $\|\phi_{z_n} - \phi\| < 1/n$, $n = 1, 2, \ldots$. Now let $\{h_k\}$ be a sequence in $H^\infty(\Omega)$ with $h_k \to 0$ weak-* in $L^\infty(\Omega, dx\,dy)$. We must show $\phi(h_k) \to 0$ as well. We know that $\{h_k\}$ tends to 0 uniformly on compact subsets of Ω and so $|h_k(z_n)| \to 0$ as $k \to \infty$ for each n. Hence,

$$\begin{aligned}|\phi(h_k)| &\leqslant |\phi(h_k) - \phi_{z_n}(h_k)| + |\phi_{z_n}(h_k)| \\ &< \frac{\|h_k\|}{n} + |\phi_{z_n}(h_k)| \\ &\leqslant \frac{M}{n} + |h_k(z_n)| \\ &< \varepsilon, \qquad k \geqslant k_0\end{aligned}$$

if n is chosen so that $M/n < \varepsilon/2$ and then k_0 is chosen so that $|h_k(z_n)| < \varepsilon/2$, $k \geqslant k_0$. Hence, (b) holds.

Suppose next that (b) is true. Then (c) follows almost by definition; there is a function $u \in L^1(\Omega, dx\,dy)$ with

$$\int_\Omega fu\,dx\,dy = \phi(f), \qquad f \in H^\infty(\Omega)$$

Finally, if (c) holds then (a) must also. For if (a) fails then there is a sequence $\{h_n\}$ of elements of $H^\infty(\Omega)$ with

$$\|h_n\| \leqslant 1 \tag{4.13a}$$

$$\phi(h_n) = 0 \tag{4.13b}$$

$$h_n(z_0) \to 1 \tag{4.13c}$$

From (4.13a) and (4.13c) we see that $h_n(z) \to 1$ uniformly on compact subsets of Ω. But by assumption

$$\phi(f) = \int f\,d\alpha$$

for some measure α supported on Ω. Thus

$$1 = \lim \int h_n \, d\alpha = \lim \phi(h_n) = 0$$

an obvious absurdity.

Definition. Suppose $\lambda \in \partial\Omega$ and $\mathbf{M}_\lambda$ is not a peak set for $H^\infty(\Omega)$. The homomorphism $\phi_\lambda \in \mathbf{M}_\lambda$ described in Theorem 4.1 and characterized in Theorem 4.2 will be called a *distinguished homomorphism*. We do know that each fiber $\mathbf{M}_\lambda$ contains at most one distinguished homomorphism.

Proposition 4.3. Suppose $\lambda \in \partial\Omega$ and $\mathbf{M}_\lambda$ is not a peak set for $H^\infty(\Omega)$; let $\phi_\lambda \in \mathbf{M}_\lambda$ be the distinguished homomorphism. Set $\Omega_\delta = \{z \in \Omega : |z - \lambda| < \delta\}$ and

$$S_\delta(h) = \frac{1}{\pi\delta^2} \iint_{\Omega_\delta} h \, dx \, dy, \qquad h \in H^\infty(\Omega) \tag{4.14}$$

$$L_\delta(h) = \frac{1}{\pi} \iint_{\Omega_\delta} \frac{f(z)}{z - \lambda} \frac{\partial v_\delta}{\partial \bar{z}} \, dx \, dy, \qquad h \in H^\infty(\Omega) \tag{4.15}$$

where $v_\delta(z) = u((z - \lambda)/\delta)$ and u is given in (3.1). Then $S_\delta \to \phi_\lambda$ as $\delta \to 0$ and $L_\delta \to -\phi_\lambda$ as $\delta \to 0$, both in the norm of $(H^\infty(\Omega))^*$.

Proof. We know that the area measure of Ω_δ when divided by $\pi\delta^2$ goes to 1 as $\delta \to 0$; see (4.12). Hence, $S_\delta(1) \to 1$ as $\delta \to 0$. Let h have norm one and lie in the kernel of ϕ_λ. Then $|h| \leqslant \varepsilon$ on $A_{\varepsilon,\delta}$ so that

$$|S_\delta(h)| \leqslant \varepsilon + \frac{1}{\pi\delta^2} \operatorname{area}(\Delta_\delta \setminus A_{\varepsilon,\delta})$$

Hence, $\lim \sup\{|S_\delta(h)|\colon \delta \to 0\} \leqslant \varepsilon$ for all h, $\|h\| \leqslant 1$, $\phi_\lambda(h) = 0$. Thus $\lim\{|S_\delta(h)|\colon \delta \to 0\} = 0$ and so $S_\delta \to \phi_\lambda$ in norm as $\delta \to 0$.

Likewise, $L_\delta(1) \to -1$ as $\delta \to 0$, so we need only check the behavior of L_δ on the kernel of ϕ_λ. Let $\|h\| \leqslant 1$ and $\delta_\lambda(h) = 0$; thus, $|h| \leqslant \varepsilon$ on $A_{\varepsilon,\delta}$ so that

$$|L_\delta(h)| \leqslant \tfrac{1}{2}\varepsilon\pi\delta^{-1} \iint_{\Omega \cap A_{\varepsilon,\delta}} |z - \lambda|^{-1} \, dx \, dy + \tfrac{1}{2}\delta^{-1} \iint_{\Omega_\delta \setminus A_{\varepsilon,\delta}} |z - \lambda|^{-1} \, dx \, dy$$

Now it is a fact about area integrals that for each measurable set E we have

$$\iint_E |z - \lambda|^{-1} \, dx \, dy \leqslant [4\pi \operatorname{area}(E)]^{1/2} \tag{4.16}$$

Hence, from above, we have

$$|L_\delta(h)| \leqslant \varepsilon 8\pi\sqrt{\pi} + 8\sqrt{\pi}\,\delta^{-1}\big[\pi\delta^2 \setminus \operatorname{area}(A_{\varepsilon,\delta})\big]^{1/2}$$

$$\leqslant 80\varepsilon + 20\left[1 - \frac{\operatorname{area}(A_{\varepsilon,\delta})}{\pi\delta^2}\right]^{1/2}$$

Thus, by (4.12), we have

$$\limsup\{|L_\delta(h)|\colon \delta \to 0\} \leqslant 80\varepsilon, \qquad \|h\| \leqslant 1, \qquad \phi_\lambda(h) = 0$$

As before, this gives the desired conclusion.

Proposition 4.4. If $\mathbf{M}_\lambda$ is not a peak set for $H^\infty(\Omega)$, then λ is a component of $\mathbf{C} \setminus \Omega$.

Proof. Suppose there is a continuum γ in $\mathbf{C} \setminus \Omega$ which contains λ. Let f be the Riemann mapping onto Δ of that component of $\mathbf{C} \setminus \gamma$ which contains Ω. The cluster set of f at λ lies within the unit circle $\mathbf{T}$ and thus the range of $\hat{f}$ on $\mathbf{M}_\lambda$ lies within $\mathbf{T}$, as well. Hence, no point of the fiber $\mathbf{M}_\lambda$ can be in the same part for $H^\infty(\Omega)$ as Ω, a contradiction. Thus, no such continuum exists and $\{\lambda\}$ is a component of $\mathbf{C} \setminus \Omega$.

Remark. Proposition 4.4 provides a partial answer to a question raised by Beck in his paper (1964a) on what we've termed in Chapter 3 as weak peak points: namely, can the peaking function be chosen to be continuous at λ? From Proposition 4.4 we see that if $\lambda \in \partial\Omega$ lies in nontrivial continuum in $\mathbf{C} \setminus \Omega$, then $\mathbf{M}_\lambda$ is a peak fiber so the answer is "yes." However, it is not always "yes" since $\mathbf{M}_\lambda$ can fail to be a peak set for $H^\infty(\Omega)$.

In addition to the conditions given in Theorems 4.1 and 4.2 there is another description of the behavior of points of Ω near a distinguished homomorphism. Let us begin by assuming that $\lambda \in \partial\Omega$ and that there is a nonconstant function $f \in H^\infty(\Omega)$ with

$$\|f\|_\infty = 1$$

$$\lim\{f(z) : z \in \Omega,\, z \to \lambda\} = 1$$

Let $\{z_n\}$ be any sequence in Ω with $z_n \to \lambda$. Then $\zeta_n = f(z_n) \to 1$ so that there is some subsequence $\{\zeta_{n_j}\}$ which is an interpolation sequence for $H^\infty(\Delta)$. Hence, $\{z_{n_j}\}$ is an interpolation sequence for $H^\infty(\Omega)$. For if $\{w_j\}$ is any bounded sequence of complex numbers and if $h \in H^\infty(\Delta)$ satisfies $h(\zeta_{n_j}) = w_j$, $1 \leqslant j < \infty$, then $(h \circ f)(z_{n_j}) = w_j$, $1 \leqslant j < \infty$, and, of course, $h \circ f$ lies in $H^\infty(\Omega)$. These remarks then set the stage for the next result.

Theorem 4.5. *Suppose* $z_n \to \lambda$, $\lambda \in \partial\Omega$, *and no subsequence of* $\{z_n\}$ *is an interpolation sequence for* $H^\infty(\Omega)$. *Then the fiber* $\mathbf{M}_\lambda$ *contains a distinguished homomorphism* ϕ_λ *and* $\|\phi_{z_n} - \phi_\lambda\| \to 0$ *in the norm of* $(H^\infty(\Omega))^*$.

Proof. The remarks preceding the statement of the theorem show that $\mathbf{M}_\lambda$ can not be a peak set for $H^\infty(\Omega)$ and so there is a distinguished homomorphism ϕ_λ in $\mathbf{M}_\lambda$. We shall assume that $\{\phi_{z_n}\}$ does not converge to ϕ_λ in norm and show that $\{z_n\}$ has an interpolation (sub)sequence.

To begin, we may assume that

$$\sup\{|f(z_n)| : \|f\|_\infty \leqslant 1, \phi_\lambda(f) = 0\} \geqslant \eta > 0$$

for some η independent of n. (This is at least true for a subsequence of $\{z_n\}$.) We next employ Proposition 3.2 to obtain a sequence $\{f_n\}$ of elements of $H^\infty(\Omega)$ with

$$f_n(z_n) = 1 \quad \text{for all } n \tag{4.17a}$$

$$f_n \text{ is holomorphic in a neighborhood of } \lambda \text{ and } f_n(\lambda) = 0 \tag{4.17b}$$

$$\|f_n\| \leqslant M \quad \text{for all } n \tag{4.17c}$$

Now let $\mathcal{U}$ be a disc centered at λ and let $\varepsilon > 0$ be given. We shall see that there is an integer N and a function F_N in $H^\infty(\Omega)$ satisfying

$$\|F_N\| \leqslant 5M \tag{4.18a}$$

$$|F_N| < \varepsilon \quad \text{on } \Omega \setminus \mathcal{U} \tag{4.18b}$$

$$F_N \text{ is analytic on a neighborhood of } \lambda \text{ and } F_N(\lambda) = 0 \tag{4.18c}$$

$$F_N(z_N) = 1 \tag{4.18d}$$

To prove (4.18) let $\{T_\delta f\}$ be the functions constructed in Proposition 3.2 for the function f; we shall take f to be one of the $\{f_k\}$ presently. Choose δ so small that $|T_\delta f| < \varepsilon/2$ on $\Omega \setminus \mathcal{U}$ for all $f \in H^\infty(\Omega)$, $\|f\| \leqslant M$. Then, by Proposition 4.3, δ can be also chosen so small that $|L_\delta(f)| < \varepsilon/2$ if $f \in H^\infty(\Omega)$, $\|f\| \leqslant M$, and $\phi_\lambda(f) = 0$. Hence, $|f_k - T_\delta f_k| < \varepsilon$ for all very large k by (3.9). Thus,

$$\varepsilon > |f_k(z_k) - (T_\delta f_k)(z_k)| = |1 - (T_\delta f_k)(z_k)|, \qquad k \geqslant N$$

Set $F_N = (T_\delta f_N + L_\delta f_N)((T_\delta f_N)(z_n) + (L_\delta f_N)(z_N))^{-1}$; this is the desired function.

Now we employ (4.18) to obtain a subsequence $\{\zeta_j\}$ of $\{z_n\}$, a sequence $\{h_j\}$ of elements of $H^\infty(\Omega)$, and a sequence $\{\mathcal{U}_j\}$ of nested discs centered at λ

having these properties:

$$\|h_j\| \leqslant 5M \tag{4.19a}$$

$$|h_j| \leqslant 2^{-j-2} \quad \text{on } \Omega \setminus \mathcal{U}_j \tag{4.19b}$$

$$h_j(\zeta_j) = 1 \tag{4.19c}$$

$$|h_j(\zeta_k)| \leqslant 2^{-j-2} \quad \text{if } k \neq j \tag{4.19d}$$

$$h_j \text{ is analytic in a neighborhood of } \lambda \text{ and } h_j(\lambda) = 0 \tag{4.19e}$$

$$|h_j| < 2^{-j-2} \quad \text{on } \mathcal{U}_{j+1}. \tag{4.19f}$$

One way to obtain such functions is this. Given $\zeta_1,\ldots,\zeta_n$, $h_1,\ldots,h_n$ and $\mathcal{U}_1 \supset \cdots \supset \mathcal{U}_n$ define $\mathcal{U}_{n+1}$ to be a disc centered at λ, inside $\mathcal{U}_n$, and so small that $|h_j| < 2^{-n-2}$ on $\mathcal{U}_{n+1}$ for $j = 1,\ldots,n$. Then employ (4.18) to find an index N and a function F_N satisfying (4.18) for $\mathcal{U} = \mathcal{U}_{n+1}$, $\varepsilon = 2^{-n-2}$. Let $\zeta_{n+1} = z_N$ and $h_{n+1} = F_N$.

Now the sum $\sum_{j=1}^{\infty}|h_j|$ converges uniformly on compact subsets of Ω and the sum is bounded above by $5M + \frac{1}{2}$. If $\{w_j\}$ is a sequence of complex numbers with $|w_j| \leqslant 1$ for all j, then

$$h(z) = \sum_{j=1}^{\infty} w_j h_j(z)$$

lies in $H^\infty(\Omega)$, and

$$\|h\| \leqslant 5M + \tfrac{1}{2} \tag{4.20a}$$

$$|h(\zeta_j) - w_j| < \tfrac{1}{2} \tag{4.20b}$$

It just remains, then, to show that (4.20) implies $\{\zeta_j\}$ is an interpolation sequence. This is an entirely separate matter, contained in the next result.

Proposition 4.6. Let $\{\zeta_j\}$ be a sequence of points in Ω. Suppose there are constants M and m, $0 < m < 1$, with this property: if $\{w_j\}$ is a sequence in l^∞, then there is an $f \in H^\infty(\Omega)$ with

$$\|f\|_\infty \leqslant M\|\{w_j\}\|_\infty \tag{4.21a}$$

$$\|\{f(\zeta_j) - w_j\}\| \leqslant m\|\{w_j\}\|_\infty \tag{4.21b}$$

Then $\{\zeta_j\}$ is an interpolation sequence for $H^\infty(\Omega)$.

Proof. Let $\{w_j\} \in l^\infty$ with $\|\{w_j\}\| \leqslant 1$. Choose $h_1 \in H^\infty(\Omega)$ with $\|h_1\| \leqslant M$ and $|w_j - h_1(\zeta_j)| \leqslant m$ for all j. Apply (4.21) to the sequence $\{w_j - h_1(\zeta_j)\}_{j=1}^\infty$; there is an $h_2 \in H^\infty(\Omega)$ with $\|h_2\| \leqslant Mm$ and $|w_j - h_1(\zeta_j) - h_2(\zeta_j)| \leqslant m^2$. Apply (4.21) to the sequence $\{w_j - h_1(\zeta_j) - h_2(\zeta_j)\}_{j=1}^\infty$ and continue in this fashion. In this way we obtain a sequence $\{h_n\}$ of elements of $H^\infty(\Omega)$ with

$$\|h_n\| \leqslant Mm^{n-1}, \qquad n = 1,2,\dots$$

$$\left| w_j - \sum_{n=1}^{N} h_n(\zeta_j) \right| \leqslant m^N, \qquad N = 1,2,\dots, \qquad j = 1,2,\dots$$

Thus, $\Sigma_1^\infty h_n$ lies in $H^\infty(\Omega)$ and $h(\zeta_j) = w_j, j = 1,2,\dots.$

The final results of this section extend to a general domain the idea from the example which opened this section; namely, the distinguished homomorphism could be realized as the limit of $f(x)$ as x approaches 0 along the negative real axis.

Theorem 4.7. *Suppose $\lambda \in \partial\Omega$ and $\mathbf{M}_\lambda$ is not a peak set for $H^\infty(\Omega)$; let $\phi_\lambda \in \mathbf{M}_\lambda$ be the distinguished homomorphism for the fiber $\mathbf{M}_\lambda$. Suppose $\{z_n\}$ is a sequence in Ω with*

$$z_n \to \lambda \tag{4.22a}$$

$$\operatorname{dist}(z_n, \partial\Omega) \geqslant c|z_n - \lambda|, \quad \text{all } n \tag{4.22b}$$

and some positive number c. Then $\|\phi_{z_n} - \phi_\lambda\| \to 0$ in the norm of $(H^\infty(\Omega))^*$.

Proof. There is no loss in assuming $\lambda = 0$. Let

$$A_{\varepsilon,\delta} = \{z \in \Omega : |z| < \delta \quad \text{and} \quad \|\phi_z - \phi_\lambda\| < \varepsilon\}$$

We know that area $(A_{\varepsilon,\delta})/\pi\delta^2$ goes to 1 as $\delta \to 0$ from (4.12). Select sequences $\{\varepsilon_n\}$ and $\{\nu_n\}$ both decreasing to zero so that

$$\frac{\operatorname{area}\left(A_{\varepsilon_n, 2|z_n|}\right)}{4\pi|z_n|^2} \geqslant 1 - \nu_n$$

Thus, there is a point w_n with

$$\|\phi_{w_n} - \phi_\lambda\| < \varepsilon_n$$

$$|w_n - z_n| < 2\sqrt{\nu_n}\,|z_n|$$

and hence the disc centered at z_n and of radius $c|z_n|$ lies in Ω and contains w_n. Let $f \in H^\infty(\Omega)$ with $\|f\| \leqslant 1$ and $f(z_n) = 0$. Then

$$|f(w_n)| \leqslant \frac{|z_n - w_n|}{c|z_n|} \leqslant 2\frac{\sqrt{\nu_n}}{c}$$

by an application of Schwarz's lemma. Thus,

$$\|\phi_{z_n} - \phi_{w_n}\| \leqslant 2\sqrt{\nu_n}/c$$

and so

$$\|\phi_{z_n} - \phi_\lambda\| \leqslant 2\sqrt{\nu_n}/c + \varepsilon_n \to 0$$

Corollary 4.9. *If γ is an arc in Ω with endpoint at λ and if $dist(z, \partial\Omega) \geqslant c|z - \lambda|$ for all $z \in \gamma$ and some $c > 0$, then*

$$\lim\{f(z) : z \in \gamma, z \to \lambda\} = \phi_\lambda(f)$$

for all $f \in H^\infty(\Omega)$.

6.5. THE SHILOV BOUNDARY OF $H^\infty(\Omega)$

The Shilov boundary $\mathbf{S}(\Omega)$ of $H^\infty(\Omega)$ is the smallest closed set in $\mathbf{M}(\Omega)$ on which each $f \in H^\infty(\Omega)$ attains its norm; that is,

$$\|\hat{f}\|_{\mathbf{S}(\Omega)} = \|f\|_{H^\infty(\Omega)}, \qquad f \in H^\infty(\Omega) \tag{5.1}$$

and

$$\mathbf{S}(\Omega) \subseteq \mathbf{X} \quad \text{if} \quad \mathbf{X} \subset \mathbf{M}(\Omega) \quad \text{and} \quad \|\hat{f}\|_{\mathbf{X}} = \|f\|_{H^\infty(\Omega)} \quad \text{for all } f \in H^\infty(\Omega) \tag{5.2}$$

The existence of such a set $\mathbf{S}(\Omega)$ is a matter of Banach algebra considerations; we refer the reader to Stout (1971, section 7) for a proof. We shall eventually make a connection between $\mathbf{S}(\Omega)$ and the maximal ideal space of a certain L^∞ space so we begin by amplifying and extending the discussion of Section 3 in Chapter 5 on such spaces.

Theorem 5.1. *Let μ be a positive finite regular Borel measure on a topological space $\mathbf{S}$ and let $\mathbf{X}$ be the maximal ideal space of $L^\infty(\mathbf{S}, \mu)$. Then the Gelfand transform maps L^∞ isometrically onto all of $\mathbf{C}(\mathbf{X})$. Further, the topology of $\mathbf{X}$ has a basis of sets which are both open and closed.*

Proof. It is apparent that $\|\hat{u}\| \leqslant \|u\|$ for each $u \in L^\infty$. Suppose that α is a complex number such that $\|u\| = \alpha$ and the set of those $y \in \mathbf{S}$ at which $|u(y) - \alpha| < \varepsilon$ has positive μ-measure for all $\varepsilon > 0$. Then $u - \alpha$ is not invertible in L^∞ and so $u - \alpha$ lies in a proper ideal in L^∞, for example, the ideal of functions v with the property that 0 lies in their essential range. Hence, $u - \alpha$ lies in a maximal ideal and so there is a $\phi \in \mathbf{X}$ with $\phi(u - \alpha) = 0$; that is, $\phi(u) = \alpha$. Hence, $\|\hat{u}\| \geqslant |\hat{u}(\phi)| = |\alpha| = \|u\|$.

Next, we note that by definition the topology of $\mathbf{X}$ has as a basis sets of the form $\{\phi \in \mathbf{X} : |\hat{u}_j(\phi)| < \varepsilon, j = 1,\ldots, N\}$ where $u_1,\ldots, u_N$ lie in L^∞ and $\varepsilon > 0$. Because the linear span of characteristic functions is dense in L^∞, the basic open sets can be defined by limiting $u_1,\ldots, u_N$ to be characteristic functions. But if u is the characteristic function of a measurable set, then $u^2 = u$ so $\hat{u}^2 = \hat{u}$ on $\mathbf{X}$ and hence $\hat{u}$ is the characteristic function of some set A in $\mathbf{X}$ which must in consequence be both open and closed. If $u = \sum_1^N c_j u_j$ where u_j is the characteristic function of a measurable set E_j, then $\bar{u} = \sum_1^N \overline{c_j} u_j$ and so for $\phi \in \mathbf{X}$ we have

$$\phi(\bar{u}) = \sum_1^N \overline{c_j}\phi(u_j) = \overline{\sum_1^N c_j\phi(u_j)} = \overline{\phi(u)}$$

Because the linear span of the characteristic functions of measurable sets is dense in L^∞ we see that

$$\phi(\bar{u}) = \overline{\phi(u)}, \qquad u \in L^\infty, \qquad \phi \in \mathbf{X}$$

Consequently, the Gelfand transform, $\hat{L}^\infty$, of L^∞ is a self-adjoint point separating subalgebra of $\mathbf{C(X)}$ and hence it is dense in $\mathbf{C(X)}$ by the Stone-Weierstrass theorem. But $\hat{L}^\infty$ is closed and so $\hat{L}^\infty = \mathbf{C(X)}$.

We begin our investigation of the Shilov boundary of $H^\infty(\Omega)$ by examining the case when $\Gamma = \partial\Omega$ consists of a finite number of analytic simple closed curves. In this setting we shall be able to identify $\mathbf{S}(\Omega)$ rather well. For a general Ω some information is available, but not as much.

Suppose that $\Gamma = \partial\Omega$ consists of $m + 1$ disjoint analytic simple closed curves. Then $H^\infty(\Omega)$ can be considered a subalgebra of $L^\infty(\Gamma, \omega)$ where ω is harmonic measure on Γ for a particular point $z_0 \in \Omega$. We shall show that the Shilov boundary of $H^\infty(\Omega)$ coincides with the maximal ideal space of $L^\infty(\Gamma, \omega)$. To set the notation let $\mathbf{M}(L^\infty)$ denote the maximal ideal space of $L^\infty(\Gamma, \omega)$. Obviously, $\mathbf{M}(L^\infty) \subseteq \mathbf{M}(\Omega)$. We also know that

$$\|f\|_{H^\infty(\Omega)} = \|f\|_{L^\infty(\Gamma,\omega)}, \qquad f \in H^\infty(\Omega) \tag{5.3}$$

and so $\mathbf{S}(\Omega) \subseteq \mathbf{M}(L^\infty)$. This leads us to the first major result on $\mathbf{S}(\Omega)$.

Theorem 5.2. *If* $\Gamma = \partial\Omega$ *consists of a finite number of disjoint analytic simple closed curves, then* $\mathbf{S}(\Omega)$ *coincides with the maximal ideal space of* $L^\infty(\Gamma, \omega)$.

Proof. Suppose, by way of obtaining a contradiction, that $\mathbf{S}(\Omega)$ is a proper subset of $\mathbf{M}(L^\infty)$, the maximal ideal space of $L^\infty(\Gamma, \omega)$. Then there is a basic open set $\mathcal{U}$ in $\mathbf{M}(L^\infty)\setminus \mathbf{S}(\Omega)$ and hence there is a measurable set E in Γ of positive measure with characteristic function u such that $\hat{u} = 0$ on $\mathbf{S}(\Omega)$ but $\hat{u} = 1$ on $\mathcal{U}$. Let $\tilde{u}$ be the harmonic extension of u to Ω and let $r_1, \ldots, r_m$ be the periods of $*\tilde{u}$ about $\Gamma_1, \ldots, \Gamma_m$, respectively. According to Theorem 4.6.3 there is a positive measure ν on Γ with a finite number of points in its support such that $*\tilde{\nu}$ has periods $r_1, \ldots, r_m$, respectively, as well. The function

$$f = \exp[\tilde{u} - \tilde{\nu} + i(*\tilde{u} - *\tilde{\nu})]$$

is then in $H^\infty(\Omega)$ and $|f| = e^u$ a.e. ω on Γ. Hence, $|\hat{f}| = \exp[\hat{u}]$ on $\mathbf{M}(L^\infty)$ so that $|\hat{f}| = 1$ on $\mathbf{S}(\Omega)$, which implies that $\|f\|_\infty = 1$. However, $|f| = e$ a.e. ω on the set E so that $\|f\|_\infty \geqslant e$. This is a contradiction.

We can further identify the elements of $\mathbf{S}(\Omega)$ from those of $\mathbf{M}(\Omega)$ by their effect on inner functions.

Theorem 5.3. *Let* $\partial\Omega = \Gamma$ *consist of* $m + 1$ *disjoint analytic simple closed curves and let* $\phi \in \mathbf{M}(\Omega)$. *The following are equivalent.*

(*a*) $\phi \in \mathbf{S}(\Omega)$

(*b*) $|\phi(f)| = 1$ *for all inner functions* f *in* $H^\infty(\Omega)$

(*c*) $\phi(f) \neq 0$ *for all inner functions* f *in* $H^\infty(\Omega)$

Proof. Clearly (a) implies (b) and (b) implies (c). Suppose (c) holds. If f is inner, $f \in H^\infty(\Omega)$, and if $a = \phi(f)$ is in Δ then $g = (f - a)(1 - \bar{a}f)^{-1}$ is inner and in $H^\infty(\Omega)$ while $\phi(g) = 0$, contradicting (c). Thus, (b) holds. Suppose (b) holds. Let $\mathbf{B}$ consist of all quotients of the form f/g where $f, g \in H^\infty(\Omega)$ and g is inner. We know that $\mathbf{B}$ is a dense subalgebra of $L^\infty(\Gamma, \omega)$ from Corollary 5.3.3. Let ϕ satisfy (b) and extend ϕ to $\mathbf{B}$ by the rule

$$\tilde{\phi}\left(\frac{f}{g}\right) = \frac{\phi(f)}{\phi(g)}$$

This extension is well defined since $\phi(g) \neq 0$ and since ϕ is multiplicative on $H^\infty(\Omega)$. Further, $\tilde{\phi}$ is both linear and multiplicative on $\mathbf{B}$ and if $u = f/g$ then $|\tilde{\phi}(u)| = |\phi(f)| \leqslant \|f\| = \|u\|$ since g is inner. Hence, $\tilde{\phi}$ extends ϕ to a multiplicative linear functional on $\mathbf{B}$ which is a dense subalgebra of $L^\infty(\Gamma, \omega)$. Thus, $\phi \in \mathbf{S}(\Omega)$ by Theorem 5.2.

We begin our look at $\mathbf{S}(\Omega)$ for a general domain Ω by the simple observation that if $\lambda \in \partial\Omega$ is essential (and we shall always assume that each point of $\partial\Omega$ is essential) then the fiber $\mathbf{M}_\lambda$ over λ must meet $\mathbf{S}(\Omega)$. This is a direct consequence of Theorem 3.7.1: there is an $f \in H^\infty(\Omega)$ with

$$\limsup\{|f(z)| : z \to \lambda\} = 1$$

$$\limsup\{|f(z)| : z \to x\} < 1, \qquad x \neq \lambda$$

For this f, $1 = \|f\|_{H^\infty(\Omega)} = \|\hat{f}\|_{\mathbf{S}(\Omega)}$ but the range of $\hat{f}$ on the fiber $\mathbf{M}_x$, $x \neq \lambda$, lies within the unit disc. Hence, the only fiber on which $|\hat{f}|$ can assume the value 1 is $\mathbf{M}_\lambda$. Thus, $\mathbf{M}_\lambda$ must meet $\mathbf{S}(\Omega)$.

We also take note of the fact that no distinguished homomorphism of $H^\infty(\Omega)$ lies in $\mathbf{S}(\Omega)$; see exercise 2 at the end of the chapter.

The major result that we have for $\mathbf{S}(\Omega)$ is stated in Theorem 5.4, which follows some preliminary remarks.

Let m denote Lebesgue area measure on Ω. As we noted in Section 3 of this chapter, $H^\infty(\Omega)$ is a weak-* closed subalgebra of $L^\infty(\Omega, m)$. Let $\mathbf{X}$ denote the maximal ideal space of $L^\infty(\Omega, m)$; then obviously $\mathbf{X}$ is a boundary for $H^\infty(\Omega)$ since

$$\|u\|_{L^\infty(\Omega, m)} = \|\hat{u}\|_{\mathbf{X}}, \qquad u \in L^\infty(\Omega, m)$$

Thus,

$$\mathbf{S}(\Omega) \subseteq \mathbf{X} \subseteq \mathbf{M}(\Omega) \tag{5.4}$$

Let $\hat{H}^\infty$ denote the lift of $H^\infty(\Omega)$ to $\mathbf{X}$ via the Gelfand transform. We then have the following result.

Theorem 5.4. *Suppose u is a continuous real-valued function on $\mathbf{M}(\Omega)$. Then there is an $F \in H^\infty(\Omega)$ with*

$$|F(z)| \leqslant e^{u(z)} \quad \text{for } z \in \Omega \tag{5.5a}$$

$$|\hat{F}| = e^u \qquad \text{on } \mathbf{S}(\Omega) \tag{5.5b}$$

Proof. Fix a point $p \in \Omega$, and define Λ to be those functions f in $H^\infty(\Omega)$ which satisfy

$$f(p) = 0, \qquad |f(z)|e^{-u(z)} \leqslant 1, \qquad z \in \Omega$$

Set

$$\gamma = \sup\{|f'(p)| : f \in \Lambda\}$$

The number γ is positive since e^u is bounded below on $\mathbf{M}(\Omega)$. Let F be any element of $H^\infty(\Omega)$ with

$$F(p) = 0; \qquad |F(z)|e^{-u(z)} \leqslant 1, \qquad z \in \Omega; \qquad F'(p) = \gamma$$

From this point on the proof is virtually a replication of that of Theorem 5.1.3.

Let Ω' consist of those points $b \in \Omega$ for which there is some function $h \in H^\infty(\Omega)$ with $|h(b)| > 1$ but yet $|h(z)F(z)e^{-u(z)}| \leqslant 1$ for all $z \in \Omega$. We shall show that Ω' is empty. Obviously, Ω' is open in Ω and p is not in Ω'. We now show that Ω' is also closed. Suppose $c \notin \Omega'$; then whenever $h \in H^\infty(\Omega)$ and

$$|h(z)F(z)e^{-u(z)}| \leqslant 1 \quad \text{on } \Omega \text{ we must have } |h(c)| \leqslant 1$$

Thus, the linear functional l acting on $\hat{F}e^{-u}\hat{H}^\infty$ by the rule

$$l(\hat{F}e^{-u}\hat{h}) = h(c), \qquad h \in H^\infty(\Omega)$$

has norm 1 and so there is a measure ν on $\mathbf{S}(\Omega)$ of norm 1 with

$$h(c) = \int \hat{F}e^{-u}\hat{h}\, d\nu, \qquad h \in H^\infty(\Omega)$$

In particular, $1 = \int \hat{F}e^{-u}\, d\nu \leqslant \int |\hat{F}|e^{-u}\, d|\nu| \leqslant \|\nu\| \leqslant 1$ so that

$$\hat{F}e^{-u}\, d\nu \geqslant 0 \quad \text{as a measure} \tag{5.6a}$$

$$|\hat{F}e^{-u}| = 1 \quad \text{a.e. } d\nu \tag{5.6b}$$

Let $d\mu = \hat{F}e^{-u}\, d\nu$. If c' is near c, then the argument in the proof of Theorem 4.1 (or better, the argument in the proof of Theorem 5.1.3) shows that c' also lies in the complement of Ω'. Thus, Ω' is both open and closed and since Ω' is not all of Ω, Ω' must be empty. Hence,

$$|h(z)F(z)e^{-u(z)}| \leqslant 1 \quad \text{on } \Omega \text{ implies } |h(z)| \leqslant 1 \text{ on } \Omega$$

Thus, $|\hat{F}|e^{-u} = 1$ identically on $\mathbf{S}(\Omega)$ and the theorem is proved.

Remark. The conclusion of the theorem, and the proof as well, remain unaltered under several different hypotheses. For example, u can be replaced by an element v of $L^\infty_r(\Omega, m)$ so that we obtain an $F \in H^\infty(\Omega)$ with

$$|F(z)| \leqslant e^{v(z)} \quad \text{a.e. } m \text{ on } \Omega$$

and

$$|\hat{F}| = e^{\hat{v}} \qquad \text{on } \mathbf{S}(\Omega)$$

Also ω and m may be replaced by $\mathcal{U}$ and $m_{\mathcal{U}}$ where $\mathcal{U}$ is any open set in Ω

satisfying

$$\|f\|_{H^\infty(\Omega)} = \|f\|_{\mathcal{U}} = \sup\{|f(z)| : z \in \mathcal{U}\} \quad \text{for all } f \in H^\infty(\Omega)$$

and $m_{\mathcal{U}}$ is the restriction to $\mathcal{U}$ of m. Another possibility is to replace Ω and m by $\{s_j\}_1^\infty$ and μ where $\{s_j\}$ is a dominating sequence for $H^\infty(\Omega)$ and μ places weight 2^{-j} at s_j for $j = 1, 2, \ldots$.

6.6. THE CORONA THEOREM

Let Ω be a domain on the sphere which supports nonconstant bounded analytic functions and suppose each point in $\partial\Omega$ is essential for $H^\infty(\Omega)$. Each point p of Ω gives an element of the maximal ideal space $\mathbf{M}(\Omega)$ of $H^\infty(\Omega)$ via "evaluation at p"

$$\phi_p(f) = f(p), \qquad f \in H^\infty(\Omega)$$

The mapping $p \to \phi_p$ is a homeomorphism of Ω into $\mathbf{M}(\Omega)$ if $\mathbf{M}(\Omega)$ is given the weak-* topology as a subset of the dual space of $H^\infty(\Omega)$; this is virtually a corollary of the definition of the weak-* topology. Further, the image of Ω is an open set in $\mathbf{M}(\Omega)$. To see this, note that the mapping ρ defined just prior to Proposition 3.3 is continuous and the complement of Ω in $\mathbf{M}(\Omega)$ is exactly $\rho^{-1}(\partial\Omega)$. Of course, as we have seen, the set $\rho^{-1}(\partial\Omega)$ is very big. However, it is sometimes the case that Ω is actually dense in $\mathbf{M}(\Omega)$. That is, if we let $\text{CL}(\Omega)$ stand for the weak-* closure of Ω in $\mathbf{M}(\Omega)$, then it is sometimes true that the "corona," $\mathbf{M}(\Omega) \setminus \text{CL}(\Omega)$, is empty. The corona theorem asserts that this is true when Ω is bounded by a finite number of disjoint analytic simple closed curves. It is possible that Ω is *always* dense in $\mathbf{M}(\Omega)$, at least when Ω is planar. At this time this remains an intriguing open question in the function theory of planar domains; see the notes at the end of the chapter for some comments. In this section we shall give a proof that Ω is dense in $\mathbf{M}(\Omega)$ when Ω is bounded by a finite number of disjoint simple closed curves. We begin, however, by recasting the problem into one of function theory.

Theorem 6.1. *Ω is dense in $\mathbf{M}(\Omega)$ if and only if whenever $f_1, \ldots, f_N$ are elements of $H^\infty(\Omega)$ with*

$$|f_1(z)|^2 + \cdots + |f_N(z)|^2 \geqslant \delta > 0, \qquad z \in \Omega \tag{6.1}$$

then there are elements $g_1, \ldots, g_N$ of $H^\infty(\Omega)$ with

$$f_1(z)g_1(z) + \cdots + f_N(z)g_N(z) = 1, \qquad z \in \Omega \tag{6.2}$$

Proof. Suppose Ω is not dense in $\mathbf{M}(\Omega)$; then there is an element ϕ_0 in $\mathbf{M}(\Omega)$ and a basic neighborhood $\mathcal{U}$ of ϕ_0 with $\mathcal{U} \cap \Omega = \phi$. Hence, there are functions $f_1,\ldots,f_N$ in $H^\infty(\Omega)$ and a $\delta > 0$ such that $\hat{f}_j(\phi_0) = 0$ and no point z of Ω lies in the set $\{\phi \in \mathbf{M}(\Omega) : |\hat{f}_j(\phi)| < \delta^{1/2}, 1 \leqslant j \leqslant N\}$. Thus, $|f_1(z)^2| + \cdots + |f_N(z)|^2 \geqslant \delta$ for all $z \in \Omega$. But if there were $g_1,\ldots, g_N$ in $H^\infty(\Omega)$ such that (6.2) holds, then

$$1 = \phi_0(1) = \phi_0\left(\sum_1^N f_j\phi_j\right) = \sum_1^N \phi_0(f_j)\phi_0(g_j) = 0$$

since $\phi_0(f_j) = 0$ for $1 \leqslant j \leqslant N$. This contradiction shows no such functions $g_1,\ldots, g_N$ exist.

Conversely, suppose Ω is dense in $\mathbf{M}(\Omega)$. If $f_1,\ldots,f_N$ satisfy (6.1), then $\Sigma_1^N|\hat{f}_j(\phi)|^2 \geqslant \delta$ for all $\phi \in \mathbf{M}(\Omega)$ and so $\hat{f}_1,\ldots,\hat{f}_N$ have no common zero $\mathbf{M}(\Omega)$. Thus, the ideal they generate is all of $H^\infty(\Omega)$. In particular, 1 lies in the ideal generated by $f_1,\ldots,f_N$, which is just statement (6.2).

We prove the corona theorem first on the unit disc Δ and later show that it holds on other domains that are locally like Δ.

Theorem 6.2. *The corona theorem is true on the open unit disc Δ; that is, Δ is dense in the maximal ideal space of $H^\infty(\Delta)$.*

Proof. We shall actually show somewhat more, namely, if (6.1) holds then (6.2) holds with

$$|g_j(z)| \leqslant \frac{129N^2}{\delta^4}, \qquad z \in \Delta \tag{6.3}$$

That is, the g_j can be chosen so that their sup norm on Δ depends only on N and δ, *not* on $f_1,\ldots,f_N$. We shall show this under the additional assumption that $f_1,\ldots,f_N$ are analytic and bounded by 1 on $|z| < 1 + \eta$ for some $\eta > 0$. A normal families argument, which makes use of (6.3), then allows us to pass to the limit and obtain the case when $f_1,\ldots,f_N$ are holomorphic and bounded by 1 on Δ. Henceforth, then, in the proof of this theorem we assume that $f_1,\ldots,f_N$ are holomorphic and bounded by 1 on the disc $\{|z| < 1 + \eta\}$.

Set $h_j = \bar{f}_j/(|f_1|^2 + \cdots + |f_N|^2)$ for $j = 1,\ldots, N$. Then h_j is C^∞ as a function of x and y, and $\Sigma f_j h_j \equiv 1$. Obviously, though, h_j is not analytic so that we must modify it. Let w_{jk} be smooth functions on $|z| < 1 + \eta$ satisfying the equation

$$\frac{\partial w_{jk}}{\partial \bar{z}} = h_j\frac{\partial h_k}{\partial \bar{z}}, \qquad j, k = 1,\ldots, N. \tag{6.4}$$

Here the notation $\partial/\partial\bar{z}$ means, as in Section 3, that

$$\frac{\partial u}{\partial \bar{z}} = \frac{\partial u}{\partial x} + i\frac{\partial u}{\partial y}$$

Hence, a smooth function h is analytic if and only if $\partial h/\partial \bar{z} = 0$. Next, set

$$g_j = h_j + \sum_{k=1}^{N} (w_{jk} - w_{kj}) f_k, \qquad 1 \leqslant j \leqslant N.$$

Then

$$\begin{aligned}\sum_1^N f_j g_j &= \sum_1^N f_j h_j + \sum_{j,k=1}^{N} f_j f_k (w_{jk} - w_{kj}) \\ &= 1\end{aligned}$$

and

$$\begin{aligned}\frac{\partial g_j}{\partial \bar{z}} &= \frac{\partial h_j}{\partial \bar{z}} + \sum_{k=1}^{N} \left(\frac{\partial w_{jk}}{\partial \bar{z}} - \frac{\partial w_{kj}}{\partial \bar{z}} \right) f_k + \sum_{k=1}^{N} (w_{jk} - w_{kj}) \frac{\partial f_k}{\partial \bar{z}} \\ &= \frac{\partial h_j}{\partial \bar{z}} + h_j \sum_{k=1}^{N} \frac{\partial h_k}{\partial \bar{z}} f_k - \frac{\partial h_j}{\partial \bar{z}} \sum_{k=1}^{N} h_k f_k \\ &= \frac{\partial h_j}{\partial \bar{z}} + h_j \frac{\partial}{\partial \bar{z}} \left(\sum_{k=1}^{N} h_k f_k \right) - \frac{\partial h_j}{\partial \bar{z}} \\ &= 0\end{aligned}$$

To finish the proof that (6.3) holds we need only show that

$$\|w_{jk}\|_\infty \leqslant c, \qquad j, k = 1, \ldots, N \tag{6.5}$$

where c depends on N and δ in the indicated fashion.

Let u be a smooth function on the disc $\{|z| < 1 + \eta\}$; we shall later take u to be $h_j(\partial h_k/\partial \bar{z})$ but this special choice is not relevant now. We shall derive an upper bound on the infimum of $\|w\|_{L^\infty(\mathbf{T}, d\theta)}$ for those w with

$$\frac{\partial w}{\partial \bar{z}} = u \tag{6.6}$$

If w_0 is some solution of (6.6) on a neighborhood $\mathfrak{D}$ of $\mathrm{CL}(\Delta)$, say,

$$w_0(z) = \frac{1}{\pi} \iint_{\mathfrak{D}} \frac{u(\zeta)}{\zeta - z} \, d\sigma \, d\tau, \qquad \zeta = \sigma + i\tau,$$

then any solution w of (6.6) on $\mathfrak{D}$ has the form

$$w = w_0 + p$$

where $\partial p/\partial \bar{z} = 0$ on $\mathfrak{D}$; that is, where p is analytic on $\mathfrak{D}$. Hence, with $\mathfrak{D}$ any neighborhood of $\{|z| \leqslant 1\}$ we have

$$\inf\left\{\|w\|_\infty : \frac{\partial w}{\partial \bar{z}} = u \text{ on } \mathfrak{D}\right\} = \inf\{\|w_0 + p\|_\infty : p \text{ analytic on } \mathfrak{D}\}$$

Since $\mathfrak{D}$ is an arbitrary neighborhood of $\{|z| \leqslant 1\}$, we can apply the basic duality result on $\mathbf{T}$,

$$L^\infty/H^\infty \text{ is the dual space of } H_0^1$$

and so

$$\inf\left\{\|w\| : \frac{\partial w}{\partial \bar{z}} = u \text{ on some neighborhood of } \operatorname{CL}(\Delta)\right\}$$

$$= \sup\left\{\left|\frac{1}{2\pi}\int_0^{2\pi} w_0 F\, dt\right| : F \in H_0^1, \|F\|_1 \leqslant 1\right\}$$

We now apply Green's formula to the integral, assuming, as we may, that F is holomorphic on a neighborhood of $\operatorname{CL}(\Delta)$

$$\int_0^{2\pi} w_0(e^{it})F(e^{it})\, dt = \iint_{|z|<1} \Delta(w_0 F)\log\frac{1}{|z|}\, dx\, dy - 2\pi(w_0 F)(0)$$

But $F(0) = 0$ and

$$\Delta(w_0 F) = 4\frac{\partial^2}{\partial z\, \partial \bar{z}}(w_0 F)$$

$$= 4\frac{\partial}{\partial z}(uF) = 4uF' + 4F\frac{\partial u}{\partial z}$$

Thus, we must derive an upper bound for the quantity

$$\left|\iint_\Delta uF' \log\frac{1}{|z|}\, dx\, dy + \iint_\Delta F\frac{\partial u}{\partial z} \log\frac{1}{|z|}\, dx\, dy\right| \tag{6.7}$$

for each $F \in H_0^1$, $\|F\|_1 \leqslant 1$, and $u = h_j(\partial h_k/\partial \bar{z})$, and show that there is a bound depending only on N and δ. We do this in a series of small steps.

We first note that if $h \in H^2$ then

$$\iint_\Delta |h'(z)|^2 \log\frac{1}{|z|}\, dx\, dy = \frac{\pi}{2}\left(\|h\|_2^2 - |h(0)|^2\right)$$

$$\leqslant \frac{\pi}{2}\|h\|_2^2 \tag{6.8}$$

This is proved by expanding h in a power series and directly computing the integral.

Next, if $f \in H^\infty$ and $g \in H^2$, then

$$\iint_\Delta |gf'|^2 \log \frac{1}{|z|} dx\, dy \leqslant 2\pi \|g\|_2^2 \|f\|_\infty^2 \tag{6.9}$$

This follows by writing $gf' = (gf)' - g'f$; then

$$|gf'|^2 \leqslant 2\{|(gf)'|^2 + |g'f|^2\}$$

$$\leqslant 2\{|(gf)'|^2 + \|f\|_\infty^2 |g'|^2\}$$

Now apply (6.8); we find

$$\iint_\Delta |gf'|^2 \log \frac{1}{|z|} dx\, dy \leqslant \pi \|gf\|_2^2 + \pi \|f\|_\infty \|g\|_2^2$$

$$\leqslant 2\pi \|f\|_\infty^2 \|g\|_2^2$$

The third step is to show for $F \in H^1$, $f_1, f_2 \in H^\infty$

$$\iint_\Delta |Ff_1'f_2'| \log \frac{1}{|z|} dx\, dy \leqslant 2 \|F\|_1 \|f_1\|_\infty \|f_2\|_\infty \tag{6.10}$$

To see this, write $F = g_1 g_2$ where $g_1, g_2 \in H^2$ and $\|g_1\|_2^2 = \|g_2\|_2^2 = \|F\|_1$; see problem 20, Chapter 3. Then use the Cauchy–Schwarz inequality on

$$\iint_\Delta |g_1 g_2 f_1' f_2'| \log \frac{1}{|z|} dx\, dy$$

and follow this by an application of (6.9) to each of the two terms obtained.

The final inequality we need is this: for $F \in H^1$ and $f \in H^\infty$

$$\iint_\Delta |F'f'| \log \frac{1}{|z|} dx\, dy \leqslant 2\pi \|F\|_1 \|f\|_\infty \tag{6.11}$$

Again write $F = g_1 g_2$ where g_1, g_2 are in H^2 and $\|g_1\|_2^2 = \|g_2\|_2^2 = \|F\|_1$. Hence,

$$\iint_\Delta |F'f'| \log \frac{1}{|z|} dx\, dy \leqslant \iint_\Delta |g_1' g_2 f'| \log \frac{1}{|z|} dx\, dy$$

$$+ \iint_\Delta |g_1 g_2' f'| \log \frac{1}{|z|} dx\, dy$$

Apply the Cauchy–Schwarz inequality to each of these terms to obtain the terms

$$\left(\iint_\Delta |g_1'|^2 \log\frac{1}{|z|}\,dx\,dy\right)^{1/2} = A_1$$

$$\left(\iint_\Delta |g_2 f'|^2 \log\frac{1}{|z|}\,dx\,dy\right)^{1/2} = B_1$$

and two similar terms A_2 and B_2 with the roles of g_1 and g_2 interchanged. A_1 is estimated by (6.8) and B_1 by (6.9). Multiplying A_1 times B_1 and then adding the corresponding product A_2B_2 we obtain (6.11).

Now we are in a position to finish the proof. To simplify notation let

$$\phi = \left(\sum_1^N |f_j|^2\right)^{-1}$$

Recall that $h_k = \bar{f}_k\phi$ for $k = 1,\ldots, N$. Hence,

$$\frac{\partial h_k}{\partial \bar{z}} = \phi\bar{f}_k' - \phi^2\bar{f}_k \sum_{m=1}^N f_m\overline{f_m'}$$

and thus

$$u = h_j\frac{\partial h_k}{\partial \bar{z}} = \phi^2\bar{f}_j\bar{f}_k' - \phi^3\bar{f}_j\bar{f}_k \sum_{m=1}^N f_m\overline{f_m'} \tag{6.12}$$

Consequently, we obtain

$$\frac{\partial u}{\partial z} = -2\phi^3\bar{f}_j\bar{f}_k' \sum_{m=1}^N f_m'\bar{f}_m - \phi^3\bar{f}_j\bar{f}_k \sum_{m=1}^N |f_m'|^2 + 3\phi^4\bar{f}_j\bar{f}_k\left|\sum_{m=1}^N f_m f_m'\right|^2 \tag{6.13}$$

Now look back at (6.7). Substituting the expression for u given in (6.12) into the first integral in (6.7) and then applying (6.11) $N+1$ times we obtain

$$\left|\iint_\Delta uF'\log\frac{1}{|z|}\,dx\,dy\right| \leqslant 2\pi\frac{1}{\delta^2} + 2\pi\frac{1}{\delta^3}N \leqslant \frac{2\pi(N+1)}{\delta^3} \tag{6.14}$$

where we have made use of the fact that $|f_k| \leqslant 1$ for all k and $|\phi| \leqslant 1/\delta$. Next take the expression for $\partial u/\partial z$ given in (6.13), substitute it for $\partial u/\partial z$ in the second integral in (6.7) and then apply (6.10). This yields the estimate

$$\left|\iint_\Delta F\frac{\partial u}{\partial z}\log\frac{1}{|z|}\,dx\,dy\right| \leqslant \frac{2\pi}{\delta^4}(6N)$$

Hence, we have established (6.5) with $c \leqslant 64N/\delta^4$. Since

$$g_j = h_j + \sum_{k=1}^{N} (w_{jk} - w_{kj}) f_k$$

we finally obtain

$$\begin{aligned} \|g_j\|_\infty &\leqslant \|h_j\|_\infty + \sum_{k=1}^{N} \left(\|w_{jk}\|_\infty + \|w_{kj}\|_\infty\right) \\ &\leqslant \frac{1}{\delta} + \frac{128N^2}{\delta^4} \\ &< \frac{129N^2}{\delta^4} \end{aligned}$$

Corollary. *If Ω is a simply connected domain, then the corona theorem is true on Ω.*

Proof. Since $H^\infty(\Omega)$ is nontrivial, Ω is conformally equivalent to the open unit disc Δ.

Theorem 6.3. *Suppose Ω is bounded by a finite number of disjoint analytic simple closed curves. Then the corona theorem is true on Ω.*

Proof. Let $\lambda \in \partial\Omega$ and let $\mathfrak{U}$ be a neighborhood of λ such that $\mathfrak{D} = \mathfrak{U} \cap \Omega$ is bounded by a piecewise smooth simple closed curve. Proposition 3.1 shows that each element ℓ of the maximal ideal space of $H^\infty(\Omega)$ which lies in the fiber over λ extends to an element of the maximal ideal space of $H^\infty(\mathfrak{D})$ which lies in the fiber over λ. Conversely, if m is an element of the maximal ideal space of $H^\infty(\mathfrak{D})$ lying in the fiber over λ, then the restriction of m to $H^\infty(\Omega)$ is an element of the maximal ideal space of $H^\infty(\Omega)$ in the fiber over λ. Thus, if $\ell \in \mathbf{M}(\Omega)$ and ℓ lies in the fiber over λ, then there is a net $\{z_\alpha\}$ of points of $\mathfrak{D}$ such that ℓ is an accumulation point of $\{\phi_{z_\alpha}\}$ in the maximal ideal space of $H^\infty(\mathfrak{D})$ and hence in the maximal ideal space of $H^\infty(\Omega)$. Thus, the closure of Ω in $\mathbf{M}(\Omega)$ contains the fiber over λ. Since such a neighborhood $\mathfrak{U}$ can be found for all $\lambda \in \partial\Omega$, we see that Ω is dense in $\mathbf{M}(\Omega)$.

There is another proof of Theorem 6.3 which is also quick but yields a little less information. Use the uniformizer T to lift $H^\infty(\Omega)$ to the subspace $H^\infty/\mathfrak{G}$ of $H^\infty(\Delta)$, as described in Chapter 2. Let $\mathfrak{P}$ be the bounded linear projection from $H^\infty(\Delta)$ onto $H^\infty/\mathfrak{G}$ detailed in Theorem 4.5.2. If $f_1, \ldots, f_N$ lie in $H^\infty/\mathfrak{G}$ and satisfy

$$|f_1|^2 + \cdots + |f_N|^2 \geqslant \delta \quad \text{on } \Delta$$

then by Theorem 6.2 there are functions $g_1, \ldots, g_N$ in $H^\infty(\Delta)$ with

$$f_1 g_1 + \cdots + f_N g_N \equiv 1 \quad \text{on } \Delta$$

Hence,

$$1 = \mathcal{P}\left(\sum_1^N f_j g_j\right) = \sum_1^N f_j \mathcal{P}(g_j)$$

However, $\mathcal{P}(g_j)$ is an element of $H^\infty/\mathfrak{G}$ and so the proof is complete.

ADDITIONAL READINGS AND NOTES

The material on peak points, peak sets, and parts contained in Section 2 is only a part of an extensive literature on the subject. Once again the books of Gamelin (1969) and Stout (1971) are good references. See Hoffman (1967) for a good look at the parts of $\mathbf{M}(\Delta)$. More on the construction found at the beginning of Section 3 is in Gamelin's book, Chapter VIII, Section 10. Distinguished homomorphisms were first discovered and named by Zalcman (1969); the example given here is from his paper. The bulk of what's done in Sections 3, 4, and 5 comes from papers of Gamelin (1970, 1974) and a paper of Gamelin and Garnett (1970). The use of the functions $\tilde{\nu}$ and $\hat{\nu}$ in the proof of Theorem 4.1 was inspired by Browder's lovely paper (1967). One of the virtues of this proof of Theorem 4.1 is that it frees the concept of a distinguished homomorphism of $H^\infty(\Omega)$ from the grip of ideas in rational approximation, in particular from Melnikov's Theorem. A proof of Theorem 4.1 quite similar to the one in the text can be found in Gamelin's La Plata Lecture Notes (1972). The proof of the corona theorem for the unit disc Δ presented in Theorem 6.2 is taken directly from Gamelin (1980). This proof is "elementary" and greatly simpler than either Carleson's original proof (1962b) or even the reasonably expository presentation of Carleson's proof like, for instance, the one in Duren (1970). Soon after the appearance of Carleson's proof, Stout (1965) and Forelli (1966), among others, gave proofs valid on finite open Riemann surfaces; Forelli's proof is second proof given here for Theorem 6.3. Efforts to extend the corona theorem to arbitrary planar domains have met with partial success. Behrens (1971) showed the result to hold for a certain class of infinitely connected domains and further, showed that if it fails, then it fails for a domain Ω obtained from $0 < |z| < 1$ by deleting a sequence $\{\Delta_j\}$ of disjoint closed discs which accumulate only at the origin. On the other hand, Cole has constructed an example, which appears in Gamelin (1978, Chapter IV), of a Riemann surface for which the corona theorem fails. Efforts to prove, or disprove, the corona theorem have lead to a number of significant new results in function theory—perhaps more will come in the future.

EXERCISES

1. Let Ω be a domain and let $\lambda \in \partial\Omega$. Prove the following three assertions to be equivalent

a. $\mathbf{M}_\lambda$ is a peak set for $H^\infty(\Omega)$

b. there is a nonconstant f in the unit ball of $H^\infty(\Omega)$ with

$$\lim\{f(z): z \to \lambda, z \in \Omega\} = 1$$

c. each sequence $\{z_n\}$ of points of Ω with $z_n \to \lambda$ has a subsequence which is an interpolation sequence for $H^\infty(\Omega)$.

2. If $\phi \in \mathbf{M}(\Omega)$ is a distinguished homomorphism, prove ϕ is not in $\mathbf{S}(\Omega)$. HINT: let $z_0 \in \Omega$ lie close to ϕ in the norm of $H^\infty(\Omega)^*$; consider the Ahlfors function for z_0.

3. Let ν be a measure on $\mathbf{C}$ with compact support and set

$$\hat{\nu}(z) = \int (z - \zeta)^{-1} d\nu(\zeta), \qquad \tilde{\nu}(z) = \int |z - \zeta|^{-1} d|\nu|(\zeta), \qquad z \in \mathbf{C}$$

Show (a) $\hat{\nu}$ and $\tilde{\nu}$ are finite a.e. $dx\,dy$. (b) $|\hat{\nu}(z)| \leqslant \tilde{\nu}(z)$ all points where $\tilde{\nu}(z) < \infty$. (c) If $\hat{\nu} = 0$ a.e. $dx\,dy$, then ν is the zero measure.

4. Let $\mathbf{A}$ and $\mathbf{X}$ be as in Section 2 and suppose ϕ and ψ are elements of $\mathbf{M}(\mathbf{A})$. Show that ϕ and ψ lie in the same part if and only if there is a constant $\rho > 0$ such that

$$\rho \operatorname{Re} \phi(u) \leqslant \operatorname{Re} \psi(u) \leqslant \frac{1}{\rho} \operatorname{Re} \phi(u) \tag{6.15}$$

for all positive functions u, $u = \operatorname{Re} f$, $f \in \mathbf{A}$.

5. Suppose ϕ, ψ are in the same part for $\mathbf{A}$. Let

$$b(\phi, \psi) = \sup\{\rho : (6.15) \text{ holds}\}$$

Show (a) $b(\phi, \psi) = b(\psi, \phi)$. (b) $b(\phi_1, \phi_2) b(\phi_2, \phi_3) \leqslant b(\phi_1, \phi_3)$. Conclude that $\log b(\phi, \psi)$ is a metric on the part.

If $\mathbf{A} = H^\infty(\Omega)$, show that this metric agrees on Ω with the usual planar metric; that is, generates the same open opens.

6. If ϕ and ψ are in the same part, show ϕ and ψ have mutually absolutely continuous representing measures. HINT: use the result of exercise 4.

7. If ϕ and ψ are in different parts, show any representing measure for ϕ is singular with respect to any representing measure for ψ.

8. If μ is a complex measure with $\int f\,d\mu = \phi(f)$, then show that there is a representing measure ν for ϕ with $\nu \ll \mu$.
9. Let m be Lebesgue area measure on Ω and define $\hat{m}$ on $\mathbf{X} = \mathbf{M}(L^\infty(\Omega, m))$ by

$$\int_{\mathbf{X}} u\,d\hat{m} = \int_{\Omega} v\,dm, \qquad \hat{v} = u, \qquad v \in L^\infty(\Omega, m)$$

Show that $\hat{H}^\infty$ is a weak-* closed subalgebra of $L^\infty(\mathbf{X}, \hat{m})$.
10. Prove the remark made after the proof of Theorem 5.4; namely, that the function u can be replaced by a function $v \in L_r^\infty(\Omega, m)$.
11. Let μ be any measure on $\mathbf{M}(\Omega)$ such that the linear functional $h \mapsto \int \hat{h}\,d\mu$, $h \in H^\infty(\Omega)$, is weak-* continuous. Show that the closed support of μ contains $\mathbf{S}(\Omega)$.
12. Let Ω be the domain cited in the example in the beginning of Section 4. (a) Let F be the Ahlfors function for Ω and some point $p \in \Omega$. Show that F is analytic across $\partial\Omega$ except at 0, that F maps each component of $\partial\Omega \setminus \{0\}$ both one-to-one and onto $\mathbf{T}$, and that F is not continuous at 0. Find $\lim\{F(x): x \to 0, x < 0\} = \phi_0(F)$. (b) The Shilov boundary of $H^\infty(\Omega)$ is just the maximal ideal space of $L^\infty(\partial\Omega, ds)$.

Problems 13 and 14 use the Shilov Idempotent Theorem.

13. The maximal ideal space of $H^\infty(\Omega)$ is connected.
14. If the domain Ω is locally connected at $\lambda \in \partial\Omega$, then the fiber $\mathbf{M}_\lambda$ is connected.
15. Let $\mathbf{K}$ be a compact set in the plane and let $\mathbf{A}(\mathbf{K})$ denote those functions which are continuous on $\mathbf{K}$ and analytic on the interior of $\mathbf{K}$. Show that each continuous linear functional m on $\mathbf{A}(\mathbf{K})$ which is multiplicative is of the form $m(f) = f(z_0)$ for some z_0 in $\mathbf{K}$. That is, the maximal ideal space of $\mathbf{A}(\mathbf{K})$ is precisely $\mathbf{K}$. Show as well that the Shilov boundary of $\mathbf{A}(\mathbf{K})$ is just the topological boundary of $\mathbf{K}$.
16. Let $\mathbf{K}_1$ and $\mathbf{K}_2$ be two compact sets in the plane and suppose L is a linear, one-to-one, mapping of $\mathbf{A}(\mathbf{K}_1)$ onto $\mathbf{A}(\mathbf{K}_2)$ with $L(fg) = L(f)L(g)$. Show that there is an element $\phi \in \mathbf{A}(\mathbf{K}_2)$, ϕ a homeomorphism of $\mathbf{K}_2$ onto $\mathbf{K}_1$ such that $L(f) = f \circ \phi$, $f \in \mathbf{A}(\mathbf{K}_1)$.
17. Let Ω be a domain. If each point $\lambda \in \partial\Omega$ has a neighborhood $\mathcal{U}$ such that the corona theorem is true on $\mathcal{U} \cap \Omega$, show that the corona theorem is true on Ω.
18. Suppose that $\infty \in \Omega$ and that the complement of Ω consists of a countable number $\mathbf{K}_1, \mathbf{K}_2, \ldots$ of compact connected sets whose diameters are bounded from below by a positive constant. Use problem 17 to show that the corona theorem is true for Ω.

19. Establish this formula. If u is smooth on an open disc $\mathfrak{D}$ and if w is defined by

$$w(z) = \frac{1}{\pi} \iint_{\mathfrak{D}} \frac{u(\zeta)}{\zeta - z} \, d\sigma \, d\tau, \qquad \zeta = \sigma + i\tau$$

then $\partial w / \partial \bar{z} = u$ on $\mathfrak{D}$.

7

LINEAR OPERATORS ON H^p SPACES

In this chapter we explore several types of linear operators whose domain and range lie in $H^p(\Omega)$. In Section 1, we discover the isometries of $H^p(\Omega)$, $p \neq 2$, when Ω is bounded by a finite number of disjoint smooth simple closed curves. Such mappings turn out to involve composition of the $H^p(\Omega)$ functions with an analytic function ϕ which maps Ω into itself. If ϕ is such a function, then the mapping given by $f \to f \circ \phi$ is a composition operator on $H^p(\Omega)$ and it is such maps that we investigate in Sections 2 through 4. Finally, in Sections 5 and 6 we examine the question of estimating, in an optimal way, functions in H^∞ or in H^2. It turns out that there are linear operators which provide such an optimal estimation in a sense made precise in these sections.

7.1. THE ISOMETRIES OF $H^p(\Omega)$

Let Ω be a domain bounded by $m + 1$ disjoint analytic simple closed curves, $m \geqslant 0$. Fix p, $0 < p < \infty$, but $p \neq 2$, and let T be a linear operator from $H^p(\Omega)$ into $H^p(\Omega)$ which is an isometry:

$$\|Tf\|_p = \|f\|_p, \qquad f \in H^p(\Omega) \tag{1.1}$$

We shall analyze T and discover its basic makeup. As a consequence, we will be able to fully describe the linear isometries of $H^p(\Omega)$ onto itself. We also can characterize the linear isometries of $H^\infty(\Omega)$ onto itself.

We begin with two propositions which are unrelated to the particular setting of $H^p(\Omega)$.

Proposition 1.1. Let μ and ν be two positive measures of mass one on measurable spaces X and Y, respectively. Suppose $f \in L^p(\mu)$, $g \in L^p(\nu)$ and

$$\int |1 + zf|^p \, d\mu = \int |1 + zg|^p \, d\nu, \qquad \text{all } z \in \mathbb{C} \tag{1.2}$$

Then

$$\int |f|^2 \, d\mu = \int |g|^2 \, d\nu$$

Proof. If neither f nor g is in L^2, there is nothing to prove, so suppose that $f \in L^2(\mu)$. For small values of $|z|$ we have

$$\frac{1}{2\pi}\int_0^{2\pi} |1 + ze^{it}|^p \, dt - 1 = \left(\frac{p^2}{4}\right)|z|^2 + \sum_{j=2}^{\infty} \binom{p/2}{j}^2 |z|^{2j} \tag{1.3}$$

so that

$$r^{-2}\left(\frac{1}{2\pi}\int_0^{2\pi} |1 + rg(y)e^{it}|^p \, dt - 1\right) \to \left(\frac{p^2}{4}\right)|g(y)|^2, \qquad r \downarrow 0 \tag{1.4}$$

However, (1.3) shows that the left-hand side in (1.4) is non-negative for all r so that by Fatou's lemma

$$\liminf_{r \downarrow 0} r^{-2}\left(\frac{1}{2\pi}\int_Y\int_0^{2\pi} |1 + rg(y)e^{it}|^p \, dt \, d\nu(y) - 1\right)$$

must exceed

$$\left(\frac{p^2}{4}\right)\int_Y |g(y)|^2 \, d\nu(y)$$

Next the left-hand side of (1.3) is bounded by $A|z|^2$ if $|z| < 1$ and by $A'|z|^p$ if $|z| \geqslant 1$ where A, A' are constants depending on p but not z. Thus, the left-hand side of (1.3) does not exceed $A''(|z|^2 + |z|^p)$ in general and, if $0 < p \leqslant 2$, it does not exceed $A'''|z|^2$. Thus, the quantity

$$r^{-2}\left(\frac{1}{2\pi}\int_0^{2\pi} |1 + rf(x)e^{it}|^p \, dt - 1\right)$$

does not exceed

$$A''(|f(x)|^2 + r^{p-2}|f(x)|^p) \quad \text{if} \quad p \geqslant 2$$

or

$$A'''|f(x)|^2 \quad \text{if} \quad 0 < p < 2$$

We may therefore apply the dominated convergence theorem to

$$r^{-2}\left(\frac{1}{2\pi}\int_0^{2\pi}|1 + rf(x)e^{it}|^p\,dt - 1\right)$$

and conclude

$$\lim_{r\to 0} r^{-2}\left(\frac{1}{2\pi}\int_X\int_0^{2\pi}|1 + rf(x)e^{it}|^p\,dt\,d\mu(x) - 1\right) = \frac{p^2}{4}\int_X |f(x)|^2\,d\mu(x)$$

However, by (1.2)

$$\int_X |1 + rf(x)e^{it}|^p\,d\mu(x) = \int_Y |1 + rg(y)e^{it}|^p\,d\nu(y)$$

so that

$$\int_Y |g(y)|^2\,d\nu(y) \leqslant \int_X |f(x)|^2\,d\mu(x)$$

Now reverse the argument to learn that

$$\int_X |f(x)|^2\,d\mu(x) \leqslant \int_Y |g(y)|^2\,d\nu(y)$$

This completes the proof of the proposition.

Proposition 1.2. Let μ and ν be as in Proposition 1.1. Suppose $\mathcal{Q}$ is a subalgebra of $L^\infty(\mu)$, $1 \in \mathcal{Q}$, which is carried by a linear transformation T into $L^\infty(\nu)$, $T(1) = 1$. If $p \neq 2$ and if

$$\int_Y |Tf|^p\,d\nu = \int_X |f|^p\,d\mu, \qquad f \in \mathcal{Q}$$

then T is multiplicative: $T(fg) = (Tf)(Tg)$.

Proof. The hypotheses immediately give

$$\int_Y |1 + zTf|^p\,d\nu = \int_X |1 + zf|^p\,d\mu, \qquad f \in \mathcal{Q} \tag{1.5}$$

Expand the left- and right-hand sides of (1.5) in powers of z and $\bar{z}$ to get

$$\sum_{j,k=0}^{\infty} \binom{p/2}{j}\binom{p/2}{k} z^j\bar{z}^k \int_Y (Tf)^j\,(\overline{Tf})^k\,d\nu$$

and

$$\sum_{j,k=0}^{\infty} \binom{p/2}{j}\binom{p/2}{k} z^j \bar{z}^k \int_X f^j \bar{f}^k \, d\mu,$$

respectively. We equate the coefficients of like powers of z, $\bar{z}$ and we learn that

$$\int_X |f|^2 \, d\mu = \int_Y |Tf|^2 \, d\nu \tag{1.6a}$$

$$\int_X |f|^4 \, d\mu = \int_Y |Tf|^4 \, d\nu \tag{1.6b}$$

$$\int_X \bar{f} f^2 \, d\mu = \int_Y \overline{Tf}(Tf)^2 \, d\nu \tag{1.6c}$$

[Of course, (1.6a) follows from Proposition 1.1 and (1.5), anyway.] Here we have used the fact that $\binom{p/2}{1}$ and $\binom{p/2}{2}$ are not zero if $p \neq 2$. In (1.6a) replace f by $f + zg$ and again equate the coefficients of equal powers of z, $\bar{z}$; we get

$$\int_Y Tg \, \overline{Tf} \, d\nu = \int_X g\bar{f} \, d\mu, \qquad f, g \in \mathcal{Q} \tag{1.7}$$

Next, replace g by f^2 in (1.7) and compare with (1.6c). We find

$$\int_Y (\overline{Tf})(Tf)^2 \, d\nu = \int_X \bar{f} f^2 \, d\mu, \qquad f \in \mathcal{Q}$$

Again replace f by $f + zg$ and equate coefficients of equal powers of z, $\bar{z}$. This gives

$$\int_Y (Tf)^2 \overline{T} g \, d\nu = \int_X f^2 \bar{g} \, d\mu, \qquad f, g \in \mathcal{Q} \tag{1.8}$$

Thus, we get

$$\begin{aligned}\int_Y |(Tf)^2 - T(f^2)|^2 \, d\nu &= \int_Y |Tf|^4 \, d\nu - 2\,\mathrm{Re} \int_Y (Tf)^2 \overline{T(f^2)} \, d\nu \\ &\quad + \int_Y |T(f^2)|^2 \, d\nu \\ &= \int_X |f|^4 \, d\mu - 2\,\mathrm{Re} \int_X |f|^4 \, d\mu + \int_X |f|^4 \, d\mu \\ &= 0\end{aligned}$$

by applying (1.6b), (1.8), and (1.7), respectively. Thus,

$$T(f^2) = (Tf)^2, \qquad f \in \mathcal{A}$$

and the proposition follows by replacing f by $f + g$.

With the two propositions in hand we can begin our analysis of a linear isometry T of $H^p(\Omega)$ into $H^p(\Omega)$. Set $F = T(1)$ and let ν be the measure on $\Gamma = \partial\Omega$ given by

$$d\nu = |F|^p \, d\omega$$

where, as usual, ω is harmonic measure on Γ for the point $z_0 \in \Omega$ at which the H^p norm is computed. The function F cannot vanish on any set of positive ω-measure and so we can define a linear transformation S of $H^p(\Omega)$ into $L^p(\nu)$ by

$$Sf = Tf/F, \qquad f \in H^p(\Omega)$$

Then $S(1) = 1$ and S is an isometry because T is an isometry. Let G be any inner function in $H^\infty(\Omega)$. Then

$$\int |1 + zG|^p \, d\omega = \int |1 + zS(G)|^p \, d\nu, \qquad \text{all } z \in \mathbb{C}$$

Hence, Proposition 1.1 implies that

$$1 = \int |G|^2 \, d\omega = \int |S(G)|^2 \, d\nu$$

But

$$1 = \int |G|^p \, d\omega = \int |S(G)|^p \, d\nu$$

as well, so that

$$1 = \int |S(G)|^2 \, d\nu = \int |S(G)|^p \, d\nu$$

Thus, $S(G)$ is unimodular a.e. on Γ. Consequently, S maps the algebra generated by the inner functions into $L^\infty(\nu)$ and we may apply Proposition 1.2 to conclude that S is multiplicative on $H^p(\Omega)$. Let ϕ be the image under S of the identity function $I(z) = z$. Then for any rational function f with no poles on $\Omega \cup \Gamma$ we have

$$S(f) = f(\phi)$$

and consequently

$$Tf = Ff \circ \phi, \qquad f \text{ rational}, \qquad f \in H^p(\Omega). \tag{1.9}$$

Since $\phi = H/F$, where $H = T(I)$, we see that the bounded function ϕ is the quotient of two $H^p(\Omega)$ functions, at least on Γ. Let us call the ratio $H(z)/F(z)$ by the name $\phi_1(z)$ for $z \in \Omega$. We shall show that (1.9) implies ϕ_1 maps Ω into itself and the boundary values of ϕ_1 are what is to be expected, namely ϕ.

Suppose $f = P/Q$ where P, Q are polynomials of degree N and M, respectively. Then by (1.9)

$$F^M(\zeta)Q\left(\frac{H(\zeta)}{F(\zeta)}\right)Tf(\zeta) = F^N(\zeta)P\left(\frac{H(\zeta)}{F(\zeta)}\right)F(\zeta)^{M-N+1}, \qquad \zeta \in \Gamma$$

Assume with no loss that $M - N + 1 \geqslant 0$ (otherwise put the term F^{M-N+1} on the other side of the equation). Then both sides of this equation are analytic functions in $H^q(\Omega)$ for some small q and they are identically equal on Γ; hence, they are also equal identically on Ω. Thus, if ϕ_1 mapped some point $a \in \Omega$ to a point b not in $\Omega \cup \Gamma$, we could choose a rational function f which was very big at b but very small on all of Ω. It would follow that $(Tf)(a)$ is very small while $f(b)$ is big; this is a contradiction since

$$f(b) = (Tf)(a)$$

Consequently, ϕ_1 maps Ω into Ω and

$$(Tf)(z) = F(z)f(\phi_1(z)), \qquad z \in \Omega, \qquad f \text{ rational} \tag{1.10}$$

Since the rational functions are dense in $H^p(\Omega)$, (1.10) holds as well for all $f \in H^p(\Omega)$ and by taking boundary values [with $f(z) \equiv z$] we see that $\phi_1 = \phi$ a.e. on Γ. This means that we have proved the first part of the following theorem.

Theorem 1.3. *Let $0 < p < \infty$, $p \neq 2$, and let T be a linear isometry of $H^p(\Omega)$ into $H^p(\Omega)$. Then there is an analytic function ϕ mapping Ω into Ω and a function $F \in H^p(\Omega)$ with*

$$Tf = Ff \circ \phi, \qquad f \in H^p(\Omega) \tag{1.11}$$

Further, ϕ maps Γ into Γ a.e. *and F and ϕ are related by*

$$\omega(E) = \int_{\phi^{-1}(E)} |F|^p \, d\omega \tag{1.12}$$

for each measurable set E in Γ.

Proof. We need to show that ϕ maps Γ into Γ. Note that

$$\int_\Gamma |f|^p \, d\omega = \int_\Gamma |Tf|^p \, d\omega = \int_\Gamma |F|^p |f \circ \phi|^p \, d\omega \tag{1.13}$$

If there were a measurable set A in Γ with $\omega(A) > 0$ and $\phi(A)$ contained within a compact set of Ω, then by taking f to be ψ^n where ψ is an Ahlfors function and then letting $n \to \infty$ we would obtain

$$\begin{aligned} 1 \leqslant \int_{\Gamma \setminus A} |F|^p \, d\omega &= \int_\Gamma |F|^p \, d\omega - \int_A |F|^p \, d\omega \\ &= \int_\Gamma |T(1)|^p \, d\omega - \int_A |F|^p \, d\omega \\ &= 1 - \int_A |F|^p \, d\omega < 1 \end{aligned}$$

which is a contradiction. Thus, the values of ϕ on Γ lie in Γ a.e. ω. Finally, if E is a measurable set in Γ there is a sequence $\{f_n\}$ of functions in $H^\infty(\Omega)$ with $|f_n| \leqslant 1$ and $|f_n| \to \chi_E$ pointwise a.e. ω on Γ; this follows easily from results in Section 6 of Chapter 4. Here, as usual, χ_E stands for the characteristic function of E. Thus, we find

$$\begin{aligned} \omega(E) = \lim \int_\Gamma |f_n|^p \, d\omega &= \lim \int_\Gamma |Tf_n|^p \, d\omega \\ &= \lim \int_\Gamma |F|^p |f_n \circ \phi|^p \, d\omega \\ &= \int_\Gamma |F|^p |\chi_E \circ \phi|^p \, d\omega \\ &= \int_{\phi^{-1}(E)} |F|^p \, d\omega \end{aligned}$$

which is the desired conclusion.

Corollary 1.4. *Let Ω be bounded by $m + 1$ disjoint analytic simple closed curves. Suppose T is an isometry of $H^p(\Omega)$ onto $H^p(\Omega)$, $1 \leqslant p < \infty$, $p \neq 2$. Then*

$$Tf = \lambda F f \circ \phi \tag{1.14}$$

where ϕ is a one-to-one analytic mapping of Ω onto Ω and F is an outer function in

$H^p(\Omega)$ with the property that

$$|F|^p = \frac{d\omega'}{d\omega} \tag{1.15}$$

and ω' is harmonic measure for the point $\phi^{-1}(z_0)$. In particular, if $T(1) = 1$, then

$$Tf = \lambda f \circ \phi$$

Proof. Since T is an onto isometry T^{-1} is also, so there is an analytic function ψ mapping Ω into Ω and Γ into Γ (a.e. ω) and a function $G \in H^p(\Omega)$ with

$$T^{-1}g = Gg \circ \psi \tag{1.16a}$$

$$\omega(E) = \int_{\psi^{-1}(E)} |G|^p \, d\omega \tag{1.16b}$$

Thus, for each $f \in H^p(\Omega)$

$$f = T^{-1}(Tf) = (G)(F \circ \psi)(f \circ \phi \circ \psi)$$

and

$$f = T(T^{-1}f) = (F)(G \circ \phi)(f \circ \psi \circ \phi)$$

Hence,

$$1 = (G)(F \circ \psi) = (F)(G \circ \phi) \tag{1.17a}$$

$$\text{and } \phi \circ \psi = \psi \circ \phi = \text{identity} \tag{1.17b}$$

We see that ϕ is a one-to-one analytic mapping of Γ onto itself which maps Γ onto Γ.

We further have (1.17a) and (1.12). The former implies immediately that F and G are outer functions. The latter implies that

$$|F|^p = \frac{d\omega'}{d\omega} \quad \text{a.e. on } \Gamma \tag{1.18}$$

where ω' is harmonic measure for the point $z_1 = \phi^{-1}(z_0)$; of course, ω is harmonic measure for z_0, as usual.

Corollary 1.5. *Suppose T is a linear isometry of $H^p(\Delta)$ onto $H^p(\Delta)$, $1 \leqslant p < \infty$, $p \neq 2$. Then there is a linear fractional transformation ϕ, $\phi(z) =$*

$\mu(z-a)(1-\bar{a}z)^{-1}$, $|\mu|=1$, $a\in\Delta$, *and a unimodular constant* λ, *such that*

$$Tf=\lambda(\phi')^{1/p}f\circ\phi \tag{1.19}$$

Proof. Each one-to-one analytic function ϕ mapping Δ onto Δ is a linear fractional transformation of the indicated form. Further, the harmonic measure for $a=\phi^{-1}(0)$ is just

$$d\omega_a=\frac{1}{2\pi}P_a\,d\theta=\frac{1}{2\pi}\frac{1-|a|^2}{|1-\bar{a}e^{i\theta}|^2}d\theta$$

Thus,

$$|F|^p=2\pi\frac{d\omega_a}{d\theta}=\frac{1-|a|^2}{|1-\bar{a}e^{i\theta}|^2}=|\phi'(e^{i\theta})|$$

Since ϕ' is analytic on a neighborhood of $\{z\colon |z|\leqslant 1\}$ and since $|\phi'(z)|>0$ for $|z|\leqslant 1$, we must have $F=\lambda(\phi')^{1/p}$ for some unimodular constant λ.

EXAMPLE.

In a problem at the end of the chapter the reader is asked to find the isometries of $\mathbf{A}(\Omega)$ onto $\mathbf{A}(\Omega)$. The *into* isometries, however, are not fully known and are considerably more complicated. In particular, let $\mathbf{A}(\Delta)$ be the disc algebra: those functions f which are analytic on Δ and continuous on $\{z : |z|\leqslant 1\}$. Let $\mathbf{C}$ be a Cantor set in $\mathbf{T}$ of Lebesgue measure zero and let $\lambda\in\mathbf{T}\setminus\mathbf{C}$. There is a continuous function u on $\mathbf{C}$ whose range is all of $\mathbf{T}$; see Newman (1961). Further, by the results of Section 3 of Chapter 4 there are two functions h_1, h_2 in $\mathbf{A}(\Delta)$ with

$$h_1=\begin{cases} u & \text{on } \mathbf{C}\\ 1 & \text{at } \lambda\end{cases},\qquad h_2=\begin{cases} u & \text{on } \mathbf{C}\\ -1 & \text{at } \lambda\end{cases}$$

and $\|h_1\|_\infty=\|h_2\|_\infty=1$. Let

$$Tf=\tfrac{1}{2}(f\circ h_1+f\circ h_2)$$

Then T is a linear norm reducing map of $\mathbf{A}(\Delta)$ into $\mathbf{A}(\Delta)$. But if $|f(p)|=\|f\|_\infty$, then there is a $q\in\mathbf{C}$ with $h_1(q)=h_2(q)=u(q)=p$ so that

$$\|Tf\|\geqslant|(Tf)(q)|=\tfrac{1}{2}|f(h_1(q))+f(h_2(q))|$$

$$=\tfrac{1}{2}|f(p)+f(p)|=|f(p)|=\|f\|$$

Thus, T is an isometry of $\mathbf{A}(\Delta)$ into $\mathbf{A}(\Delta)$ and T is not of the form $Tf = \mu f \circ \phi$ where $|\mu| = 1$ and $\phi \in \mathbf{A}(\Delta)$, $\|\phi\| = 1$.

We conclude this section by describing the linear isometries of $H^\infty(\Omega)$ *onto* $H^\infty(\Omega)$ for an arbitrary region Ω (for which $H^\infty(\Omega)$ is nontrivial, of course). The description involves showing that such an isometry is "almost" multiplicative and then using Theorem 3.6.1, which describes the algebra automorphisms of $H^\infty(\Omega)$. The description of the onto isometry depends only on the fact that $H^\infty(\Omega)$ may be viewed as a closed subalgebra of $\mathbf{C(M)}$ where $\mathbf{M}$ is the maximal ideal space of $H^\infty(\Omega)$. The result will be given first in that abstract context and then specialized to $H^\infty(\Omega)$.

Theorem 1.6. *Let* $\mathbf{M}$ *be a compact Hausdorff space and* $\mathbf{B}$ *a closed subalgebra of* $\mathbf{C(M)}$ *with* $1 \in \mathbf{M}$. *Suppose* T *is a linear isometry of* $\mathbf{B}$ *onto* $\mathbf{B}$. *Then*

$$T(h) = \lambda A(h), \qquad h \in \mathbf{B} \tag{1.20}$$

where λ *is an element of* $\mathbf{B}$ *of unit modulus on* $\mathbf{M}$, $1/\lambda$ *is in* $\mathbf{B}$, *and* A *is an algebra automorphism of* $\mathbf{B}$ *onto* $\mathbf{B}$.

Proof. There is no loss of generality in assuming initially that $\mathbf{B}$ separates the points of $\mathbf{M}$ since we can always form the quotient space of $\mathbf{M}$ modulo the relation $x \sim y$ if $f(x) = f(y)$ for all $f \in \mathbf{B}$.

Since T is an isometry of $\mathbf{B}$ onto $\mathbf{B}$, we know that its adjoint T^* is an isometry of $\mathbf{B}^*$ onto $\mathbf{B}^*$. Hence, T^* carries the set $\mathfrak{E}$ of extreme points of the unit ball of $\mathbf{B}^*$ onto itself. We shall now spend a moment to determine what these extreme points are. Let $\ell \in \mathbf{B}^*$ be an extreme point in the unit ball of $\mathbf{B}^*$; let S be the set of all linear functionals L on $\mathbf{C(M)}$ with $\|L\| = 1$ and $L = \ell$ on $\mathbf{B}$. Then S is a compact, convex set in $\mathbf{C(M)}^*$ so that S has an extreme point L_0. But then L_0 must also be an extreme point of the unit ball of $\mathbf{C(M)}^*$ since if $L_0 = \frac{1}{2}L_1 + \frac{1}{2}L_2$ as linear functionals on $\mathbf{C(M)}$, $\|L_1\| = \|L_2\| = 1$, then $L_j = \ell$ on $\mathbf{B}$ for $j = 1, 2$ so that $L_j \in S$ for $j = 1, 2$. Hence, $L_1 = L_2 = L_0$. Consequently, L is an extreme point of the unit ball of $\mathbf{C(M)}^*$ and so is easily identified: there is a point $y_0 \in \mathbf{M}$ and a unimodular constant λ such that

$$L_0(u) = \lambda u(y_0), \qquad u \in \mathbf{C(M)} \tag{1.21}$$

Let $\mathbf{Y}$ consist of all points $y_0 \in \mathbf{M}$ which arise in (1.21); that is, $y_0 \in \mathbf{Y}$ if and only if there is an extreme point ℓ in the unit ball of $\mathbf{B}^*$ with

$$\ell(h) = \lambda h(y_0), \qquad h \in \mathbf{B}$$

If $f_0 \in \mathbf{B}$, then there is a point $y_0 \in \mathbf{M}$ at which $\|f_0\| = |f_0(y_0)|$. Set $\lambda =$

$|f_0(y_0)|/f_0(y_0)$ and let P consist of all linear functionals ℓ on **B** with

$$\ell(f_0) = \lambda, \qquad \|\ell\| = 1$$

P is convex and weak-* compact so that P contains an extreme point ℓ_0. This ℓ_0 is an extreme point of the unit ball of **B*** since if $\ell_0 = \frac{1}{2}\ell_1 + \frac{1}{2}\ell_2$, then

$$\lambda = \ell_0(f_0) = \tfrac{1}{2}\ell_1(f_0) + \tfrac{1}{2}\ell_2(f_0)$$

and so $\ell_j(f_0) = \lambda, j = 1, 2$, because $\|\ell_j\| = 1$. Hence, $\ell_0(h) = \mu h(y)$ for some $y \in \mathbf{Y}$ and some $\mu, |\mu| = 1$. In particular,

$$\|f_0\| = |\ell_0(f_0)| = |f_0(y)|$$

Thus,

$$\|f\| = \sup\{|f(y)| : y \in \mathbf{Y}\}, \qquad \text{for all } f \in \mathbf{B} \tag{1.22}$$

Now let $y \in \mathbf{Y}$ and let $L_y(h) = h(y)$. Then L_y is an extreme point of the unit ball of **B*** so that $T^*(L_y)$ is the same. Hence,

$$T^*(L_y) = \lambda(y) L_{\tau(y)}$$

for some point $\tau(y) \in \mathbf{Y}$, and some unimodular constant $\lambda(y)$. Thus,

$$(Th)(y) = \lambda(y) h(\tau(y)), \qquad h \in \mathbf{B}, \qquad y \in \mathbf{Y}. \tag{1.23}$$

If g, h both lie in **B**, then

$$T(gh)(y) = \lambda(y) g(\tau(y)) h(\tau(y))$$

so

$$\lambda(y) T(gh)(y) = (Tg)(y)(Th)(y)$$

Note that by taking $h \equiv 1$ in (1.23) we see that $\lambda(y)$ is the restriction to Y of a function in **B**. Hence, the function

$$\lambda T(gh) - T(g)T(h)$$

is the restriction to **Y** of a function in **B** and since this function is zero on **Y** we know from (1.22) that it is zero on all of **M**; thus,

$$\lambda T(gh) = T(g)T(h)$$

Choosing $g = h = T^{-1}(1)$, we see that $1/\lambda$ lies in **B**. Thus, λ is a function in **B** whose reciprocal is in **B** and λ has unit modulus on **Y**, which is larger than the

Shilov boundary of **B**. Hence, λ is a unimodular function in **B** and the rule

$$A(h) = \bar{\lambda}Th$$

is an algebra automorphism of **B**.

Theorem 1.7. *Let Ω be a domain maximal for $H^\infty(\Omega)$ and let T be a linear isometry of $H^\infty(\Omega)$ onto itself. Then*

$$Th = \lambda h \circ \phi$$

where λ is a unimodular constant and ϕ is a one-to-one analytic function mapping Ω onto Ω.

Proof. We know already from Theorem 1.6 that

$$Th = \lambda A(h)$$

where λ is an invertible element of $H^\infty(\Omega)$ which has unit modulus on **M**, the maximal ideal space of $H^\infty(\Omega)$, and A is an algebra automorphism of $H^\infty(\Omega)$ onto itself. It follows that λ is a constant and that $A(h) = h \circ \phi$ where ϕ is as described; see Theorem 3.6.1.

7.2. SELF-MAPPINGS OF A DOMAIN

Suppose Ω is a domain on the sphere which omits at least three points and ϕ is an analytic mapping of Ω into itself. The properties of the linear operator $f \mapsto f \circ \phi$ depend in large part on the behavior of sequence $\{\phi_n\}$ of iterates of ϕ defined by

$$\phi_0(z) \equiv z, \qquad \phi_{n+1}(z) = \phi_n(\phi(z)), \qquad n = 0, 1, 2, \ldots \tag{2.1}$$

It is this question that we investigate in this section. The first result which we shall obtain is the following.

Theorem 2.1. *Let Ω be a domain whose boundary has at least three points and let ϕ be an analytic function mapping Ω into itself, $\phi(z) \not\equiv z$. Then one, and only one, of the following possibilities can occur:*

(*a*) *ϕ is a one-to-one conformal map of Ω onto itself and is either periodic in the sense that there is an integer N with $\phi_N(z) \equiv z$ or there is a subsequence $\{\phi_{n_k}\}$ which converges uniformly on compact subsets of Ω to the function $G(z) = z$. If the latter occurs, then for each integer $m \geqslant 2$ there is a one-to-one conformal map ψ of Ω onto Ω which is not the identity for which $\psi_m(z) \equiv z$.*

(*b*) *There is a point* $\zeta_0 \in \Omega$ *such that* $\phi(\zeta_0) = \zeta_0$, $|\phi'(\zeta_0)| < 1$, *and the sequence* $\{\phi_n\}$ *converges uniformly on compact subsets of* Ω *to the constant function* $F(z) \equiv \zeta_0$.

(*c*) *There is a continuum* $\mathbf{C}$ *in* $\Gamma = \partial\Omega$ *with the property that every subsequence* $\{\phi_{n_k}\}$ *of* $\{\phi_n\}$ *which converges uniformly on compact subsets of* Ω *has as its limit a constant function* $F(z) \equiv \lambda$ *for some* $\lambda \in \mathbf{C}$; *each point* λ *of* $\mathbf{C}$ *so appears.*

Proof. Suppose there is a point $\zeta_0 \in \Omega$ which ϕ fixes: $\phi(\zeta_0) = \zeta_0$. Let $a = \phi'(\zeta_0)$; we shall first show that $|a| \leqslant 1$ and that the possibility $|a| = 1$ implies (a) holds and that the case $|a| < 1$ implies that (b) holds. Near ζ_0 we have

$$\phi(z) = \zeta_0 + a(z - \zeta_0) + c_2(z - \zeta_0)^2 + \cdots$$

Then a computation shows that

$$\phi_n(z) = \zeta_0 + a^n(z - \zeta_0) + c_{n,2}(z - \zeta_0)^2 + \cdots$$

Since the sequence $\{\phi_n\}$ forms a normal family we know at least some subsequence $\{\phi_{n_k}\}$ of $\{\phi_n\}$ converges uniformly on compact sets in Ω to an analytic function $f(z)$. Thus, $\{\phi'_{n_k}\}$ also converges uniformly on compact sets in Ω of f' and so $a^{n_k} \to f'(\zeta_0)$; clearly, then, $|a| \leqslant 1$ is the only possibility. Suppose first that $|a| < 1$. Then for $|h| < \delta$ we have

$$\begin{aligned} |\phi(\zeta_0 + h) - \phi(\zeta_0)| &\leqslant |a||h| + o(|h|) \\ &< r|h| \end{aligned}$$

where $r < 1$. Hence, ϕ maps the disc $\mathfrak{D} = \{z : |z - \zeta_0| < \delta\}$ into the disc $\{|w - \zeta_0| \leqslant r\delta\}$ and so ϕ_2 maps $\mathfrak{D}$ into $\{w : |w - \zeta_0| \leqslant r^2\delta\}$ and, in general, ϕ_n maps $\mathfrak{D}$ into $\{w : |w - \zeta_0| \leqslant r^n\delta\}$. Hence, $\{\phi_n\}$ converges uniformly on $\mathfrak{D}$ to the constant ζ_0 and so the same is true (by normal families) on any compact set in Ω.

Next suppose that $|a| = 1$ and that $a = \exp[2\pi i t]$ where t is rational, $t = M/N$ in lowest terms. Then

$$\begin{aligned} \phi_N(z) &= \zeta_0 + (z - \zeta_0) + c_{2,N}(z - \zeta_0)^2 + \cdots \\ &= z + c_{2,n}(z - \zeta_0)^2 + \cdots \end{aligned}$$

Let $c = c_{k,N}$ be the first nonzero coefficient past z in the expansion of ϕ_N. Then it is easily established that

$$\phi_{mN}(z) = z + mc(z - \zeta_0)^k + \cdots, \qquad m = 1, 2, \ldots$$

A normal families argument shows that this is an impossibility and we conclude that $c = 0$. Hence, if $a = \exp[2\pi iM/N]$, then $\phi_N(z) \equiv z$ and so ϕ itself must be a one-to-one conformal map of Ω onto Ω. Suppose next that $a = \exp[2\pi it]$ where t is irrational. Given a positive integer m let $\{n_k(m)\}$ be a sequence of integers tending to ∞ such that

$$\exp[2\pi itn_k(m)] \to \exp[2\pi i/m] \quad \text{as } k \to \infty$$

Then at least a subsequence of $\{\phi_{n_k(m)}\}$ converges uniformly on compact subsets of Ω to an analytic function ψ which maps Ω into itself, which fixes ζ_0, and which satisfies $\psi'(\zeta_0) = \exp[2\pi i/m]$. Such a ψ must be a one-to-one conformal mapping of Ω onto itself with

$$\psi \circ \cdots \circ \psi(z) \equiv z, \qquad m \text{ factors}$$

as we just showed. If $\phi(w_1) = \phi(w_2)$, then the same would hold for $\phi_{n_k(m)}$ and so also for ψ. Thus, ϕ is one-to-one. The same sort of argument shows ϕ maps Ω onto itself.

Finally, we have the possibility that ϕ has no fixed points at all within Ω. We must show that either conclusion (a) or (c) holds. Suppose that there is some integer M and some point ζ_1 in Ω for which

$$\phi_M(\zeta_1) = \zeta_1$$

If $|\phi_M'(\zeta_1)| = 1$, then, as before, ϕ_M must be a one-to-one conformal map of Ω onto itself and so the same is true for ϕ; this puts us back into conclusion (a). If, on the other hand, $|\phi_M'(\zeta_1)| < 1$, then we know that $\{\phi_{mM}(z)\}_{m-1}^{\infty}$ converges uniformly on compact sets in Ω to the function which is identically equal to ζ_1. Thus, because

$$\phi(\phi_{mM}(z)) \equiv \phi_{mN}(\phi(z)), \qquad z \in \Omega$$

we find $\phi(\zeta_1) = \zeta_1$ and $1 > |\phi_M'(\zeta_1)| = |\phi'(\zeta_1)|^M$ so that $|\phi'(\zeta_1)| < 1$, as well. Thus, $\{\phi_n\}$ converges uniformly on compact sets in Ω to the function which is identically equal to ζ_1.

Suppose next that no ϕ_n has a fixed point but that there are points $\zeta_1, \zeta_2 \in \Omega$ and a sequence $n_k \to \infty$ with

$$\phi_{n_k}(\zeta_1) \to \zeta_2$$

There is no loss in assuming that the integers satisfy $m_k = n_{k+1} - n_k \to \infty$, and further, that

$$\phi_{m_k} \to f, \qquad \text{uniformly on compact sets}$$

Then $\phi_{n_{k+1}} = \phi_{m_k} \circ \phi_{n_k}$ so that at ζ_1 we find

$$\zeta_2 = \lim_{k \to \infty} \phi_{n_{k+1}}(\zeta_1)$$
$$= \lim_{k \to \infty} \phi_{m_k}(\phi_{n_k}(\zeta_1))$$
$$= f(\zeta_2)$$

However, we have hypothesized that ϕ_n has no fixed point so that $\phi_n(z) - z \neq 0$ in Ω. Hence, either $f(z) - z \neq 0$ in Ω or else $f(z) - z \equiv 0$ in Ω. Since $f(\zeta_2) = \zeta_2$ we must have $f(z) \equiv z$. It follows as above that ϕ must be a one-to-one conformal mapping of Ω onto itself.

We have thus reduced the possibilities to this: ϕ is not a one-to-one mapping of Ω onto itself and for every $\zeta \in \Omega$, all limit points of the sequence $\{\phi_n(\zeta)\}$ lie in $\Gamma = \partial\Omega$.

Let $\mathbf{C}$ denote the totality of limit points of all the sequences $\{\phi_n(\zeta)\}_{n=1}^{\infty}$ for $\zeta \in \Omega$. This set is independent of ζ and could just as well be described as the set of limit points of the sequence $\{\phi_n(\zeta)\}_{n=1}^{\infty}$ as ζ ranges over some fixed subset D of Ω. Let D be a subdomain of Ω whose closure lies in Ω and which satisfies $\phi(D) \cap D \neq \varnothing$. Then

$$\phi_{n+1}(D) \cap \phi_n(D) \neq \varnothing \quad \text{for } n = 0, 1, 2, \ldots$$

and hence the set $\cup_{n=N}^{\infty} \phi_n(D)$ and the set

$$\mathbf{C}_N = \mathrm{CL}\left[\bigcup_{n=N}^{\infty} \phi_n(D)\right]$$

are both connected. Thus, $\cap_{N=1}^{\infty} \mathbf{C}_N$ is also connected. But $\cap_{N=1}^{\infty} \mathbf{C}_N$ is exactly the set of limit points of $\{\phi_n(D)\}_{n=1}^{\infty}$ which in turn is exactly the set $\mathbf{C}$. Thus, the set $\mathbf{C}$ is connected, as we wished to show.

Definition. In the event that conclusion (b) holds, we say that ϕ has an *attractive fixed point* at ζ_0.

When the domain Ω is the open unit disc Δ more precise information is available.

Theorem 2.2. *Suppose that ϕ is a holomorphic function mapping Δ into Δ which is not a linear fractional transformation of the form $\lambda(z - a)(1 - \bar{a}z)^{-1}$, $\lambda \in \mathbf{T}$, $a \in \Delta$; suppose as well that ϕ does not have an attractive fixed point in Δ. Then there is a unique point $p \in \mathbf{T}$ such that $\phi_n(z) \to p$ as $n \to \infty$ for each $z \in \Delta$. That is, the set $\mathbf{C}$ described in (c) of Theorem 2.1 is a single point.*

Proof. We know from Theorem 2.1 that ϕ must either be a one-to-one conformal mapping of Δ onto itself and hence of the form $\lambda(z-a)(1-\bar{a}z)^{-1}$ or, all the limit points of the sequence $\{\phi_n(0)\}$ form a continuum **C** in the boundary of Δ. The first possibility has been specifically ruled out so that we need only show that the arc **C** is, in fact, a single point. For $z \in \Delta$ and $a \in \Delta$ the function

$$\frac{\phi_n(z)-\phi_{n+1}(a)}{1-\bar{\phi}_{n+1}(a)\phi_n(z)}$$

is holomorphic, bounded by 1, and vanishes at $z = \phi(a)$. Hence,

$$\left|\frac{\phi_n(z)-\phi_{n+1}(a)}{1-\overline{\phi_{n+1}}(a)\phi_n(z)}\right| \leqslant \left|\frac{z-\phi(a)}{1-\overline{\phi(a)}z}\right|$$

Set $z = a$; this yields

$$\left|\frac{\phi_n(a)-\phi_{n+1}(a)}{1-\overline{\phi_{n+1}}(a)\phi_n(a)}\right| \leqslant \left|\frac{a-\phi(a)}{1-\overline{\phi(a)}a}\right| \tag{2.2}$$

Suppose that $\{n_k\}$ is a sequence such that $\phi_{n_k}(0) \to \lambda$, $\lambda \in \mathbf{T}$. We shall show that $\phi_{n_k+1}(0) \to \lambda$ also. If this is not the case, then by passing to a subsequence if necessary we can assume $\phi_{n_k}(0) \to \lambda$ and $\phi_{n_k+1}(0) \to \mu$, $\mu \neq \lambda$. Thus,

$$1 = \left|\frac{\mu-\lambda}{1-\bar{\mu}\lambda}\right| \leqslant \left|\frac{a-\phi(a)}{1-a\bar{\phi}(a)}\right| < 1$$

from (2.2); this contradiction shows that if $a \in \Delta$ and

$$\text{if } \phi_{n_k}(0) \to \lambda, \quad \text{then} \quad \phi_{n_k+1}(0) \to \lambda \tag{2.3}$$

Let us assume that the arc **C** is not a single point. Let C_0 consist of those points λ in **C** with the property that there is some triangle $D = D(\lambda)$ in Δ with vertex at λ and some point $a \in \Delta$ such that infinitely many of the points $\phi_n(a)$ lie in D. We now show that $\mathbf{C} \setminus C_0$ has at most three points. If not we can find distinct points α and β in the interior of **C** but not in C_0 such that the arc in **T** from α to β lies in **C**.

Let L be the chord from α to β; L divides Δ into disjoint open sets which we call G and H. Fix any $a \in \Delta$; we note that only a finite number of the points $\zeta_n = \phi_n(a)$ can lie on L so that the indices larger than some N can be divided into disjoint sets: P and Q with $\zeta_p \in G$, $p \in P$ and $\zeta_q \in H$, $q \in Q$. We assume that H is that portion of Δ which has the arc from α to β as a part of its boundary. There must be infinitely many indices $\rho \in P$ such that $\zeta_{\rho+1} \in H$.

For otherwise, no point in the arc from α to β would lie in **C**. We also note that (2.2) implies that $|\zeta_\rho - \zeta_{\rho+1}| \to 0$. For if $\{\sigma\}$ is a subsequence of $\{\rho\}$ with $|\zeta_\sigma - \zeta_{\sigma+1}| \geqslant \varepsilon > 0$ for all σ, then we could pass to a further subsequence and conclude $\zeta_\sigma \to \lambda \in \mathbf{C}$. But this would yield $\zeta_{\sigma+1} \to \lambda$ by (2.2), contradicting $|\zeta_\sigma - \zeta_{\sigma+1}| \geqslant \varepsilon$. We conclude, therefore, that $\{\zeta_\rho\}$ converges to either α or β; let us assume that the limit is α. Note that if L' is the chord obtained by reflecting L across the radius from 0 to α, then all but finitely many of the points $\{\zeta_\rho\}$ lie outside the triangle determined by L, L' and the chord joining the ends of L and L'. This is because α is not in C_0. Indeed, if D is any triangle in Δ with vertex at λ, then at most a finite number of the points $\{\phi_n(a)\}$ lie in D. Thus,

$$\lim_{\rho\to\infty} \arg(\zeta_\rho - \alpha) = \frac{\pi}{2} \tag{2.4a}$$

$$\lim_{\rho\to\infty} \arg(\zeta_{\rho+1} - \alpha) = \frac{-\pi}{2} \tag{2.4b}$$

Let $u_\rho(z) = \arg(\phi_\rho(z) - \alpha)$; then $\{u_\rho\}$ is a sequence of bounded harmonic functions on Δ and we may pass to a subsequence and conclude that there is a harmonic function $u(z)$ on Δ with $u_\rho(z) \to u(z)$ uniformly on compact subsets of Δ. However, by (2.4a) and (2.4b) we have

$$u_\rho(a) = \arg(\zeta_\rho - \alpha) \to \frac{\pi}{2}, \qquad a \in \Delta \tag{2.5a}$$

$$u_\rho(\phi(a)) = \arg(\zeta_{\rho+1} - \alpha) \to \frac{-\pi}{2}, \qquad a \in \Delta \tag{2.5b}$$

From (2.5a) we conclude $u \equiv \pi/2$; this contradicts (2.5b). This contradiction establishes that the interior of **C** has at most one point which is not in C_0. Hence, C_0 has measure equal to that of **C**. Since ϕ is a bounded holomorphic function on Δ, ϕ has nontangential limits at almost all points of C_0. Let $\lambda \in C_0$ be such a point. Then there is a point $a \in \Delta$ and a triangle D in Δ such that $\phi_{n_k}(a) \in D$, $k = 1, 2, \ldots$ and $\phi_{n_k}(a) \to \lambda$. Then from (2.2), $\phi_{n_k+1}(a) \to \lambda$ as well. But $\phi(\phi_{n_k}(a)) = \phi_{n_k+1}(a)$, so that the boundary value of ϕ at λ is exactly λ. Thus, $\phi(\lambda) = \lambda$ a.e. on **C** which implies $\phi(z) \equiv z$, a contradiction. This final contradiction implies that the limit set **C** is a single point.

There is another facet of the behavior of holomorphic functions which map Δ into itself: the angular derivative. We now investigate this; the definition of angular derivative follows the discussion.

Suppose again that ϕ is a nonconstant analytic function mapping Δ into itself. Then the function $(1 + \phi(z))(1 - \phi(z))^{-1}$ has positive real part in Δ and hence there is a unique positive measure μ on **T** and a real scalar a with

$$\frac{1 + \phi(z)}{1 - \phi(z)} = \int \frac{e^{i\theta} + z}{e^{i\theta} - z}\, d\mu(\theta) + ia$$

Let us suppose that $\|\phi\|_\infty = 1$. Thus, there is a sequence $\{z_n\}$ in Δ with $z_n \to \lambda \in \mathbf{T}$ and $\phi(z_n) \to \mu$, $\mu \in \mathbf{T}$. By replacing ϕ by $\bar{\mu}\phi(\bar{\lambda}z)$, we may suppose that

$$z_n \to 1, \qquad \phi(z_n) \to 1$$

Let c_0 be the mass of μ at $\{1\}$ so that $c_0 \geqslant 0$ and let μ_0 be the rest of μ; that is, the restriction of μ to $\mathbf{T} \setminus \{1\}$. Then

$$\frac{1+\phi(z)}{1-\phi(z)} = c_0 \frac{1+z}{1-z} + \int \frac{e^{i\theta}+z}{e^{i\theta}-z} d\mu_0(\theta) + ia$$

so

$$\frac{1+\phi(z)}{1-\phi(z)} = c_0 \frac{1+z}{1-z} + \frac{1+f(z)}{1-f(z)} + ia \tag{2.6}$$

where f is holomorphic in Δ and bounded by 1, $f(0)$ real. Taking real parts we find

$$\frac{1-|\phi(z)|^2}{|1-\phi(z)|^2} = c_0 \frac{1-|z|^2}{|1-z|^2} + \frac{1-|f(z)|^2}{|1-f(z)|^2}$$

Now

$$\frac{1-|f(z)|^2}{|1-f(z)|^2} = \int \frac{1-|z|^2}{|e^{i\theta}-z|^2} d\mu_0(\theta)$$

so that

$$\frac{1-|f(z)|^2}{1-|z|^2} \cdot \frac{|1-z|^2}{|1-f(z)|^2} = \int \frac{|1-z|^2}{|e^{i\theta}-z|^2} d\mu_0(\theta)$$

The quantity $|1-z|^2/|e^{i\theta}-z|^2$ remains uniformly bounded as $z \to 1$ from within a triangle D in Δ with vertex at 1. Hence, because μ_0 has no mass at 1, we have

$$\lim_{\substack{z\to 1\\ z\in D}} \int \frac{|1-z|^2}{|e^{i\theta}-z|^2} d\mu_0(\theta) = 0 \tag{2.7}$$

and thus

$$\lim_{\substack{z\to 1\\ z\in D}} \frac{1-|\phi(z)|^2}{1-|z|^2} \frac{|1-z|^2}{|1-\phi(z)|^2} = c_0$$

Set

$$\gamma = \sup_{z \in \Delta} \frac{|1 - \phi(z)|^2}{|1 - z|^2} \frac{1 - |z|^2}{1 - |\phi(z)|^2} \tag{2.8}$$

Then clearly $\gamma \geqslant 1/c_0$ so that if $c_0 = 0$ we must have $\gamma = \infty$. Suppose next that $\gamma < \infty$. Then c_0 must be positive and

$$\frac{|1 - \phi(z)|^2}{|1 - z|^2} \leqslant \gamma \frac{1 - |\phi(z)|^2}{1 - |z|^2}, \qquad z \in \Delta$$

However, from (2.6) we also have

$$\begin{aligned} \frac{|1 - z|^2}{|1 - \phi(z)|^2} \frac{1 - |\phi(z)|^2}{1 - |z|^2} &= c_0 + \frac{1 - |f(z)|^2}{1 - |z|^2} \frac{|1 - z|^2}{|1 - f(z)|^2} \\ &\geqslant c_0 \end{aligned}$$

so that γ must equal $1/c_0$ and further,

$$\lim_{\substack{z \to 1 \\ z \in D}} \frac{1 - |\phi(z)|^2}{1 - |z|^2} \frac{|1 - z|^2}{|1 - \phi(z)|^2} = c_0 = \frac{1}{\gamma} \tag{2.9}$$

Let f continue to be as in (2.6). Then

$$\frac{1 - z}{1 - f(z)}(1 + f(z)) = \int (e^{i\theta} + z) \frac{1 - z}{e^{i\theta} - z} d\mu_0(\theta)$$

However, if $z \in D$, then

$$|e^{i\theta} + z| \left| \frac{1 - z}{e^{i\theta} - z} \right| \leqslant 2 \frac{|1 - z|}{1 - |z|} \frac{1 - |z|}{|e^{i\theta} - z|} \leqslant 2M$$

so that

$$\lim_{\substack{z \to 1 \\ z \in D}} \frac{1 - z}{1 - f(z)}(1 + f(z)) = 0 \tag{2.10}$$

We also know that

$$|1 - \phi(z)|^2 < \gamma |1 - z| \frac{|1 - z|}{1 - |z|} \leqslant \gamma M |1 - z|$$

Let us henceforth assume that $\gamma < \infty$; then $\phi(z) \to 1$ as $z \to 1$, $z \in D$, and from (2.6) and (2.10), we obtain

$$\lim_{\substack{z \to 1 \\ z \in D}} \frac{1 - z}{1 - \phi(z)} = c_0 = \frac{1}{\gamma} \tag{2.11}$$

Further, from (2.9) we find

$$\lim_{\substack{z \to 1 \\ z \in D}} \frac{1 - |\phi(z)|}{1 - |z|} = \frac{1}{c_0} = \gamma \tag{2.12}$$

However we also know that $\|f\|_\infty \leqslant 1$, so that

$$\left| \frac{f(z) - f(w)}{1 - \overline{f(w)}f(z)} \right| \leqslant \left| \frac{z - w}{1 - \bar{w}z} \right|$$

for $z, w \in \Delta$. By letting $w \to z$ this yields

$$|f'(z)| \leqslant \frac{1 - |f(z)|^2}{1 - |z|^2} \tag{2.13}$$

Thus, with f as in (2.6),

$$\gamma\phi'(z)\left[\frac{1 - z}{1 - \phi(z)}\right]^2 - 1 = \gamma f'(z)\left(\frac{1 - z}{1 - f(z)}\right)^2$$

and so if $z \in D$ we have

$$\left|\gamma\phi'(z)\left[\frac{1 - z}{1 - \phi(z)}\right]^2 - 1\right| \leqslant \gamma \frac{1 - |f(z)|^2}{1 - |z|^2} \frac{|1 - z|^2}{|1 - f(z)|^2}$$

$$\to 0 \quad \text{as } z \to 1, z \in D$$

In combination with (2.11) this shows that

$$\lim\{\phi'(z) \colon z \to 1, z \in D\} = \frac{1}{c_0} = \gamma$$

We summarize and extend the preceding in the next theorem.

Theorem 2.3. *Let ϕ be holomorphic in Δ, $\|\phi\|_\infty = 1$, and let D be any triangle in Δ with vertex at 1. Then $(1 - \phi(z))(1 - z)^{-1}$ always has a limit as $z \to 1$, $z \in D$. The limit is either ∞ or a positive real number γ where γ is given by* (2.8). *In the latter case it is also true that*

$$\lim\{\phi'(z) : z \in D, z \to 1\} = \gamma \tag{2.14}$$

Furthermore, if there is any sequence $\{z_n\}$ *in* Δ, $z_n \to 1$, *such that*

$$\lim_{n\to\infty} \frac{1 - |\phi(z_n)|}{1 - |z_n|} = L < \infty \tag{2.15}$$

then (2.11), (2.12), *and* (2.14) *hold with* $\gamma \leqslant L$.

Proof. Assume (2.15) holds. We start with the observation, established by a simple computation, that for z, $a \in \Delta$ and $0 < r < 1$,

$$\left|\frac{z-a}{1-\bar{a}z}\right| < r \quad \text{if and only if} \quad \frac{|1-\bar{a}z|^2}{1-|z|^2} < \frac{1-|a|^2}{1-r^2} \tag{2.16}$$

Now fix any $z \in \Delta$ and let s be any number greater than $|1 - z|^2(1 - |z|^2)^{-1}$. Choose r_n such that

$$\frac{1 - |z_n|}{1 - r_n} = s$$

Then for large values of n we have

$$\frac{1-|z_n|^2}{1-r_n^2} > \frac{|1-\bar{z}_n z|^2}{1-|z|^2}$$

since $|z_n|$ and r_n both converge to 1. But by (2.16) this implies

$$\left|\frac{z - z_n}{1-\bar{z}_n z}\right| < r_n \tag{2.17}$$

Since ϕ is holomorphic and bounded by 1 in Δ, (2.17) yields

$$\left|\frac{\phi(z) - \phi(z_n)}{1 - \overline{\phi(z_n)}\,\phi(z)}\right| < r_n \tag{2.18}$$

Another application of (2.16), this time to (2.18), shows that we must have

$$\frac{|1-\overline{\phi(z_n)}\,\phi(z)|^2}{1-|\phi(z)|^2} < \frac{1-|\phi(z_n)|^2}{1-r_n^2}$$

$$= \left(\frac{1-|\phi(z_n)|}{1-|z_n|}\right)\left(\frac{1-|z_n|}{1-r_n}\right)\left(\frac{1+|\phi(z_n)|}{1+r_n}\right)$$

Hence, as $n \to \infty$, we find that

$$\frac{|1-\phi(z)|^2}{1-|\phi(z)|^2} \leqslant Ls \quad \text{if} \quad \frac{|1-z|^2}{1-|z|^2} < s$$

Equivalently,

$$\sup_{z \in \Delta} \frac{|1 - \phi(z)|^2}{|1 - z|^2} \frac{1 - |z|^2}{1 - |\phi(z)|^2} \leqslant L$$

This implies that $L \geqslant \gamma$ and we're now back in the setting of (2.8).

Definition. If any of (2.11), (2.12), or (2.14) hold, then we say that ϕ has an *angular derivative* at 1 with value γ.

Theorem 2.4. *Suppose ϕ is holomorphic and bounded by 1 on Δ and $\phi_n(a) \to 1$ as $n \to \infty$ for each $a \in \Delta$. Then ϕ has an angular derivative at 1 and $|\phi'(1)| \leqslant 1$. Conversely, if ϕ has an angular derivative at 1 with modulus less than 1, then $\phi_n(a) \to 1$ as $n \to \infty$ for all $a \in \Delta$.*

Proof. There is a sequence $n_j \to \infty$ such that

$$|\phi_{n_j+1}(0)| > |\phi_{n_j}(0)| \quad \text{and} \quad \phi_{n_j}(0) \to 1$$

Set $z_j = \phi_{n_j}(0)$ so that $z_j \to 1$ and

$$\frac{1 - |\phi(z_j)|}{1 - |z_j|} < 1$$

Now apply Theorem 2.3.

Conversely, suppose ϕ has an angular derivative at 1 with modulus ρ strictly less than 1. Then

$$\frac{1 - |z|^2}{1 - |\phi(z)|^2} \frac{|1 - \phi(z)|^2}{|1 - z|^2} \leqslant \rho, \qquad z \in \Delta$$

Replace z by $\phi_n(a)$, $n = 0, 1, 2, \dots, N - 1$, and multiply the resulting inequalities. We obtain

$$\frac{1 - |a|^2}{1 - |\phi_N(a)|^2} \frac{|1 - \phi_N(a)|^2}{|1 - a|^2} \leqslant \rho^N$$

Hence,

$$\frac{1}{2} \frac{1 - |a|^2}{|1 - a|^2} |1 - \phi_n(z)| \leqslant \rho^n$$

which shows that $\phi_n(a) \to 1$ exponentially fast as $n \to \infty$.

7.3. GENERAL PROPERTIES OF COMPOSITION OPERATORS

Throughout this section Ω is a domain in the plane for which $H^p(\Omega)$ is nontrivial and ϕ is a nonconstant analytic mapping of Ω into itself. We set forth some basic properties of the composition operator C_ϕ on $H^p(\Omega)$ defined by

$$(C_\phi f)(z) = f(\phi(z)) \tag{3.1}$$

It is first worthwhile to note that C_ϕ maps $H^p(\Omega)$ into itself; for if u is the least harmonic majorant of $|f|^p$ $(0 < p < \infty)$ or if $|f|$ is bounded $(p = \infty)$, then $u \circ \phi$ is an harmonic majorant of $|f \circ \phi|^p$ $(0 < p < \infty)$ or $|f \circ \phi|$ is obviously again bounded $(p = \infty)$.

Note next that C_ϕ is always one-to-one since $\phi(\Omega)$ is a nonempty open subset of Ω and hence $C_\phi f = C_\phi g$ implies f and g agree on $\phi(\Omega)$ and thus on all of Ω. Further, C_ϕ is onto $H^p(\Omega)$ if and only if ϕ is a one-to-one mapping of Ω onto itself and C_ϕ is an isometry of $H^p(\Omega)$ onto $H^p(\Omega)$ if and only if ϕ is both one-to-one, onto, and ϕ fixes the point z_0 which is used to compute the $H^p(\Omega)$ norm.

Proposition 3.1. C_ϕ is compact on $H^p(\Omega)$, $1 \leqslant p \leqslant \infty$, if and only if whenever $\{f_n\}$ is a sequence of functions in the unit ball of $H^p(\Omega)$ with $f_n \to 0$ uniformly on compact subsets of Ω, then $f_n \circ \phi \to 0$ in $H^p(\Omega)$.

Proof. If C_ϕ is compact, then the second condition follows easily. Suppose, then, that the second condition holds. Let $\mathfrak{B}$ be any bounded set in $H^p(\Omega)$ and let $\{f_n \circ \phi\}$ be any sequence in $C_\phi(\mathfrak{B})$. A normal families argument allows us to assume that there is a subsequence $\{f_{n_j}\}$ and an $f \in H^p(\Omega)$ with $f_{n_j} \to f$ uniformly on compact sets in Ω. Hence, our hypothesis implies that $f_{n_j} \circ \phi \to f \circ \phi$ in $H^p(\Omega)$. Thus, $C_\phi(\mathfrak{B})$ is precompact so C_ϕ is a compact operator.

We shall now establish some basic properties of the operator C_ϕ under the twin assumptions that C_ϕ is compact on some $H^p(\Omega)$, $1 \leqslant p < \infty$, and that Ω is finitely connected. (The case when C_ϕ is compact on $H^\infty(\Omega)$ is explored thoroughly in Section 4.) We note first that it is a standard fact that the spectrum of a compact operator mapping a Banach space into itself consists only of isolated eigenvalues $\lambda \neq 0$ and possibly of $\{0\}$, which may not be an eigenvalue; see Dunford and Schwartz (1957).

Theorem 3.2. *Let $\partial\Omega$ consist of $m + 1$ disjoint analytic simple closed curves, let ϕ be an analytic function mapping Ω into itself, and suppose C_ϕ is compact on some $H^p(\Omega)$, $1 \leqslant p < \infty$. Then: (a) C_ϕ is compact on every $H^p(\Omega)$, $1 \leqslant p < \infty$; (b) ϕ has an attractive fixed point at some point $a \in \Omega$.*

Proof. Let $q \neq p$, $1 \leqslant q < \infty$, and suppose that $\{f_n\}$ is a sequence of elements of the unit ball of $H^q(\Omega)$ with $f_n \to 0$ uniformly on compact subsets of Ω. We must show that $f_n \circ \phi \to 0$ in $H^q(\Omega)$. Let $f_n = F_n I_n$ be the inner-outer factorization of f_n according to Theorem 4.7.3; the function I_n incorporates both the singular factor and the Blaschke product in f_n. The function $(F_n)^{q/p}$ lies in the unit ball of $MH^p(\Omega)$. By Corollary 4.6.2 there is a sequence $\{w_n\}$ of non-negative functions on Γ such that $0 \leqslant w_n \leqslant M$ for all n and

$$g_n = I_n(F_n)^{q/p}\exp(w_n + i^* w_n)$$

is single-valued on Ω. Then g_n lies in $H^p(\Omega)$, $\|g_n\|_p \leqslant e^M$, and $g_n \to 0$ uniformly on compact subsets of Ω. Hence, because C_ϕ is compact on $H^p(\Omega)$ we have

$$\int_\Gamma |F_n(\phi)|^q |I_n(\phi)|^p \, d\omega \leqslant \int_\Gamma |g_n(\phi)|^p \, d\omega \to 0 \tag{3.2}$$

as $n \to \infty$. If $q > p$, then

$$\int_\Gamma |f_n(\phi)|^q \, d\omega \leqslant \int_\Gamma |F_n(\phi)|^q |I_n(\phi)|^p \, d\omega \to 0$$

as $n \to \infty$ by (3.2) and so C_ϕ is compact on $H^q(\Omega)$ for $q > p$. If $1 \leqslant q < p$, then given $\varepsilon > 0$ we let

$$A_n = \{x \in \Gamma : |I_n(\phi(x))| \leqslant \varepsilon\}.$$

Thus,

$$\int_\Gamma |f_n(\phi)|^q \, d\omega = \int_{A_n} + \int_{\Gamma \setminus A_n}$$
$$\leqslant \varepsilon^q \int_\Gamma |F_n(\phi)|^q \, d\omega + \varepsilon^{-p} \int_\Gamma |F_n(\phi)|^q |I_n(\phi)|^p \, d\omega$$

However,

$$\int_\Gamma |F_n(\phi)|^q \, d\omega = \|C_\phi(F_n)\|_q^q$$
$$\leqslant \|C_\phi\|^q$$

and therefore

$$\|C_\phi(f_n)\|_q^q \leqslant \varepsilon^q \|C_\phi\|^q + \varepsilon^{-p} \nu_n$$

where $\nu_n \to 0$ as $n \to \infty$. Consequently, $\|C_\phi(f_n)\|_q < B\varepsilon$ if n is sufficiently large where B is a constant independent of n and ε. Thus, C_ϕ is also compact on $H^q(\Omega)$ for $1 \leqslant q < p$.

The proof of (b) is based on Theorem 2.1. If ϕ fails to have an attractive fixed point in Ω, then it must follow that all the limit points of $\{\phi_n(\zeta)\}$ form some continuum **C** in $\partial\Omega$. (If C_ϕ is compact, ϕ cannot be a one-to-one mapping of Ω onto Ω.) Let ξ be any point of **C** and let $\{n_k\}$ be an increasing sequence of integers such that

$$\xi_k = \phi_{n_k}(z_0) \to \xi$$

Let h be an element of $\mathbf{A}(\Omega)$ with $h(\xi) = 1$ and $|h(\zeta)| < 1$ for all $\zeta \in \partial\Omega$,

$\zeta \neq \xi$; such an h exists by the discussions in Chapter 4, Section 3, for example. There is no loss in assuming that $\{\xi_k\}$ satisfies

$$\sum_1^\infty (1 - |h(\xi_k)|) < \infty$$

Let $w_k = h(\xi_k)$ and then put

$$B_N(w) = \prod_{k=N}^\infty \left(\frac{-\overline{w}_k}{|w_k|}\right)\left(\frac{w - w_k}{1 - \overline{w}_k w}\right), \qquad |w| < 1$$

If $g \in H^q(\Omega)$, then $g_N(z) = g(z)B_N(h(z))$ is also in $H^q(\Omega)$ and

$$\|g_N\|_{H^q} \leqslant \|g\|_{H^q} \tag{3.3a}$$

$$g_N \to g \quad \text{uniformly on compact sets in } \Omega \tag{3.3b}$$

$$g_N(\xi_k) = 0 \quad \text{if} \quad k \geqslant N \tag{3.3c}$$

It follows from (3.3a) and (3.3b) that $g_N \to g$ in $H^q(\Omega)$; see problem 14 in Chapter 3. Thus (3.3c) leads us to the conclusion that the set of those elements in $H^q(\Omega)$ which vanish at all but a finite number of the points $\{\xi_k\}$ is dense in $H^q(\Omega)$.

We know from problem 2 at the end of this chapter that the norm of the linear functional L_ζ given by $L_\zeta(g) = g(\zeta)$, $g \in H^q(\Omega)$, goes to ∞ as $\zeta \to \partial\Omega$. Thus, we may assume that initially the subsequence $\{n_k\}$ was chosen so that

$$\|L_{\zeta_k}\| \leqslant \|L_{\xi_k}\|$$

where $\zeta_k = \phi_{n_k - 1}(z_0)$, $\xi_k = \phi_{n_k}(z_0)$. Let ℓ_k be the linear functional $\ell_k = L_{\zeta_k}/\|L_{\zeta_k}\|$. The adjoint C_ϕ^* of C_ϕ is compact and so at least a subsequence of $\{C_\phi^* \ell_k\}$ converges in norm to a linear functional l on $H^q(\Omega)$. This gives

$$\begin{aligned} |l| &= \lim \|C_\phi^*(\ell_k)\| \\ &= \lim \frac{\|L_{\xi_k}\|}{\|L_{\zeta_k}\|} \\ &\geqslant 1 \end{aligned}$$

But, if g vanishes at all but finitely many of the points $\{\xi_k\}$ we also have

$$\begin{aligned} l(g) &= \lim C_\phi^*(\ell_k)(g) \\ &= \lim \frac{g(\xi_k)}{\|L_{\zeta_k}\|} \\ &= 0 \end{aligned}$$

Thus, l vanishes on a dense set in $H^q(\Omega)$ and so is identically zero, contradicting $\|l\| \geqslant 1$. This leads us to the conclusion that ϕ must have an attractive fixed point in Ω.

The final results on the compact operator C_ϕ are contained in the next theorem. We still assume Ω is bounded by a finite number of disjoint analytic simple closed curves.

Theorem 3.3. *Let C_ϕ be compact on $H^p(\Omega)$, $1 \leqslant p < \infty$, and let $a \in \Omega$ be the attractive fixed point of ϕ. Then the spectrum of C_ϕ is $\{0, 1\} \cup \{\mu^m\}_{m=1}^{\infty}$ where*

$$\mu = \phi'(a) \tag{3.4}$$

and the eigenspace corresponding to the eigenvalue μ^m is one-dimensional and is spanned by the function h^m where h is the unique function (up to scalar multiples) satisfying

$$\mu h(z) = h(\phi(z)), \qquad z \in \Omega \tag{3.5}$$

h lies in $H^q(\Omega)$ for all $q < \infty$.

Proof. Suppose λ is an eigenvalue of C_ϕ, $\lambda \neq 1, 0$, with eigenvector F:

$$\lambda F(z) = F(\phi(z)), \qquad z \in \Omega \tag{3.6}$$

Hence, $F(a) = 0$ since $\lambda \neq 1$; let m be the order of the zero of F at a and write $F(z) = (z - a)^m G(z)$ where $G(a) \neq 0$. Then from (3.6) we have

$$\lambda(z - a)^m G(z) = (\phi(z) - a)^m G(\phi(z))$$

so

$$\lambda G(z) = \left(\frac{\phi(z) - a}{z - a}\right)^m G(\phi(z))$$

which yields $\lambda = (\phi'(a))^m = \mu^m$, as $z \to a$. Thus, the spectrum of C_ϕ lies in $\{0, 1\} \cup \{\mu^m\}_{m=1}^{\infty}$.

The one-dimensional nature of the eigenspace is based on this simple observation: if f is analytic in a neighborhood $\mathfrak{D}$ of a and if

$$f(z) = f(\phi(z)), \qquad z \in \mathfrak{D} \tag{3.7}$$

then f is constant in $\mathfrak{D}$. Indeed, there is no loss in assuming $f(a) = 0$. Let N be the order of the zero of f at a and write $f(z) = (z - a)^N g(z)$, $g(a) \neq 0$. Thus, from (3.7) we find

$$(z - a)^N g(z) = (\phi(z) - a)^N g(\phi(z))$$

so that, as above, $1 = \mu^N$. But $|\mu| < 1$ and so g must vanish identically and f must be identically equal to $f(a)$.

If F and H are both eigenfunctions for the eigenvalue $\lambda = \mu^m$ then both F and H have a zero at a of order m so that $f = F/H$ is holomorphic near a and satisfies $f = f \circ \phi$ there. Thus, f is constant near a which implies F is a multiple of H near a and so on all of Ω. Thus, the eigenspaces are one-dimensional.

Next, if C_ϕ is compact on some $H^p(\Omega)$ then we know that C_ϕ is compact on every $H^p(\Omega)$, $1 \leqslant p < \infty$, and, in particular, on $H^2(\Omega)$. Let $\{f_m\}$ be the orthonormal basis for $H^2(\Omega)$ described in problem 14 at the end of the chapter with a the attractive fixed point of ϕ. The matrix representing C_ϕ^* on $H^2(\Omega)$ with respect to this basis is upper-triangular with diagonal entries $\{1, \bar{\mu}, \bar{\mu}^2, \dots\}$. Thus, these numbers are all eigenvalues for C_ϕ^* and so $\{1, \mu, \mu^2, \dots\}$ are all eigenvalues for C_ϕ.

Let h be an element of $H^2(\Omega)$ with

$$\mu h(z) = h(\phi(z)), \qquad z \in \Omega. \tag{3.8}$$

Then h^m satisfies

$$\mu^m h^m(z) - h^m(\phi(z))$$

and so h^m must agree near a with the eigenfunction (in $H^2(\Omega)$) for the eigenvalue μ^m. Hence, h^m must be the eigenfunction for μ^m. Consequently, h lies in $H^q(\Omega)$ for all $q < \infty$ and a moment's thought shows that h^m must be (a multiple of) the eigenfunction for μ^m on $H^p(\Omega)$, as well, for each p, $1 \leqslant p < \infty$.

7.4. COMPACT COMPOSITION OPERATORS ON $H^\infty(\Omega)$

Let Ω be a domain on $\mathbb{C}$ for which $H^\infty(\Omega)$ is nontrivial and let ϕ be an analytic function mapping Ω into itself. It is possible to actually characterize those ϕ for which the composition operator C_ϕ on $H^\infty(\Omega)$ is compact and that is what we shall do in this section.

We begin with two simple but useful facts about ϕ if C_ϕ is compact. Recall that $\mathbf{M}_\lambda$ stands for the fiber over λ in the maximal ideal space of $H^\infty(\Omega)$; see Section 3, Chapter 6.

Proposition 4.1. (a) If for each $\lambda \in \partial\Omega$, the fiber $\mathbf{M}_\lambda$ is a peak set for $H^\infty(\Omega)$, then C_ϕ is compact if and only if the closure of $\phi(\Omega)$ contains no point of $\partial\Omega$.

(b) If C_ϕ is compact on $H^\infty(\Omega)$, then $\phi(\Omega)$ contains no interpolating sequence for $H^\infty(\Omega)$.

Proof. (a) Suppose C_ϕ is compact but $\lambda \in \partial\Omega$ is the limit of the sequence $\{\phi(z_n)\}$. Let $h \in H^\infty(\Omega)$ peak on the fiber $\mathbf{M}_\lambda$; then h is continuous at λ, with $h(\lambda) = 1$, and $\lim\sup\{|h(z)|: z \to \zeta\} < 1$ for all $\zeta \in \partial\Omega$, $\zeta \neq \lambda$. The sequence

$\{h^k\}$ goes to zero uniformly on compact subsets of Ω and hence $\{C_\phi(h^k)\} = \{(h \circ \phi)^k\}$ must converge uniformly to zero on Ω. But

$$\lim_{n\to\infty} (h\circ\phi)^k(z_n) = 1$$

for each k. Conversely, if the closure of $\phi(\Omega)$ contains no point of $\partial\Omega$, then there is a compact set $\mathbf{K}$ of Ω which contains $\phi(\Omega)$. Clearly, then, C_ϕ is compact on $H^\infty(\Omega)$.

To see (b), recall from Section 4 of Chapter 6 that if $\{w_n\}$ is an interpolation sequence in Ω and $w_n \to \lambda \in \partial\Omega$, then the fiber $\mathbf{M}_\lambda$ is a peak set for $H^\infty(\Omega)$. Hence, if $\{\phi(z_n)\}$ is an interpolation sequence and $\phi(z_n) \to \lambda$, then there is an $h \in H^\infty(\Omega)$ which peaks on $\mathbf{M}_\lambda$. As above, we see that although the sequence $\{h^k\}$ converges to zero uniformly on compact subsets of Ω, each element in the sequence $\{h^k \circ \phi\} = \{(h \circ \phi)^k\}$ has sup norm at least one.

One of the equivalent properties that characterize those ϕ with C_ϕ compact on $H^\infty(\Omega)$ involves the distinguished homomorphisms introduced and studied in Chapter 6, Section 4. Let Λ be the set of all distinguished homomorphisms in $\mathbf{M}(\Omega)$, the maximal ideal space of $H^\infty(\Omega)$; recall that if a distinguished homomorphism exists it must lie in a fiber $\mathbf{M}_\lambda$ over some $\lambda \in \partial\Omega$ and there is at most one distinguished homomorphism in any fiber (there may be none). If the fiber $\mathbf{M}_\lambda$ has a distinguished homomorphism we denote it by ϕ_λ.

Let ϕ_λ be a distinguished homomorphism in $\mathbf{M}(\Omega)$ and let $P(\phi_\lambda, \varepsilon)$ be the open ball of radius ε centered at ϕ_λ in the dual space of $H^\infty(\Omega)$. We know from Section 4 of Chapter 6 that $P(\phi_\lambda, \varepsilon) \cap \Omega$ is nonempty; indeed, we showed

$$\lim_{\delta\to 0} \frac{\operatorname{area}(P(\phi_\lambda, \varepsilon) \cap \{z \in \Omega : |z - \lambda| \leqslant \delta\})}{\pi\delta^2} = 1 \tag{4.1}$$

The open set $\phi(\Omega)$ is a subset of Ω and hence in turn a subset of the dual space of $H^\infty(\Omega)$. Let

$$\Omega_1 = \text{norm closure of } \phi(\Omega) \text{ in } H^\infty(\Omega)^* \tag{4.2}$$

and

$$\Omega_2 = \text{weak-* closure of } \phi(\Omega) \text{ in } H^\infty(\Omega)^* \tag{4.3}$$

so that, as a matter of fact, we have

$$\Omega_1 \subset \Omega_2 \subset \mathbf{M}(\Omega)$$

Finally, for each $\varepsilon > 0$, set

$$K_\varepsilon = \Omega_2 \Big\backslash \bigcup_{\phi_\lambda \in \Lambda} P(\phi_\lambda, \varepsilon) \tag{4.4}$$

The major result which characterizes compact composition operators is this.

Theorem 4.2. *The following are equivalent:*

(*a*) C_ϕ *is compact on* $H^\infty(\Omega)$
(*b*) $\Omega_1 = \Omega_2$
(*c*) $\Omega_2 \setminus \Omega$ *is a subset of* Λ
(*d*) K_ε *is a compact subset of* Ω *for each* $\varepsilon > 0$.

Proof. To show that (a) implies (b) we suppose that C_ϕ is compact and that $m \in \mathbf{M}(\Omega)$ is a weak-* cluster point of $\phi(\Omega)$. Thus, there is a net $\{\phi(z_\alpha)\}$ converging weak-* to m; that is, for each $f \in H^\infty(\Omega)$ there is a sequence $\{\alpha_j\}_{j=1}^\infty$ such that

$$\lim_{j \to \infty} f\big(\phi(z_{\alpha_j})\big) = m(f)$$

The set $\{z_\alpha\}$ is an infinite subset of the (weak-*) compact set $\mathbf{M}(\Omega)$ and so there is a subset $\{z_\beta\}$ of $\{z_\alpha\}$ and a point m_* in $\mathbf{M}(\Omega)$ with $\{z_\beta\}$ converging weak-* to m_*; that is, for each $g \in H^\infty(\Omega)$ there is a sequence $\{\beta_k\}_{k=1}^\infty$ such that

$$\lim_{k \to \infty} g(z_{\beta_k}) = m_*(g)$$

However, C_ϕ is compact so its adjoint C_ϕ^* defined by

$$(C_\phi^* \ell)(f) = \ell(f \circ \phi), \qquad \ell \in H^\infty(\Omega)^*, \qquad f \in H^\infty(\Omega) \tag{4.5}$$

is also compact and so some subsequence of $C_\phi^*(z_\beta)$ converges in the norm of $H^\infty(\Omega)^*$ to $C_\phi^* m_*$. Hence, for any $f \in H^\infty(\Omega)$,

$$\begin{aligned} f\big(\phi(z_{\beta_j})\big) &= \hat{z}_{\beta_j}(f \circ \phi) \\ &= \big(C_\phi^* z_{\beta_j}\big)(f) \\ &\to \big(C_\phi^* m_*\big)(f) = m_*(f \circ \phi) \end{aligned}$$

Hence, $m(f) = m_*(f \circ \phi)$ for all $f \in H^\infty(\Omega)$ so that m is in the norm closure of $\phi(\Omega)$.

To see that (b) implies (c) recall from Section 4 of Chapter 6 that the distinguished homomorphisms are not only norm limits of point evaluations from Ω but that property actually characterizes them. Thus, if $\Omega_1 = \Omega_2$, then $\Omega_2 \setminus \Omega$ consists of nothing but distinguished homomorphisms.

Suppose next that (c) holds. Then K_ε is a subset of Ω. Suppose that $\{\phi(z_j)\}$ is a sequence in K_ε and $m \in \mathbf{M}(\Omega)$ is a weak-* cluster point of $\{\phi(z_j)\}$. According to (c), m must be a distinguished homomorphism or a point of Ω. If the latter is not true, then we know again from Section 4 of Chapter 6 that $\{\phi(z_j)\}$ converges in norm to m. But then $\phi(z_j)$ fails to be in K_ε for large values of j, a contradiction. Hence, all limit points of K_ε lie within Ω and so K_ε is compact.

Finally, we show that (d) implies (a). There is no loss in choosing ε so small that K_ε is nonempty. Let $\{f_n\}$ be a sequence from the unit ball of $H^\infty(\Omega)$ with $f_n \to 0$ uniformly on compact subsets of Ω. Choose N so big that $|f_n(z)| < \varepsilon$ for $z \in K_\varepsilon$ and all $n \geqslant N$. Each point w in $\phi(\Omega)\setminus K_\varepsilon$ lies in some $P(\phi_\lambda, \varepsilon)$ and $\phi(\Omega)$ is connected, so there must be a point w in both K_ε and $P(\phi_\lambda, 2\varepsilon)$. Let ζ lie in $P(\phi_\lambda, \varepsilon)$. Then $|f_n(w)| < \varepsilon$ and

$$\begin{aligned} |f_n(w) - f_n(\zeta)| &\leqslant |f_n(\zeta) - \phi_\lambda(f_n)| + |\phi_\lambda(f_n) - f_n(\zeta)| \\ &< 2\varepsilon + \varepsilon \end{aligned}$$

Thus, $|f_n(\zeta)| < 4\varepsilon$ if $n \geqslant N$ and $\zeta \in P(\phi_\lambda, \varepsilon)$. Consequently, $|f_n| < 4\varepsilon$ on all of $\phi(\Omega)$ if $n \geqslant N$ or, equivalently, $|f_n \circ \phi| < 4\varepsilon$ on all of Ω. Thus, C_ϕ is compact.

The next result provides a further characterization of those ϕ which produce compact composition operators on $H^\infty(\Omega)$.

Theorem 4.3. *C_ϕ is compact on $H^\infty(\Omega)$ if and only if whenever the Euclidean closure of $\phi(\Omega)$ contains a point $\lambda \in \partial\Omega$ then there is a distinguished homomorphism ϕ_λ in $\mathbf{M}_\lambda$ and each $f \in H^\infty(\Omega)$ has a weak-* continuous extension from $\phi(\Omega)$ to λ by the rule $f(\lambda) = \phi_\lambda(f)$.*

Proof. Suppose C_ϕ is compact on $H^\infty(\Omega)$. If $\lambda \in \partial\Omega$ lies in the Euclidean closure of $\phi(\Omega)$, then $\mathbf{M}_\lambda$ cannot be a peak set for $H^\infty(\Omega)$; let ϕ_λ be the distinguished homomorphism in $\mathbf{M}_\lambda$. Suppose $\{z_n\}$ is a sequence in Ω with $\phi(z_n) \to \lambda$. We know from Proposition 4.1 that $\{\phi(z_n)\}$ cannot contain any subsequence which is an interpolation sequence for $H^\infty(\Omega)$ so that $\phi(z_n) \to \phi_\lambda$ in the norm of $H^\infty(\Omega)^*$. Thus, ϕ_λ is the only point of $\mathbf{M}_\lambda$ which lies in the weak-* closure of $\phi(\Omega)$ and each $f \in H^\infty(\Omega)$ has a weak-* continuous extension from $\phi(\Omega)$ to λ by the rule $f(\lambda) = \phi_\lambda(f)$.

Conversely, suppose C_ϕ is not compact. Then Theorem 4.2 implies there is a point L in $\mathbf{M}(\Omega)\setminus\Omega$ which is not a distinguished homomorphism but which does lie in the weak-* closure of $\phi(\Omega)$. Thus, $L \in \mathbf{M}_\lambda$ for some $\lambda \in \partial\Omega$. If $\mathbf{M}_\lambda$ contains no distinguished homomorphism there is nothing more to do; if $\mathbf{M}_\lambda$ has a distinguished homomorphism ϕ_λ then $L \neq \phi_\lambda$. If it happens that ϕ_λ is a weak-* cluster point of $\phi(\Omega)$, then there are nets $\{z_\alpha\}$ and $\{w_\beta\}$ in Ω with $\{\phi(z_\alpha)\}$ and $\{\phi(w_\beta)\}$ converging to L and ϕ_λ, respectively. Let h be an element of $H^\infty(\Omega)$ with $L(h) \neq \phi_\lambda(h)$. Then h has at least two different weak-* limits at λ. If, on the other hand, ϕ_λ is not a weak-* cluster point of $\phi(\Omega)$, then any sequence $\{\phi(z_n)\}$ which converges to λ has a subsequence which is an interpolation sequence for $H^\infty(\Omega)$. Thus, there is an $h \in H^\infty(\Omega)$ with $h(\phi(z_n)) = (-1)^n$ and this h does not have a weak-* continuous extension from $\phi(\Omega)$ to λ. This concludes the proof.

The mapping $\phi\colon \Omega \to \Omega$ induces a mapping Φ of $\mathbf{M}(\Omega)$ into $\mathbf{M}(\Omega)$ given by

$$\Phi(m)(f) = m(f \circ \phi), \qquad m \in \mathbf{M}(\Omega) \tag{4.6}$$

Note that the restriction of Φ to Ω is nothing but ϕ; that is,

$$\Phi(L_\zeta) = L_{\phi(\zeta)}, \qquad \zeta \in \Omega$$

With this definition of Φ we can now show that a compact composition operator C_ϕ has associated with it a fixed point, either of ϕ or of Φ.

Theorem 4.4. *Suppose C_ϕ is compact on $H^\infty(\Omega)$. Then either ϕ has an attractive fixed point $a \in \Omega$ or else there is a unique $\lambda \in \partial\Omega$ with distinguished homomorphism ϕ_λ such that $\phi_n(z) \to \lambda$ for each $z \in \Omega$ and $\Phi(\phi_\lambda) = \phi_\lambda$.*

Proof. Suppose ϕ does not have an attractive fixed point in Ω. Then Theorem 2.1 implies there is a compact connected set $\mathbf{C}$ in $\partial\Omega$ such that $\mathbf{C}$ is the totality of limit points of $\{\phi_n(z)\}$. We know from Proposition 4.1, however, that $\phi(\Omega)$ contains no interpolation sequences so that $\mathbf{C}$ can contain only points λ whose fibers $\mathbf{M}_\lambda$ are not peak sets for $H^\infty(\Omega)$. However, each such point λ is a singleton component in $\partial\Omega$ by Proposition 6.4.4 so that there is a unique $\lambda \in \partial\Omega$ and a unique distinguished homomorphism ϕ_λ in the fiber $\mathbf{M}_\lambda$ such that $z_n = \phi_n(z) \to \lambda$, $z \in \Omega$. Further, $\phi(z_n) \to \phi_\lambda$ in the norm of $H^\infty(\Omega)^*$. However, Φ, defined by (4.6) is norm continuous so that $\Phi(\phi(z_n)) \to \Phi(\phi_\lambda)$ and so

$$\phi(\phi(z_n)) = \Phi(\phi(z_n)) \to \Phi(\phi_\lambda)$$

and

$$\phi(z_n) \to \phi_\lambda$$

Thus, $\Phi(\phi_\lambda)$ is in the closure of $\phi(\Omega)$ and so must either be a point of Ω or a distinguished homomorphism. If it is a distinguished homomorphism it must be ϕ_λ since ϕ_λ is the only distinguished homomorphism in the closure of $\phi(\Omega)$. Otherwise, $\Phi(\phi_\lambda) = b \in \Omega$. Iterating Φ the way we iterate ϕ we find that

$$\Phi_n(\zeta) = \phi_n(\zeta), \qquad \zeta \in \Omega$$

and

$$\Phi_{n+1}(\phi_\lambda) = \phi_n(b)$$

But $\phi_n(b) \to \phi_\lambda$ in norm as $n \to \infty$ so that $\Phi_{n+1}(\phi_\lambda) \to \phi_\lambda$. However,

$$\Phi_{n+1}(\phi_\lambda) = \Phi(\Phi_n(\phi_\lambda)) \to \Phi(\phi_\lambda)$$

Thus, $\phi_\lambda = \Phi(\phi_\lambda)$, in contradiction to $\Phi(\phi_\lambda) = b$.

7.5. OPTIMAL ESTIMATION AND WIDTHS OF SPACES OF HOLOMORPHIC FUNCTIONS: PART I. THE H^∞ CASE

To set the stage for what follows we begin with an example. Let $z_1,\ldots, z_n$ be distinct points in the open unit disc Δ and let a be some other point in Δ. Our problem is to estimate the value $f(a)$ from the values $f(z_1),\ldots, f(z_n)$ where f is a function holomorphic in Δ and bounded there by 1, but otherwise unknown. What is the optimal way to do this and how much error can we expect, at the most? Thus, for each choice $c_1,\ldots, c_n$ of complex scalars we wish to compute

$$E(a; c_1,\ldots, c_n) = \sup\left\{\left|f(a) - \sum_{j=1}^{n} c_j f(z_j)\right| : \|f\|_\infty \leqslant 1\right\} \tag{5.1}$$

and then

$$E(a) = \inf\{E(a; c_1,\ldots, c_n) : c_1,\ldots, c_n \in \mathbb{C}\} \tag{5.2}$$

We begin by noting that

$$E(a; c_1,\ldots, c_n) = \sup\left\{\mathrm{Re}\left(f(a) - \sum_{j=1}^{n} c_j f(z_j)\right) : \|f\|_\infty \leqslant 1\right\}$$

so that

$$\begin{aligned} E(a) &= \inf_{c_1,\ldots,c_n} \sup_f \mathrm{Re}\left\{f(a) - \sum_1^n c_j f(z_j)\right\} \\ &= \sup_f \inf_{c_1,\ldots,c_n} \mathrm{Re}\left\{f(a) - \sum_1^n c_j f(z_j)\right\} \end{aligned}$$

by an application of von Neumann's "mini-max" principle; see Nikaido (1954). However,

$$\inf_{c_1,\ldots,c_n} \mathrm{Re}\left\{f(a) - \sum_1^n c_j f(z_j)\right\} = \begin{cases} -\infty & \text{if some } f(z_j) \neq 0 \\ \mathrm{Re}\, f(a) & \text{if all } f(z_j) = 0 \end{cases}$$

so that

$$\begin{aligned} E(a) &= \sup\{\mathrm{Re}\, f(a) : \|f\|_\infty \leqslant 1, f(z_j) = 0, j = 1,\ldots, n\} \\ &= \sup\{|f(a)| : \|f\|_\infty \leqslant 1, f(z_j) = 0, j = 1,\ldots, n\} \end{aligned}$$

If $f \in H^\infty$ is bounded by 1 and vanishes at $z_1, \ldots, z_n$ then $f = Bg$ where g is in H^∞ and is bounded by 1 and

$$B(z) = \prod_1^n \frac{z - z_j}{1 - \bar{z}_j z} \tag{5.3}$$

Thus,

$$E(a) = |B(a)| \tag{5.4}$$

Of course, the maximum error is attained when $f = B$. Thus, our problem is completely solved. But there's more.

Let's continue this line of reasoning a little longer. Suppose that $\mathbf{K}$ is a compact set in Δ; we denote by $\|u\|_{\mathbf{K}}$ the maximum of $|u(z)|$ as z varies over $\mathbf{K}$ for a continuous function u on $\mathbf{K}$. Let X_n be some subspace of $\mathbf{C(K)}$ of (complex) dimension n. We can then attempt to estimate each element f of the unit ball of $H^\infty(\Delta)$ by elements of X_n:

$$\inf_{u \in X_n} \|f - u\|_{\mathbf{K}}$$

and then to find the maximum deviation of the unit ball of $H^\infty(\Delta)$ from X_n:

$$d(A_\infty, X_n; \mathbf{C(K)}) = \sup_{f \in A_\infty} \inf_{u \in X_n} \|f - u\|_{\mathbf{K}} \tag{5.5}$$

where we have adopted the notation

$$A_\infty = \{f \in H^\infty(\Delta) : \|f\|_\infty \leqslant 1\}$$

For example, X_n might be the polynomials of degree $n - 1$ so that (5.5) then measures the maximum error made in estimating, uniformly on $\mathbf{K}$, a function f holomorphic and bounded by 1 on Δ by a polynomial of degree $n - 1$. Finally, we can seek an "optimal" subspace X_n; that is, we set

$$d_n(A_\infty, \mathbf{C(K)}) = \inf_{X_n} d(A_\infty, X_n; \mathbf{C(K)}) \tag{5.6}$$

where X_n is allowed to vary over all n dimensional subspaces of $\mathbf{C(K)}$. This number $d_n(A_\infty, \mathbf{C(K)})$ is called the n-width of A_∞ in $\mathbf{C(K)}$. In the context of a general Banach space $\mathbf{X}$ and a subset A of $\mathbf{X}$, the *n-width of A in* $\mathbf{X}$ is defined by

$$d_n(A; \mathbf{X}) = \inf_{X_n} \sup_{f \in \mathbf{A}} \inf_{u \in X_n} \|f - u\|_{\mathbf{X}} \tag{5.7}$$

In this section we shall investigate the n-width of A_∞ in $\mathbf{C(K)}$, as just indicated, both in the unit disc and in more general domains. In the next section we investigate the n-width of the unit ball of $H^2(\Delta)$ in certain L^2 spaces. All the investigations make use of properties of these Hardy spaces that have been derived in earlier chapters, and especially results on the role of Blaschke products as solutions of many extremal problems. We shall see as well that there are linear operators which associate with each $f \in A_\infty$ its "optimal" estimator, a result which is not at all obvious in the H^∞ setting.

To begin, let Ω be a domain bounded by $m + 1$ disjoint analytic simple closed curves, $m \geqslant 0$, and let $\mathbf{K}$ be a compact subset of Ω. Further, let μ be a positive measure on $\mathbf{K}$ of total mass 1. As before we let A_∞ denote the restriction to $\mathbf{K}$ of the unit ball of $H^\infty(\Omega)$. We shall compute the value of the n-width of A_∞ in $L^s(\mu)$ for $1 < s < \infty$; the value of the n-width of A_∞ in $\mathbf{C(K)}$ will follow by a limiting process.

We start with the upper bound since it is easier. It is also easier to do the investigation in $H^p(\Omega)$ for $1 < p < \infty$ since that space and the space $L^p(\Gamma, \omega)$ are strictly convex. Let A_p denote the closed unit ball of $H^p(\Omega)$ and let ζ be a point of $\mathbf{K}$. Finally, let $z_1, \ldots, z_n$ be any n distinct points of Ω. Then

$$E(\zeta) = \inf_{c_1, \ldots, c_n} \sup_{f \in A_p} \left| f(\zeta) - \sum_1^n c_j f(z_j) \right|$$

$$= \inf_{c_1, \ldots, c_n} \sup_{f \in A_p} \left| \int_\Gamma f \left[P_\zeta - \sum_1^n c_j P_{z_j} \right] d\omega \right|$$

where ω is harmonic measure on $\Gamma = \partial\Omega$ for some (fixed) point z_0 and P_z is the Radon-Nikodym derivative of ω_z with respect to ω. Thus,

$$E(\zeta) = \inf_{c_1, \ldots, c_n} \inf_{g \perp H^p} \left\| P_\zeta - \sum_1^n c_j P_{z_j} - g \right\|_{L^q(\Gamma, \omega)}$$

where q is the conjugate exponent to p. Thus,

$$E(\zeta) = \inf_{g \perp H^p} \inf_{c_1, \ldots, c_n} \left\| P_\zeta - \sum_1^n c_j P_{z_j} - g \right\|_{L^q(\Gamma, \omega)}$$

and this exactly the distance in $L^q/(H^p)^\perp$ from P_ζ to the finite-dimensional space spanned by $\{P_{z_1}, \ldots, P_{z_n}\}$. Since the Banach space $L^q/(H^p)^\perp$ is strictly convex there is a unique choice $c_1(\zeta), \ldots, c_n(\zeta)$ of scalars which gives the minimum distance and, further, $c_1(\zeta), \ldots, c_n(\zeta)$ vary continuously with ζ. Hence, there are n continuous functions $c_1(\zeta), \ldots, c_n(\zeta)$ on $\mathbf{K}$ such that

$$\sup_{f \in A_p} \left| f(\zeta) - \sum_1^n c_j(\zeta) f(z_j) \right| = E(\zeta)$$

However, as in the opening paragraph of this section, we also have

$$E(\zeta) = \sup\{|f(\zeta)| : f \in A_p, f(z_j) = 0, j = 1, \ldots, n\}$$

Let $g(z; a)$ be the Green's function for Ω with pole at a and set

$$B(z) = \exp\left[-\sum_1^n g(z; z_j) - i\left(\sum_1^n h(z; z_j)\right)\right]$$

Then B is an element of $MH^\infty(\Omega)$, in the notation of Chapter 5, and each $f \in A_p$ which vanishes at $z_1, \ldots, z_n$ has the form $f = Bg$ where $g \in MH^p(\Omega)$ and $\|g\|_p = \|f\|_p \leqslant 1$. Hence,

$$E(\zeta) \leqslant |B(\zeta)| \sup\{|g(\zeta)| : g \in MH^p(\Omega), \|g\|_p \leqslant 1\}$$

Let $\tilde{\mathbf{K}}$ be a compact set in Δ such that $T(\tilde{\mathbf{K}}) = \mathbf{K}$ where T is the uniformizer of Ω. If $h \in H^p(\Delta)$ and $\|h\|_p \leqslant 1$ then $|h(b)| \leqslant (1 - |b|^2)^{-1/p}$; see problem 1 at the end of this chapter. Hence,

$$|f(b)| \leqslant \left(1 - |b_0|^2\right)^{-1/p}$$

for all $f \in H^p(\Delta)$, $\|f\| \leqslant 1$, and all $b \in \tilde{\mathbf{K}}$ where b_0 is any point of $\tilde{\mathbf{K}}$ with the largest possible modulus. In particular, this estimate holds when $f = g \circ T$, $g \in MH^p(\Omega)$. Thus,

$$E(\zeta) \leqslant \left(1 - |b_0|^2\right)^{-1/p} |B(\zeta)|$$

Now the distance in $L^s(\mu)$ from A_p to the subspace W_n spanned by $c_1(\zeta), \ldots, c_n(\zeta)$ is at most the $L^s(\mu)$ norm of $E(\zeta)$. Hence,

$$\begin{aligned} \operatorname{dist}(A_p, W_n) &\leqslant \left(\int_K E(\zeta)^s \, d\mu(\zeta)\right)^{1/s} \\ &\leqslant \left(1 - |b_0|^2\right)^{-1/p} \left(\int_K |B(\zeta)|^s \, d\mu(\zeta)\right)^{1/s} \end{aligned}$$

However, A_∞ is a subset of A_p so that

$$\operatorname{dist}(A_\infty, W_n) \leqslant \left(1 - |b_0|^2\right)^{-1/p} \|B\|_{L^s(\mu)}$$

Let $p \to \infty$ in this inequality. We conclude that the distance in $L^s(\mu)$ from A_∞ to W_n is no larger than $\|B\|_s$. But B is an arbitrary Blaschke product of degree n (with distinct zeros) and the n-width of A_∞ in $L^s(\mu)$ is less than or equal to the distance of A_∞ to W_n. Thus, we arrive at the relation that the n-width of A_∞ in $L^s(\mu)$ is no larger than

$$\inf\{\|B\|_{L^s(\mu)} : B \in \mathcal{B}_n\} \tag{5.8}$$

where we introduce the notation $\mathcal{B}_n$

$$\mathcal{B}_n \text{ is the set of all Blaschke products of degree } n \text{ or less.} \tag{5.9}$$

This is the upper bound. The lower bound for the n-width of A_∞ in $L^s(\mu)$ is arrived at by employing the "antipodality" theorem of Borsuk which we state here in the form we need.

Theorem 5.1. Borsuk (1933). *Let F be a continuous mapping from the sphere $\mathbf{S}^m$ into $\mathbb{R}^m$ which is odd: $F(-x) = -F(x)$. Then F has a zero; that is, there is at least one point $x_0 \in \mathbf{S}^m$ with $F(x_0) = 0 \in \mathbb{R}^m$.*

For a proof of this theorem the reader is referred to (Krasnosel'ski 1964, chapter II, section 2).

If X_n is an n-dimensional subspace of $L^s(\mu)$ then each $f \in A_\infty$ has a unique element from X_n which is its best approximation from X_n. That is, if X_n has a basis $v_1, \ldots, v_n$, then there is a unique choice $c_1(f), \ldots, c_n(f)$ of scalars such that

$$\inf_{c_1, \ldots, c_n} \left\| f - \sum_1^n c_j v_j \right\|_{L^s(\mu)} = \left\| f - \sum_1^n c_j(f) v_j \right\|_{L^s(\mu)}$$

Further, the mapping $f \to (c_1(f), \ldots, c_n(f))$ is easily seen to be both continuous and odd, in the sense that $-f \to -(c_1(f), \ldots, c_n(f))$. We now construct another odd mapping, this one from a certain sphere into the set of single-valued Blaschke products of degree at most $n + m$.

Theorem 5.2. *There is a continuous odd mapping from $\mathbf{S}^{2n+1}$ into $\mathbf{C}(\mathbf{K})$ which takes values in the set of single-valued Blaschke products of degree at most $n + m$.*

Proof. Fix $n + 1$ distinct points $z_0, \ldots, z_n$ in Ω. Let $\mathbf{w} = (w_0, \ldots, w_n)$ be any $(n + 1)$-tuple of complex numbers with $\Sigma_0^n |w_j|^2 = 1$. Then we know from problem 5 of Chapter 5 that there is a unique (single-valued) Blaschke product $B_\mathbf{w}$ of degree at most $n + m$ such that

$$B_\mathbf{w}(z_j) = \frac{w_j}{\sigma(\mathbf{w})}$$

where

$$\sigma(\mathbf{w}) = \inf\{\|f\|_\infty : f \in H^\infty(\Omega), f(z_j) = w_j, 0 \leqslant j \leqslant n\}$$

and, further, $B_\mathbf{w}$ varies continuously on $\mathbf{K}$ with $\mathbf{w}$. Since $\sigma(-\mathbf{w}) = \sigma(\mathbf{w})$, we clearly have

$$B_{-\mathbf{w}} = -B_{\mathbf{w}}$$

so that the mapping $\mathbf{w} \to B_{\mathbf{w}}$ is both odd and continuous.

Consider now the mapping

$$F(\mathbf{w}) = \left(c_1(B_{\mathbf{w}}), \ldots, c_n(B_{\mathbf{w}})\right)$$

This mapping is both continuous and odd as the composition of two such mappings and so by Theorem 5.1, it has a zero. That is, there is a Blaschke product B_0 of degree at most $m + n$ whose best approximation from the subspace X_n is zero. Thus,

$$\begin{aligned} \|B_0\|_{L^s(\mu)} &= \operatorname{dist}(B_0, X_n) \\ &\leqslant \operatorname{dist}(A_\infty, X_n) \end{aligned}$$

Hence,

$$\inf\{\|B\|_{L^s(\mu)} : B \text{ a single-valued element of } \mathfrak{B}_{n+m}\} \tag{5.10}$$

is no larger than the distance from A_∞ to X_n in $L^s(\mu)$. Thus, the number in (5.10) is a lower bound for the n-width of A_∞ in $L^s(\mu)$.

We collect all the preceding in the next theorem.

Theorem 5.3. *The n-width of A_∞ in $L^s(\mu)$, $1 \leqslant s < \infty$, lies between the two numbers given in* (5.8) *and* (5.10), *respectively. This inequality holds as well if $L^s(\mu)$ is replaced throughout by* **C(K)**. *If $\Omega = \Delta$, then all the inequalities become equalities.*

Proof. All that remains to be shown are the cases $s = 1$ and **C(K)**. The former follows easily by letting s decrease to 1. The latter follows by making a special choice of μ so that

$$\lim_{s \to \infty} \|u\|_{L^s(\mu)} = \|u\|_{\mathbf{K}}$$

for each $u \in \mathbf{C(K)}$. For example, let $\{a_j\}$ be a countable dense set in **K** and let $\mu = \Sigma_1^\infty 2^{-j}\delta_j$ where δ_j is the unit point mass at a_j.

In the context of the open unit disc Δ we can obtain an explicit optimal estimation formula. We let **X** stand for either of $L^s(\mu)$ or **C(K)**. Let B^* be any Blaschke product of degree n with

$$\|B^*\|_{\mathbf{X}} \leqslant \|B\|_{\mathbf{X}}, \qquad B \in \mathfrak{B}_n$$

and let $z_1^*, \ldots, z_n^*$ be the zeros of B^*, listed according to multiplicity; we note that B^* has degree exactly n since otherwise $zB^*(z)$ lies in $\mathfrak{B}_n$ and has norm on **K** strictly less than that of B^*. Let $f \in H^\infty$; the expression

$$\frac{1}{2\pi i}\int \frac{B^*(z)}{B^*(\zeta)}\frac{f(\zeta)}{\zeta - z}\frac{1-|z|^2}{1-\zeta\bar{z}}\,d\zeta$$

can be evaluated by the residue theorem with the result that it equals

$$f(z) - \sum_{j=1}^{n} c_j(z) f(z_j^*)$$

for some functions $c_1, \ldots, c_n$ which are continuous (but not analytic) on Δ. Hence,

$$\left| f(z) - \sum_{1}^{n} c_j(z) f(z_j^*) \right| \leqslant |B^*(z)|$$

at each $z \in \Delta$ if we know that $\|f\|_\infty \leqslant 1$. Thus,

$$\sup_{f \in A_\infty} \left\| f - \sum_{1}^{n} c_j f(z_j^*) \right\|_{\mathbf{X}} \leqslant \|B^*\|_{\mathbf{X}}$$

But $\|B^*\|_{\mathbf{X}}$ is the value of the n-width of A_∞ in **X**. Thus, the rule

$$f \to \sum_{j=1}^{n} c_j(z) f(z_j^*), \qquad z \in \mathbf{K}$$

is an optimal method of estimating $f \in A_\infty$. Specially, the functions $c_j(z)$ are given

$$c_j(z) = B^*(z)\left(z_j^* - z\right)^{-1}\left(1 - z_j^*\bar{z}\right)^{-1}\left(B'(z_j^*)\right)^{-1}\left(1 - |z|^2\right)$$

if the zero at z_j^* is of order 1 and a similar, but more complicated expression if the zero at z_j^* has higher order. It is worthwhile making explicit here the point that if the zero z_j^* has multiplicity exactly r, $r > 1$, then

$$f\left(z_{j+s}^*\right) = f^{(s)}\left(z_j^*\right), \qquad s = 0, \ldots, r-1$$

We also make note of two more points here. If $\mathbf{K} = \{z\colon |B_0(z)| \leqslant \rho\}$ for some B_0 of exact degree N and some ρ, $0 < \rho < 1$, then B_0^r is the *unique* minimal Blaschke product of degree rN, at least in the sup norm; see problem 13 in this chapter. Further, in the unit disc Δ for either the sup norm or the $L^s(\mu)$ norm, the zeros of B^* lie within the non-Euclidean hull of **K**; see Walsh (1952).

The issue remains as to just how large the n-width is, now that we have bracketed it by two similar expressions. To answer this, we introduce another definition.

Definition. Let Ω be a domain with Green's function g and let $\mathbf{K}$ be a compact set in Ω. A non-negative measure μ on $\mathbf{K}$ is *admissible* if

$$\int_{\mathbf{K}} g(z,\zeta)\, d\mu(\zeta) \leqslant 1, \qquad z \in \Omega \tag{5.11}$$

Set

$$\mathrm{c}(\mathbf{K},\Omega) = \sup\{\mu(\mathbf{K}) : \mu \text{ admissible}\} \tag{5.12}$$

$\mathrm{c}(\mathbf{K},\Omega)$ is the *Green's capacity* of K relative to Ω.

The Green's capacity is very similar to the usual logarithmic capacity covered in Chapter 1. If μ_0 is a positive measure on $\mathbf{K}$ which is admissible and which satisfies $\mu_0(\mathbf{K}) = \mathrm{c}(\mathbf{K},\Omega)$, then the measure $\lambda_0 = \mu_0/\mathrm{c}(\mathbf{K},\Omega)$ is called an *equilibrium potential* for $\mathbf{K}$. We summarize some of the properties of $\mathrm{c}(\mathbf{K},\Omega)$ and λ_0 in the next proposition.

Proposition 5.4. (a) If $\{\mathbf{K}_n\}$ is a sequence of compact sets with $\mathbf{K}_1 \supset \mathbf{K}_2 \supset \cdots$ and $\mathbf{K} = \cap \mathbf{K}_n$, then

$$\lim \mathrm{c}(\mathbf{K}_n,\Omega) = \mathrm{c}(\mathbf{K},\Omega) \tag{5.13}$$

(b) Any equilibrium potential λ_0 is supported on the boundary of $\mathbf{K}$.

(c) If $\mathbf{K}$ is bounded by a finite number of disjoint analytic simple closed curves and if λ_0 is an equilibrium potential for $\mathbf{K}$, then

$$\int_{\mathbf{K}} g(z,\zeta)\, d\lambda_0(\zeta) = \frac{1}{\mathrm{c}(\mathbf{K},\Omega)} \quad \text{for all } z \in \operatorname{INT}\mathbf{K} \tag{5.14}$$

Proof. Only a sketch of the proof will be given since these results parallel those of Chapter 1, Section 7. (a) is obtained by considering the sequence $\{\mu_n\}$ of admissible measures with $\mu_n(\mathbf{K}_n) = \mathrm{c}(\mathbf{K}_n,\Omega)$. The measures have a weak-* cluster point μ which is necessarily an admissible measure on $\mathbf{K}$. Hence,

$$\lim \mathrm{c}(\mathbf{K}_n,\Omega) = \|\mu\| \leqslant \mathrm{c}(\mathbf{K},\Omega)$$

But clearly $\mathrm{c}(\mathbf{K}_n,\Omega) \geqslant \mathrm{c}(\mathbf{K},\Omega)$ for all n so that (a) must follow.

To prove (b) and (c) we assume that $\mathrm{c} = \mathrm{c}(\mathbf{K},\Omega)$ is positive. Let

$$\alpha = \inf\Big\{ \iint g(z,\zeta)\, d\lambda(z)\, d\lambda(\zeta) : \lambda \in \mathcal{P}(\mathbf{K})\Big\} \tag{5.15}$$

where $\mathcal{P}(\mathbf{K})$ is the set of probability measures on $\mathbf{K}$. The reader will have little difficulty establishing that $\alpha = 1/c$ and that a measure λ is an equilibrium potential if and only if λ satisfies

$$\frac{1}{c} = \alpha = \iint g(z,\zeta)\,d\lambda(z)\,d\lambda(\zeta) \tag{5.16}$$

If $\lambda \in \mathcal{P}(K)$, set

$$\mu_\lambda(z) = \int g(z,\zeta)\,d\lambda(\zeta)$$

Then u_λ is lower semicontinuous on Ω and harmonic on that part of Ω which is not in the closed support of λ. Further, we know that if λ satisfies (5.16) then $u_\lambda(z) \leqslant 1/c$ on Ω, and $u_\lambda = 1/c$ a.e. $d\lambda$.

Suppose **K** has interior and λ is an equilibrium potential with mass on some small disc $\mathcal{D}$ in the interior of **K**; let λ_0 be the restriction to $\mathcal{D}$ of λ and let λ_1 be the measure on $\partial\mathcal{D}$ defined by

$$\int_{\partial\mathcal{D}} v\,d\lambda_1 = \int_{\mathcal{D}} \tilde{v}\,d\lambda_0, \qquad v \in \mathbf{C}(\partial\mathcal{D}) \tag{5.17}$$

where $\tilde{v}$ is the harmonic extension to $\mathcal{D}$ of v. Then λ_1 is a non-negative measure on $\partial\mathcal{D}$ of the same mass as λ_0; further, if $z \notin \mathcal{D} \cup \partial\mathcal{D}$, then

$$\int_{\partial\mathcal{D}} g(z,\zeta)\,d\lambda_1(\zeta) = \int_{\mathcal{D}} g(z,\zeta)\,d\lambda_0(\zeta) \tag{5.18}$$

If $z \in \mathcal{D}$, then $g(z,\zeta) = -\log|z-\zeta| + w(\zeta)$ where w is harmonic on Ω. If the disc $\mathcal{D}$ is $\{z\colon |z-z_0| < r\}$ then for $\zeta \in \partial\mathcal{D}$ we have

$$|z-\zeta| = |r^2 - (z-z_0)(\bar{\zeta}-\bar{z}_0)|$$

and so

$$\log|z-\zeta| = \log|r^2 - (z-z_0)(\bar{\zeta}-\bar{z}_0)|$$

But $s(z) = -\log|r^2 - (z-z_0)(\bar{\zeta}-\bar{z}_0)|$ is harmonic on $\mathcal{D}$. Consequently,

$$\begin{aligned}\int_{\partial\mathcal{D}} g(z,\zeta)\,d\lambda_1(\zeta) &= \int_{\mathcal{D}} w(\zeta)\,d\lambda_0(\zeta) + \int_{\mathcal{D}} s(z)\,d\lambda_0(\zeta)\\ &< \int_{\mathcal{D}} w(\zeta)\,d\lambda_0(\zeta) - \int_{\mathcal{D}} \log|z-\zeta|\,d\lambda_0(\zeta)\\ &= \int_{\mathcal{D}} g(z,\zeta)\,d\lambda_0(\zeta)\end{aligned}$$

Let β be the measure which is λ on $\mathbf{K} \setminus \mathcal{D}$ and which is λ_1 on $\partial\mathcal{D}$. Then β is an equilibrium potential and

$$u_\beta(z) = u_\lambda(z), \qquad z \notin \mathcal{D} \tag{5.19a}$$

$$u_\beta(z) < u_\lambda(z), \qquad z \in \mathcal{D} \tag{5.19b}$$

Now consider the measure $\rho = \varepsilon\beta + (1 - \varepsilon)\lambda$. We have

$$\alpha \leqslant \iint g(z, \zeta)\, d\rho(z)\, d\rho(\zeta) \tag{5.20}$$

Expand the right-hand side of (5.20) as a function of ε and use the fact that both β and λ are equilibrium potentials. This yields

$$2\alpha \leqslant \int u_\beta(z)\, d\lambda(z) + \int u_\lambda(z)\, d\beta(z)$$

Hence,

$$\alpha = \int u_\beta(z)\, d\lambda(z) = \int u_\lambda(z)\, d\beta(z)$$

But according to (5.19b) $u_\beta(z) < \alpha$ on a set of positive λ-measure, a contradiction. Hence, λ has no mass on the interior of $\mathbf{K}$.

We now show that (5.14) holds if $\mathbf{K}$ is bounded by a finite number of disjoint analytic simple closed curves (it is not required that $\mathbf{K}$ be connected). Let μ be a positive measure on $\partial\mathbf{K}$ with

$$u(z) = \int g(z, \zeta)\, d\mu(\zeta) \leqslant 1, \qquad z \in \Omega$$

$$\|\mu\| = c = c(K, \Omega)$$

Let v be the function which is harmonic on $\Omega \setminus \mathbf{K}$ and continuous on $\partial(\Omega \setminus \mathbf{K}) = \partial\Omega \cup \partial\mathbf{K}$ with

$$v = \begin{cases} 1 & \text{on } \partial\mathbf{K} \\ 0 & \text{on } \partial\Omega \end{cases}$$

Then v is greater than or equal to u throughout the open set $\Omega \setminus \mathbf{K}$. Furthermore, if $z \in \Omega \setminus \mathbf{K}$, then by Green's formula

$$\int_{\partial\mathbf{K}} g(z, \zeta) \frac{\partial v}{\partial n}\, ds = \int_{\partial\mathbf{K}} \frac{\partial g(z, \zeta)}{\partial n}\, ds + v(z)$$

However, another application of Green's formula, this time in $\mathbf{K}$, gives

$$\int_{\partial\mathbf{K}} \frac{\partial g(z, \zeta)}{\partial n}\, ds = 0, \qquad z \in \Omega \setminus \mathbf{K}$$

Hence,

$$\int_{\partial\mathbf{K}} g(z,\zeta)\,d\mu(\zeta) = u(z) \leqslant v(z) = \int_{\partial\mathbf{K}} g(z,\zeta)\,d\beta(\zeta)$$

where

$$d\beta = \frac{\partial v}{\partial n}\,ds$$

Further, if $z \in \text{INT}\,\mathbf{K}$, then again by Green's formula

$$\int_{\partial\mathbf{K}} g(z,\zeta)\,d\beta(\zeta) = \int_{\partial\mathbf{K}} \frac{\partial g(z,\zeta)}{\partial n}\,ds = 1$$

so that β is an admissible measure. We must establish that β is an equilibrium potential. Let $\mathbf{K}_\varepsilon$ be the set on which v is no more than $1-\varepsilon$, and let the measure β_ε be defined on $\partial\mathbf{K}_\varepsilon$ by

$$d\beta_\varepsilon = \frac{\partial v}{\partial n}\,ds \quad \text{on } \partial\mathbf{K}_\varepsilon$$

Then, as above,

$$\int_{\partial\mathbf{K}_\varepsilon} g(z,\zeta)\,d\beta_\varepsilon(z) = \begin{cases} v(\zeta), & \zeta \in \text{INT}\,\mathbf{K}_\varepsilon \\ 1-\varepsilon, & \zeta \in \Omega\setminus\mathbf{K}_\varepsilon \end{cases}$$

Further, if h is a function continuous on a neighborhood of $\partial\mathbf{K}$, then

$$\int h\,d\beta_\varepsilon \to \int h\,d\beta, \qquad \varepsilon \to 0$$

This is true because v is actually harmonic across $\partial\mathbf{K}$. Thus,

$$\begin{aligned} \mathrm{c}(1-\varepsilon) &= \int_{\partial\mathbf{K}} (1-\varepsilon)\,d\mu(\zeta) = \int_{\partial\mathbf{K}}\int_{\partial\mathbf{K}_\varepsilon} g(z,\zeta)\,d\beta_\varepsilon(z)\,d\mu(\zeta) \\ &= \int_{\partial\mathbf{K}_\varepsilon}\int_{\partial\mathbf{K}} g(z,\zeta)\,d\mu(\zeta)\,d\beta_\varepsilon(\zeta) = \int_{\partial\mathbf{K}_\varepsilon} u(\zeta)\,d\beta_\varepsilon(\zeta) \\ &< \int_{\partial\mathbf{K}_\varepsilon} d\beta_\varepsilon(\zeta) \to \beta(\partial\mathbf{K}), \qquad \text{as } \varepsilon \to 0 \end{aligned}$$

Thus, β has mass c or more on $\partial\mathbf{K}$. Since it cannot have mass more than c it must be an equilibrium potential. Furthermore, the variational argument given earlier in the proof implies that

$$\int_{\partial \mathbf{K}} \int_{\partial \mathbf{K}} g(z, \zeta)\, d\mu(\zeta)\, d\beta(z) = \mathrm{c}$$

so that

$$\int_{\partial \mathbf{K}} g(z, \zeta)\, d\mu(\zeta) = 1 \quad \text{a.e. } d\beta$$

Hence, the boundary values of u are 1 a.e. ds on $\partial\mathbf{K}$ and so $u = v$ on $\Omega \setminus \mathbf{K}$ and $u \equiv 1$ in the set INT $\mathbf{K}$. This proves (5.14).

With the aid of Proposition 5.4 we can now establish the asymptotic (and more) behavior of the n-width of A_∞ in $\mathbf{C(K)}$.

Theorem 5.5. *Suppose Ω is bounded by $m + 1$ disjoint analytic simple closed curves. Then*

$$\lim_{n \to \infty} \left(d_n(A_\infty, \mathbf{C(K)})\right)^{1/n} = \exp\left[\frac{-1}{\mathrm{c}(\mathbf{K}, \Omega)}\right] \tag{5.21}$$

If $\mathbf{K}$ is of the special form

$$\mathbf{K} = \{z \in \Omega : |B_0(z)| \leqslant \rho\}$$

for some $B_0 \in \mathfrak{B}_N$ and some ρ, $0 < \rho < 1$, then there are constants a and b with

$$a \leqslant d_n(A_\infty, \mathbf{C(K)}) \exp\left[\frac{n}{\mathrm{c}(\mathbf{K}, \Omega)}\right] \leqslant b \tag{5.22}$$

In particular, if $\Omega = \Delta$ and $\mathbf{K} = \{z : |z| \leqslant \rho\}$, then

$$d_n(A_\infty, \mathbf{C(K)}) = \rho^n \tag{5.23}$$

Proof. Let B be any Blaschke product of degree r and let λ_0 be an equilibrium potential for $\mathbf{K}$. Then

$$\begin{aligned}
\log \|B\|_{\mathbf{K}} &\geqslant \log \|B\|_{L^1(\lambda_0)} \\
&\geqslant \int_{\mathbf{K}} \log |B(\zeta)|\, d\lambda_0(\zeta) \\
&= -\sum_{j=1}^{r} \int_{\mathbf{K}} g(\zeta, z_j)\, d\lambda_0(\zeta) \\
&\geqslant \frac{-r}{\mathrm{c}(\mathbf{K}, \Omega)}
\end{aligned}$$

by the inequality in (5.15). Thus, $\|B\|_{\mathbf{K}} \geqslant \exp[-r/c(\mathbf{K},\Omega)]$ and so we find that

$$d_n(A_\infty, C(\mathbf{K})) \geqslant \exp\left[\frac{-(m+n)}{c(\mathbf{K},\Omega)}\right] \tag{5.24}$$

by Theorem 5.3. To obtain an upper bound we must construct a Blaschke product B which is not too far away from $\exp[-n/c(\mathbf{K},\Omega)]$ on all of $\mathbf{K}$.

Let $\mathbf{K}'$ be a compact set bounded by a finite number of disjoint analytic simple closed curves which contains $\mathbf{K}$ in its interior and for which $c(\mathbf{K}',\Omega) < c(\mathbf{K},\Omega) + \delta$. Let λ'_0 be an equilibrium potential for $\mathbf{K}'$ so that

$$\int_{\partial \mathbf{K}'} g(z,\zeta)\, d\lambda'_0(\zeta) = \frac{1}{c(\mathbf{K}',\Omega)} \qquad \text{for all } z \in \operatorname{INT} \mathbf{K}'$$

Let $\{\sigma_n\}$ be a sequence of measures on $\partial\mathbf{K}'$ which converge weak-* in the space of measures to λ'_0, each σ_n of the form

$$\sigma_n = \frac{1}{n}\sum_{j=1}^{n} \delta_{jn}$$

where δ_{jn} is the unit point mass at the point $\zeta_{jn} \in \partial\mathbf{K}'$, $j = 1,\dots,n$ and $n = 1,2,\dots$. Then the functions

$$u_n(z) = \int_{\partial \mathbf{K}'} g(z,\zeta)\, d\sigma_n(\zeta)$$

are harmonic on $\operatorname{INT}\mathbf{K}'$ and converge uniformly on each compact set in $\operatorname{INT}\mathbf{K}'$ to the constant function $1/c(\mathbf{K}',\Omega)$. In particular, this happens on $\mathbf{K}$ so that

$$\int_{\partial \mathbf{K}'} g(z,\zeta)\, d\sigma_n(\zeta) - \frac{1}{c(\mathbf{K}',\Omega)}$$

lies in the interval $(-\varepsilon, \varepsilon)$ for all $n \geqslant N$. However,

$$n\int_{\partial \mathbf{K}'} g(z,\zeta)\, d\sigma_n(\zeta) = \sum_{j=1}^{n} g(z,\zeta_{jn})$$

so that if we let B_n be the Blaschke product with zeros at $\{\zeta_{1n},\dots,\zeta_{nn}\}$ then we find that

$$\log|B_n(z)| \leqslant -\frac{n}{c(\mathbf{K}',\Omega)} + n\varepsilon, \qquad z \in \mathbf{K}$$

Thus,

$$\|B_n\|_{\mathbf{K}}^{1/n} \leqslant \exp\left[-\frac{1}{c(\mathbf{K}',\Omega)} + \varepsilon\right]$$

so that Theorem 5.3 implies

$$\limsup_{n\to\infty} \left(d_n(A_\infty, \mathbf{C}(\mathbf{K}))\right)^{1/n} \leqslant \exp\left[-\frac{1}{c(\mathbf{K}', \Omega)} + \varepsilon\right]$$

Now let ε decrease to 0 and $\mathbf{K}'$ shrink down to $\mathbf{K}$ to obtain (5.21).

To obtain (5.22) we note that (5.24) shows

$$d_n(A_\infty, \mathbf{C}(\mathbf{K}))\exp\left[\frac{n}{c(\mathbf{K}, \Omega)}\right] \geqslant \exp\left[-\frac{m}{c(\mathbf{K}, \Omega)}\right] \tag{5.25}$$

For the upper bound let λ_0 be an equilibrium potential for $\mathbf{K}$; recall that λ_0 is concentrated on $\partial\mathbf{K}$ which here is the set where $|B_0| = \rho$. Hence, for any integer r we have

$$\begin{aligned} d_{rN}(A_\infty, \mathbf{C}(\mathbf{K})) &\leqslant \inf\{\|B\|_{\mathbf{K}} : B \in \mathfrak{B}_{rN}\} \\ &\leqslant \|B_0\|_{\mathbf{K}}^r = \rho^r \end{aligned}$$

and

$$\begin{aligned} \log\rho = \int_{\partial\mathbf{K}} \log|B_0|\, d\lambda_0 &= -\sum_{j=1}^{N} \int g(\zeta, z_j)\, d\lambda_0(\zeta) \\ &= -\frac{N}{c(\mathbf{K}, \Omega)} \end{aligned}$$

since all the zeros of B_0 clearly lie in the interior of $\mathbf{K} = \{z : |B_0(z)| \leqslant \rho\}$. Thus,

$$d_{rN}(A_\infty, \mathbf{C}(\mathbf{K})) \leqslant \exp\left[-\frac{rN}{c(\mathbf{K}, \Omega)}\right]$$

If n is any positive integer, write $n = rN + q$ where $0 \leqslant q < N$. Thus,

$$\begin{aligned} d_n(A_\infty, \mathbf{C}(\mathbf{K}))\exp\left[\frac{n}{c(\mathbf{K}, \Omega)}\right] &\leqslant d_{rN}(A_\infty, \mathbf{C}(\mathbf{K}))\exp\left[\frac{n}{c(\mathbf{K}, \Omega)}\right] \\ &\leqslant \exp\left[\frac{q}{c(\mathbf{K}, \Omega)}\right] \\ &< \exp\left[\frac{N}{c(\mathbf{K}, \Omega)}\right] \end{aligned}$$

The last conclusion of Theorem 5.5 follows directly from (5.22) since for $\Omega = \Delta$ and $\mathbf{K} = \{|z| \leqslant \rho\}$ we have $m = 0$ and $B_0(z) = z$, respectively so that

$a = 1$ and $q = 0$, in the notation of the preceding paragraph. Thus, $b = 1$, as well.

7.6. OPTIMAL ESTIMATION AND WIDTHS OF SPACES OF HOLOMORPHIC FUNCTIONS: PART 2. THE H^2 CASE

This section is devoted to results on the n-width of the unit ball of $H^2(\Delta)$ in the space $L^2(\mu)$. Here the results resemble the results obtained in Section 5 for H^∞ but are more detailed in some ways since both $H^2(\Delta)$ and $L^2(\mu)$ are Hilbert spaces. We begin with a very general result which shows that certain n-width problems in a Hilbert space context can be reduced to eigenvalue problems.

Theorem 6.1. *Let $\mathcal{H}_1$ and $\mathcal{H}_2$ be two Hilbert spaces over the complex numbers and let L be a compact linear operator mapping $\mathcal{H}_1$ into $\mathcal{H}_2$ with adjoint L^*. Let $\mathcal{U}$ be the closed unit ball of $\mathcal{H}_1$. Then the n-width of $L(\mathcal{U})$ in $\mathcal{H}_2$ is the square root of the $n+1$st eigenvalue of the self-adjoint, compact operator LL^*.*

Proof. We have

$$\begin{aligned}
\left(d_n(L(\mathcal{U}), \mathcal{H}_2)\right)^2 &= \inf_{X_n} \sup_{y \in L(\mathcal{U})} \inf_{x \in X_n} \|x - y\|^2 \\
&= \inf_{X_n} \sup_{z \in \mathcal{U}} \sup_{\substack{w \perp X_n \\ \|w\| \leqslant 1}} |\langle w, L(z)\rangle|^2 \\
&= \inf_{X_n} \sup_{\substack{w \perp X_n \\ \|w\| \leqslant 1}} \sup_{z \in \mathcal{U}} |\langle L^* w, z\rangle|^2 \\
&= \inf_{X_n} \sup_{\substack{w \perp X_n \\ \|w\| \leqslant 1}} \|L^* w\|^2 \\
&= \inf_{X_n} \sup_{\substack{w \perp X_n \\ \|w\| \leqslant 1}} \langle L^* w, L^* w\rangle \\
&= \inf_{X_n} \sup_{\substack{w \perp X_n \\ \|w\| \leqslant 1}} \langle w, LL^* w\rangle \\
&= n + 1\text{st eigenvalue of } LL^*
\end{aligned}$$

The last equality is called the Courant-Hilbert mini-max principle and is very simple. Let $\lambda_1 \geqslant \cdots \geqslant \lambda_n \geqslant \lambda_{n+1} \geqslant \cdots > 0$ be the eigenvalues of the compact self-adjoint operator LL^* and let $v_1, v_2, \ldots$ be associated eigenvectors. If we take X_n to be the span of the first n eigenvectors $v_1, \ldots, v_n$, then for any unit vector w orthogonal to X_n we have

$$w = \sum_{n+1}^{\infty} c_j v_j, \qquad \sum_{n+1}^{\infty} |c_j|^2 = 1$$

thus

$$\langle w, LL^*w \rangle = \sum_{n+1}^{\infty} |c_j|^2 \lambda_j \leqslant \lambda_{n+1}$$

Conversely, if X_n is any n-dimensional subspace, then there is a choice of scalars $c_1, \ldots, c_{n+1}$ such that $w = \Sigma_1^{n+1} c_j v_j$ has norm 1 and is orthogonal to X_n. Thus,

$$\langle w, LL^*w \rangle = \sum_{1}^{n+1} |c_j|^2 \lambda_j \geqslant \lambda_{n+1}$$

We now employ Theorem 6.1 to begin to get a grip on the n-width problem when the Hilbert space $\mathcal{H}_1$ is $H^2(\Delta)$ and the Hilbert space $\mathcal{H}_2$ is $L^2(\mu)$, μ a positive measure of mass 1 on a compact set $\mathbf{K}$ within Δ. The operator L is given by restricting an element $h \in H^2(\Delta)$ to $\mathbf{K}$ or, more specifically, if we view $H^2(\Delta)$ as a subspace of $L^2(\mathbf{T}, d\theta)$, then

$$(Lh)(w) = \frac{1}{2\pi i} \int_{\mathbf{T}} \frac{h(\zeta)}{\zeta - w} d\zeta$$

$$= \frac{1}{2\pi} \int_0^{2\pi} h(e^{i\theta})(1 - we^{-i\theta})^{-1} d\theta$$

Hence, L^*, which maps $L^2(\mu)$ into H^2, is given by

$$(L^*u)(e^{i\theta}) = \int_{\mathbf{K}} u(w)(1 - \bar{w}e^{i\theta})^{-1} d\mu(w), \qquad u \in L^2(\mu)$$

Consequently, we seek solutions of the eigenvalue problem

$$\lambda f(z) = \int_{\mathbf{K}} \frac{f(w)}{1 - \bar{w}z} d\mu(w), \qquad z \in \Delta \tag{6.1}$$

The major results on solutions of (6.1) are contained in the following theorem.

Theorem 6.2. *The eigenvalues for the eigenvalue problem* (6.1) *form a strictly decreasing sequence* $\lambda_0 > \lambda_1 > \lambda_2 > \cdots > 0$. *The eigenspace corresponding to the eigenvalue* λ_n *is* 1*-dimensional and is spanned by a function* f_n *that has exactly* *n zeros in Δ, $n = 0, 1, 2, \ldots$. Furthermore, there is a constant $\nu > 0$ depending only on* **K** *such that*

$$|f_n(e^{i\theta})| \geqslant \nu, \qquad 0 \leqslant \theta \leqslant 2\pi, \qquad n = 0, 1, 2, \ldots \tag{6.2}$$

if f_n is normalized to have $H^2(\Delta)$ norm 1.

Proof. Let f satisfy (6.1) and let g be any element of $H^2(\Delta)$. Then

$$\frac{\lambda}{2\pi}\int_{\mathbf{T}} f\bar{g}\, d\theta = \int_{\mathbf{K}} f(w)\bar{g}(w)\, d\mu(w) \tag{6.3}$$

In (6.3) set $g(z) = f(z)(1 + \bar{a}z)(1 - \bar{a}z)^{-1}$ for $a \in \Delta$; then take the real part of both sides of the resulting equation. The conclusion is

$$\frac{\lambda}{2\pi}\int_{\mathbf{T}} |f(e^{i\theta})|^2 \frac{1 - |a|^2}{|e^{i\theta} - a|^2}\, d\theta = \int_{\mathbf{K}} |f(w)|^2 \operatorname{Re}\left(\frac{1 + \bar{a}w}{1 - \bar{a}w}\right) d\mu(w) \tag{6.4}$$

In (6.4) let a approach e^{it} nontangentially; we find that

$$\lambda |f(e^{it})|^2 = \int_{\mathbf{K}} |f(w)|^2 \operatorname{Re}\left(\frac{e^{it} + w}{e^{it} - w}\right) d\mu(w)$$

However, there is constant δ depending only on **K** such that

$$\operatorname{Re}\left(\frac{e^{it} + w}{e^{it} - w}\right) \geqslant \delta, \qquad w \in \mathbf{K}, \qquad 0 \leqslant t \leqslant 2\pi$$

Hence,

$$\lambda |f(e^{it})|^2 \geqslant \delta \int_{\mathbf{K}} |f(w)|^2\, d\mu(w)$$

$$= \delta\lambda$$

if f is normalized to have $H^2(\Delta)$ norm 1. This proves (6.2) with $\nu = (\delta)^{1/2}$. Furthermore, if g is another (normalized) eigenfunction for the eigenvalue λ then $f + cg$ is an eigenfunction for each scalar c and clearly this new function can be made to vanish on **T** at some point if g is not a multiple of f. Consequently, the eigenspaces are one-dimensional.

To show that the eigenfunction f_n corresponding to the eigenvalue λ_n, $n = 0, 1, \ldots$, has exactly n zeros in Δ we begin by establishing it for a special choice of **K** and μ; namely **K** is the circle $\mathbf{C} = \{z : |z| = r\}$ and μ is Lebesgue measure on **C**.

The eigenvalue equation (6.1) becomes

$$\lambda_n \sum_{j=0}^{\infty} a_j^{(n)} z^j = \sum_{j=0}^{\infty} z^j r^{2j} a_j^{(n)}$$

where we have written $f_n(z) = \sum_{j=0}^{\infty} a_j^{(n)} z^j$. It follows that

$$\lambda_n = r^{2n} \qquad\qquad n = 0, 1, 2, \ldots \tag{6.5a}$$

$$f_n(z) = z^n \tag{6.5b}$$

and so the result is established here. Now let μ be an arbitrary positive measure of mass 1 on the compact set **K** and let μ_0 be Lebesgue measure on **C**, a circle of radius r centered at the origin. Set $\mu_t = (1-t)\mu_0 + t\mu$ and let $f_{n,t}$ be the normalized nth eigenfunction for μ_t with associated eigenvalue $\lambda_{n,t}$. Let L_t be the linear operator

$$(L_t f)(z) = \int_{\mathbf{K}} \frac{f(\zeta)}{1 - \bar{\zeta} z} \, d\mu_t(\zeta)$$

Then

$$\langle (L_t - L_s) f, f \rangle_{H^2} = (t - s) \int_{\mathbf{K}} |f(\zeta)|^2 \, d\nu(\zeta)$$

where $\nu = \mu - \mu_0$. By the "principle of majorization" (see Weinstein and Stenger, 1972) we have

$$|\lambda_{n,t} - \lambda_{n,s}| \leqslant C|t - s|, \quad C = \text{constant}$$

and so $\lambda_{n,t}$ is a continuous function of t. It is a straightforward matter to prove that $f_{n,t} \to f_{n,s}$ uniformly on compact sets in Δ as $t \to s$. Since $|f_{n,t}(e^{i\theta})| \geqslant \nu$ for θ, $0 \leqslant \theta \leqslant 2\pi$, and all t, $0 \leqslant t \leqslant 1$, we may apply Rouche's Theorem and conclude that each $f_{n,t}$ has exactly n zeros in Δ; in particular, $f_{n,1}$ has this property.

Definition. For a positive measure β on **K** let

$$\lambda_0(\beta) > \lambda_1(\beta) > \cdots > 0 \tag{6.6}$$

denote the eigenvalues described in Theorem 6.2.

Theorem 6.3. *Let μ be a positive measure on* **K**. *Then*

$$\lambda_n(\mu) = \min\{\lambda_0(|B|^2 \, d\mu) : B \in \mathcal{B}_n\} \tag{6.7}$$

Furthermore, the minimum is attained when B is the Blaschke product corre-

sponding to the zeros $z_1^*, \ldots, z_n^*$ of the function f_n described in Theorem 6.2. Finally, there are functions $M_1, \ldots, M_n$ in $H^2(\Delta)$ such that the rule

$$g \to \sum_{j=1}^{n} M_j(\zeta) g(z_j^*), \qquad g \in H^2(\Delta)$$

is an optimal method for estimating g in $L^2(\mu)$:

$$\sup_{\|g\| \leqslant 1} \left\| g - \sum_1^n M_j g(z_j^*) \right\|_{L^2(\mu)} = d_n(H^2; L^2(\mathbf{K}, \mu)) \tag{6.8}$$

Proof. We begin by showing that the left-hand side in (6.7) exceeds the right-hand side. Recall from Theorem 5.2 that there is a continuous odd map from the sphere $\mathbf{S}^{2n+1}$ into the set $\mathfrak{B}_n$; let us denote this map by $\mathbf{w} \to B_{\mathbf{w}}$. By Theorem 6.2 applied to the measure $|B_{\mathbf{w}}|^2 \, d\mu$, there is a unique zero-free function $g_{\mathbf{w}}$ in the unit ball of $H^2(\Delta)$ such that

$$\int_{\mathbf{K}} |g_{\mathbf{w}}|^2 |B_{\mathbf{w}}|^2 \, d\mu = \lambda_0\left(|B_{\mathbf{w}}|^2 \, d\mu\right) \tag{6.9a}$$

$$g_{\mathbf{w}}(0) > 0 \tag{6.9b}$$

Let $\ell_1, \ldots, \ell_n$ be any n continuous linear functionals on $H^2(\Delta)$. Then the mapping

$$\mathbf{w} \to \left(\ell_1(B_{\mathbf{w}} g_{\mathbf{w}}), \ldots, \ell_n(B_{\mathbf{w}} g_{\mathbf{w}})\right)$$

is both odd and continuous from $\mathbf{S}^{2n+1}$ into $\mathbf{C}^n$ and so has a zero, by Borsuk's antipodality theorem. Hence, there is a Blaschke product $B_0 \in \mathfrak{B}_n$ and an associated function g_0 of unit norm in $H^2(\Delta)$ with $\ell_j(B_0 g_0) = 0$ for $j = 1, \ldots, n$. Thus,

$$\inf\left\{\lambda_0\left(|B|^2 \, d\mu\right) : B \in \mathfrak{B}_n\right\} \leqslant \lambda_0\left(|B_0|^2 \, d\mu\right)$$

$$= \int_{\mathbf{K}} |g_0|^2 |B_0|^2 \, d\mu$$

But this last number is less than or equal to

$$\sup\left\{\|h\|^2_{L^2(\mathbf{K}, \mu)} : \|h\|_{H^2} \leqslant 1, \ell_j(h) = 0, \; j = 1, \ldots, n\right\}$$

Since $\ell_1, \ldots, \ell_n$ are arbitrary we conclude that the square of the Gelfand n-width of H^2 in $L^2(\mathbf{K}, \mu)$ is larger than $\inf\{\lambda_0(|B|^2 \, d\mu) : B \in \mathfrak{B}_n\}$. But the Gelfand width is the same as the Kolomorgorov width in a Hilbert space; see problem 16. This establishes the fact that the left-hand side in (6.7) exceeds the right.

On the other hand, if $B \in \mathfrak{B}_n$ has zeros at $z_1, \ldots, z_n$ then

$$\lambda_0(|B|^2\,d\mu) = \sup\left\{\int_{\mathbf{K}} |h|^2|B|^2\,d\mu : \|h\|_{H^2} \leqslant 1\right\}$$

$$= \sup\left\{\int_{\mathbf{K}} |g|^2\,d\mu : \|g\|_{H^2} \leqslant 1,\ g(z_j) = 0,\ j = 1, \ldots, n\right\}$$

$$\geqslant \inf_{\ell_1, \ldots, \ell_n} \sup\{\|g\|^2_{L^2(\mathbf{K},\mu)} : \|g\|_{H^2} \leqslant 1, \ell_j(g) = 0,\ j = 1, \ldots, n\}$$

$$= d_n^2(H^2; L^2(\mu)) = \lambda_n$$

(Again, see problem 16.) This establishes the equality in (6.7). Indeed, it does more for it shows that the equality

$$\lambda_n = \sup\left\{\int_{\mathbf{K}} |h|^2\,d\mu : \|h\|_{H^2} \leqslant 1,\ h(z_j^*) = 0,\ j = 1, \ldots, n\right\} \tag{6.10}$$

is valid.

Let $\ell_1, \ldots, \ell_n$ be the linear functionals defined by $\ell_j(h) = h(z_j^*), j = 1, \ldots, n$, and let Y be the intersection of their kernels; that is, Y consists of all functions f in H^2 which vanish at $z_1^*, \ldots, z_n^*$, counting multiplicities. Let $M_1, \ldots, M_n$ be a basis for the orthogonal complement in H^2 of Y. $M_1, \ldots, M_n$ may be chosen so that $M_k(z_j^*) = \delta_{jk}$, for $j, k = 1, \ldots, n$. Then

$$h \to \sum_1^n h(z_j^*)M_j$$

is the orthogonal projection of H^2 onto the subspace X spanned by $M_1, \ldots, M_n$ and so if $\|h\| \leqslant 1$ we have by (6.10)

$$\lambda_n \geqslant \lambda_n \left\| h - \sum_1^n h(z_j^*)M_j \right\|_{H^2}^2$$

$$\geqslant \int_{\mathbf{K}} \left| h(\zeta) - \sum_{j=1}^n h(z_j^*)M_j(\zeta) \right|^2 d\mu(\zeta)$$

which gives (6.8).

If the zeros $z_1^*, \ldots, z_n^*$ of B are distinct, then we can take

$$M_j(\zeta) = \prod_{\substack{k=1\\k\neq j}}^{n} \frac{(\zeta - z_k^*)(1 - \bar{z}_k^* z_j)}{(1 - \bar{z}_k^*\zeta)(z_j - z_k^*)},$$

$$= \left(\frac{B(\zeta)}{\zeta - z_j^*}\right)\left(B'(z_j^*)\right)^{-1} \qquad j = 1, \ldots, n$$

and similar expressions if the zeros are of higher multiplicity.

The last topic to be covered is the asymptotic behavior of the n-width, equivalently of λ_n, as $n \to \infty$. Here the result is valid for measures μ which are closely related to the equilibrium potential discussed in Section 5.

Theorem 6.4. *Let μ_0 be an equilibrium potential for the set $\mathbf{K}$ and let $d\mu = w\, d\mu_0$ where w is a non-negative function in $L^1(\mu_0)$ and* log w *also is in $L^1(\mu_0)$. Let c be the Green's capacity of $\mathbf{K}$ relative to Δ. Then*

$$\lim_{n\to\infty} \left[\lambda_n(w\, d\mu_0)\right]^{1/n} = \exp\left[\frac{-2}{\mathrm{c}}\right]$$

Proof. We assume with no loss of generality that the mass of μ is precisely 1. If B is any Blaschke product of degree n, then

$$\log\left(\int |B(\zeta)|^2\, d\mu(\zeta)\right) \geq 2\int \log|B|\, d\mu_0 + \int \log|w|\, d\mu_0$$

$$\geq -\frac{2n}{\mathrm{c}} + A$$

where $A = \int \log|w|\, d\mu_0$. Thus,

$$\int |B|^2\, d\mu \geq C\exp\left[-\frac{2n}{\mathrm{c}}\right]$$

for a constant C independent of B and n. By Theorem 6.3 we find

$$\lambda_n \geq C\exp\left[-\frac{2n}{\mathrm{c}}\right]$$

and so we obtain the desired lower bound.

As for the upper bound, we make use of the function B_n constructed in the course of the proof of Theorem 5.5. This Blaschke product B_n has degree n and satisfies, for n large enough,

$$|B_n(z)| \leq \exp\left[-n\left((\mathrm{c} + \delta)^{-1} - \varepsilon\right)\right], \qquad z \in \mathbf{K},$$

for arbitrary positive ε and δ. Again applying Theorem 6.3 we find

$$\lambda_n \leqslant \exp\left[-n\left((c+\delta)^{-1}-\varepsilon\right)\right]$$

so that

$$\limsup_{n\to\infty} \lambda_n^{1/n} \leqslant \exp\left[-(c+\delta)^{-1}+\varepsilon\right]$$

and by letting ε and δ decrease to zero we get the desired upper bound.

ADDITIONAL READINGS AND NOTES

The material in Section 1 through Corollary 1.5 is from Forelli (1964), adapted when necessary to the context of a finitely connected domain. Theorems 1.6 and 1.7 are from Hoffman (1962). Iterates of self-mappings of a domain received attention from Julia (1918) and Wolff (1926); Theorem 2.1 is from Heins (1941) and Theorem 2.2 is Wolff's result (1926). The material that leads up to Theorem 2.3 and the proof of Theorem 2.3 are modeled on the presentation in Ahlfors (1973) and Carathéodory (1960). Early results on compact composition operators are in Caughran and Schwartz (1975), Caughran (1971), and Shapiro and Taylor (1973). Theorem 3.2 is from Fisher (1983) and extends results of Caughran. Also see Kamowitz (1975). Everything in Section 4 is from the paper of Swanton (1976). Likewise, the material in Sections 5 and 6 comes from two papers of Fisher and Micchelli (1980) and (1983). Earlier work, covering many of the results in Section 5, was done by Widom (1972). Further information on n-widths and optimal estimation can be found in the book edited by Micchelli and Rivlin (1977) and the references therein, especially in the lead article.

EXERCISES

1. Let Δ be the open unit disc and let $1 \leqslant p \leqslant \infty$. Show that the norm of the linear functional

$$h \to h(a), \qquad h \in H^p(\Delta), \qquad a \in \Delta$$

is precisely $(1-|a|^2)^{-1/p}$. HINT: let $F(z) = (1-|a|^2)^{1/p}(1-\bar{a}z)^{-2/p}$; then $\|F\|_p = 1$ while $F(a) = (1-|a|^2)^{-1/p}$. Thus, the norm is at least $(1-|a|^2)^{-1/p}$. Reduce the general case to the case $p = 2$ by noting that any extremal is outer and then by taking roots. For $p = 2$, the extremal is just the Cauchy kernel appropriately normalized.

2. Let Ω be finitely connected and let $1 \leqslant p < \infty$. For a point $\zeta \in \Omega$ let L_ζ be the linear functional

$$L_\zeta(h) = h(\zeta), \qquad h \in H^p(\Omega)$$

Show $\|L_\zeta\| \to \infty$ as $\zeta \to \Gamma = \partial\Omega$.

3. Let $\phi : \Delta \to \Delta$. Show that the norm of C_ϕ as a linear operator mapping $H^p(\Delta)$ into $H^p(\Delta)$ is no more than

$$\left[\frac{1+|a|}{1-|a|}\right]^{1/p}, \qquad a = \phi(0); \text{ see Ryff (1966)}$$

Show equality holds if ϕ is inner; see Nordgren (1968).

4. Suppose Ω is bounded by a finite number of disjoint analytic simple closed curves. Let ϕ be an analytic function mapping Ω into itself and let ϕ denote as well, the boundary values of ϕ on $\Gamma = \partial\Omega$. If C_ϕ is compact on some $H^p(\Omega)$, $1 \leqslant p < \infty$, show that $\phi(\zeta) \in \Omega$ for almost all $\zeta \in \Gamma$ (with respect to harmonic measure).

5. A linear operator T mapping a Hilbert space $\mathcal{H}$ into itself is *Hilbert-Schmidt* if $\mathcal{H}$ has an orthonormal basis $(e_n\}$ such that

$$\sum_{j=1}^{\infty} \|Te_n\|^2 < \infty$$

Let ϕ be analytic function mapping Δ into Δ. Show that the composition operator C_ϕ is Hilbert-Schmidt on $H^2(\Delta)$ if and only if

$$\int_0^{2\pi} (1 - |\phi(e^{it})|)^{-1}\, dt < \infty$$

(HINT: set $e_n = z^n$ and compute $\Sigma_{n=0}^{\infty}\|Te_n\|^2$); see Shapiro and Taylor (1973).

6. Let ϕ be an analytic function mapping Δ into itself. Show that if ϕ has a (finite) angular derivative at some point of the unit circle, then C_ϕ is not compact on $H^2(\Delta)$ and so is not compact on any $H^p(\Delta)$; see Shapiro and Taylor (1973).

7. Let ϕ be a nonconstant analytic function mapping a domain Ω into itself and let T be the uniformizer of Ω, $T: \Delta \to \Omega$. Show that there is an analytic function ψ mapping Δ into itself so that $T \circ \psi = \phi \circ T$. A point $a \in \Omega$ is an attractive fixed point of ϕ if and only if each $b \in \Delta$ with $Tb = a$ is an attractive fixed point of ψ.

8. Let ϕ, T, and ψ be as in problem 7. Show that the composition operator C_ϕ induced by ϕ on $H^p(\Omega)$ can be lifted to the disc to act on the space

$H^p/\mathfrak{G}$ in the sense that C_ψ maps $H^p/\mathfrak{G}$ into $H^p/\mathfrak{G}$ and

$$(C_\phi(g))(Tz) = C_\psi(g(Tz)), \qquad z \in \Delta, \qquad g \in H^p(\Omega)$$

Show C_ϕ is compact on $H^p(\Omega)$ if and only if C_ψ is compact on $H^p/\mathfrak{G}$.

9. Let $\phi(z) = \frac{1}{2}(1 + z)$. Show that C_ϕ is not compact as an operator from $H^2(\Delta)$ into $H^2(\Delta)$.
10. Let D be the triangle in Δ with vertices at 1 and $\pm i/2$. Let ϕ be the Riemann mapping of Δ onto D. Show that C_ϕ is a compact operator on $H^2(\Delta)$; indeed, C_ϕ is a Hilbert-Schmidt operator on $H^2(\Delta)$.
11. Let Ω be bounded by a finite number of disjoint analytic simple closed curves and let ϕ be a nonconstant analytic function mapping Ω into Ω. Show C_ϕ is compact on $H^\infty(\Omega)$ if and only if the range of ϕ lies in some compact subset of Ω.
12. Let Ω be bounded by a finite number of disjoint analytic simple closed curves and let $\mathbf{A}(\Omega)$ denote those functions continuous on $\Omega \cup \partial\Omega$ and analytic on Ω. Find all the isometries of $\mathbf{A}(\Omega)$ *onto* $\mathbf{A}(\Omega)$. HINT: use Theorem 1.6 and problem 16 from Chapter 6.
13. Let Ω be bounded by a finite number of disjoint analytic simple closed curves and let B_0 be a single-valued Blaschke product on Ω of degree N. Set $\mathbf{K} = \{z : |B_0(z)| \leqslant \rho\}$ for $\rho \in (0, 1)$. If B is a single-valued Blaschke product of degree N or less and $|B(z)| \leqslant |B_0(z)|$ for all $z \in \mathbf{K}$, show that $B = \lambda B_0$ for some unimodular constant λ.
14. Let Ω be a domain for which $H^2(\Omega)$ is nontrivial; fix a point $a \in \Omega$. Let

$$\gamma_m = \sup\{|f^{(m)}(a)| : f \in H^2(\Omega), \|f\|_2 \leqslant 1, f^{(k)}(a) = 0, 0 \leqslant k < m\}$$

a. Show there is a unique $f_m \in H^2(\Omega)$ with $\|f_m\|_2 = 1$ and $f_m^{(m)}(a) = \gamma_m$.
b. Show the functions 1 and $\{f_m\}_{m=1}^\infty$ form an orthonormal basis for $H^2(\Omega)$.

15. Let $S(z) = \exp[(z + 1)(z - 1)^{-1}]$, $z \in \Delta$, and let ϕ be holomorphic and map Δ into Δ. Prove $S \circ \phi$ has a singular factor with weight $c > 0$ at 1 if and only if ϕ has an angular derivative at 1 with $\phi'(1) = c$.
16. Let A be a balanced symmetric subset of a Banach space $\mathbf{X}$. The *Gelfand n-width of A in* $\mathbf{X}$ is defined by

$$d^n(A; \mathbf{X}) = \inf_{\mathbf{Y}_n} \sup\{\|x\| : x \in A \cap \mathbf{Y}_n\}$$

where $\mathbf{Y}_n$ runs over all closed subspaces of $\mathbf{X}$ of codimension n. The n-width defined in (5.7) is the *Kolmogorov n-width of A in* $\mathbf{X}$. If $\mathcal{H}_1, \mathcal{H}_2$ are Hilbert spaces, if L is a compact operator from $\mathcal{H}_1$ into $\mathcal{H}_2$, and if A

is the image under L of the closed unit ball U of $\mathcal{H}_1$, then show that

$$d^n(A; \mathcal{H}_2) = d_n(A; \mathcal{H}_2)$$

17. Let A and $\mathbf{X}$ be as in problem 16. The *linear n-width of A in* $\mathbf{X}$ is defined by

$$s_n(A; \mathbf{X}) = \inf_P \sup\{\|x - Px\| : x \in A, \quad \|x\| \leqslant 1\}$$

where P varies over all linear operators of $\mathbf{X}$ into itself whose range has dimension n or less. Show that in general

$$s_n(A; \mathbf{X}) \geqslant \max\{d_n(A; \mathbf{X}), d^n(A; \mathbf{X})\}$$

and if $\mathcal{H}_1$, $\mathcal{H}_2$, L, U, and A are as in problem 16, then equality holds.

18. Let $\mathbf{K} = \{z : |z| \leqslant \rho\}$ and μ be Lebesgue measure on $|z| = \rho$. Find the optimal subspace for approximating $h \in H^2(\Delta)$ within $L^2(\mu)$.

BIBLIOGRAPHY

Abrahamse, M. B. (1979). The Pick interpolation theorem for finitely connected domains, *Mich. Math. J.* **26**, 195–203.

Abrahamse, M. B. and S. D. Fisher (1980). Mapping intervals to intervals, *Pac. J. Math.* **91**, 13–27.

Ahern, P. (1969). On the geometry of the unit ball in the space of real annihilating measures, *Pac. J. Math* **28**, 1–7.

Ahlfors L. (1947). Bounded analytic functions, *Duke Math. J.* **14**, 1–11.

Ahlfors L. (1950). Open Reimann surfaces and extremal problems on compact subregions, *Comment. Math. Helv.* **24**, 100–134.

Ahlfors, L. (1973). *Conformal Invariants*, McGraw-Hill, N. Y.

Ahlfors, L. and A. Beurling (1950). Conformal invariants and function theoretic null sets, *Acta Math.* **83**, 101–129.

Akutowitz, E. J. (1956). A quantitative characterization of Blaschke products in a half-plane, *Amer. J. Math.* **78**, 677–684.

Beck, A. (1964a). A theorem on maximum modulus, *Proc. Amer. Math Soc.* **15**, 345–349.

Beck, A. (1964b). On rings on rings, *Proc. Amer. Math. Soc.* **15**, 350–353.

Behrens, M. (1971). The maximal ideal space of algebras of bounded analytic functions on infinitely connected domains, *Trans. Amer. Math. Soc.* **161**, 359–380.

Bernard, A., J. B. Garnett, and D. E. Marshall (1977). Algebras generated by inner functions, *J. Funct'l. Anal.* **25**, 275–285.

Borsuk, K. (1933). Drei Sätze über die n-dimensionale euklidische Sphär., *Fund. Math.* **20**, 177–190.

Browder, A. (1967). Point derivations on function algebras, *J. Funct'l. Anal.* **1**, 22–27.

Carathéodory, C. (1960). *Theory of Functions*, II, Chelsea Publishing Company, N.Y.

Carathéodory, C. and L. Fejer (1911). Über den Zusammenhang der Extremen von harmonischen Funktionen mit ihren Koeffizienten und über den Picard-Landau'schen Satz, *Rend. Circ. Mat. Palermo* **32**, 218–239.

Carleson, L. (1958). An interpolation problem for bounded analytic functions, *Amer. J. Math* **80**, 921–930.

Carleson, A. (1962a). Interpolations by bounded analytic functions and the corona problem, Proc. International Congress of Mathematicians 1962, 314–316.

Carleson, A. (1962b). Interpolation by bounded analytic functions and the corona problem, *Ann. Math.* **76** (2), 547–559.

Carleson, A. (1967). *Selected Problems on Exceptional Sets*, Van Nostrand Mathematical Studies # 13, D. Van Nostrand Co., Princeton, N.J.

Caughran, J. (1971). Polynomial approximation and spectral properties of composition operators on H^2, *Ind. Univ. Math. J.* **21**, 81–84.

Caughran, J. and H. J. Schwartz (1975). Spectra of compact composition operators, *Proc. Amer. Math. Soc.* **51**, 127–129.

Denjoy, A. (1909). Sur les fonctions analytiques uniformes à singularitiés discontinues, *C.r. Acad. Sci. Paris*, **149**.

Douglas, R. G. and W. Rudin (1969). Approximation by inner functions, *Pac. J. Math.* **31**, 313–320.

Dunford, N. and J. T. Schwartz (1957). *Linear Operators*, Part I, Wiley-Interscience, N.Y.

Duren, P. L. (1970). *The Theory of H^p Spaces*, Academic Press, N.Y.

Earl, J. P. (1970). On the interpolation of bounded sequences by bounded functions, *J. London Math. Soc.* **2** (2), 544–548.

Earle, C. J. and A. Marden (1969). On Poincare series with application to H^p spaces on bordered Riemann surfaces, *Ill. J. Math* **13**, 202–219.

Farkas, H. M. and I. Kra (1980). *Riemann Surfaces*, Springer Graduate Texts in Mathematics #71, Springer-Verlag, Berlin.

Fekete, M. (1923). Uber die Verteilung der Wurzeln bei gewissen algebraishen Gleichungen mit ganzzahligen Koeffizienten, *Math. Zeit.* **17**, 228–249.

Fisher, S. D. (1969a). Another theorem on convex combinations of unimodular functions, *Bull. Amer. Math. Soc.* **75**, 1037–1039.

Fisher, S. D. (1969b). On Schwarz's lemma and inner functions, *Trans. Amer. Math. Soc.* **138**, 229–240.

Fisher, S. D. (1972). The moduli of extremal functions, *Mich. Math. J.* **19**, 179–183.

Fisher, S. D. (1983). Eigen-values and eigen-vectors of compact composition operators on $H^p(\Omega)$, *Ind. Univ. Math. J.*

Fisher, S. D., and C. A. Micchelli (1980). The *n* width of sets of analytic functions, *Duke Math. J.* **47**, 789–801.

Fisher, S. D. and C. A. Micchelli (1983). Optimal sampling of holomorphic functions, *Amer. J. Math.*

Forelli, F. (1964). The isometries of H^p, *Can. J. Math.* **16**, 721–728.

Forelli, F. (1966). Bounded holomorphic functions and projections. *Ill. J. Math.* **10**, 367–380.

Frostman, O. (1935). Potentiel d'equilibre et capacité des ensembles avec quelques applications à la theórie des fonctions, *Medd. Lunds Univ. Math. Sem.* **3**, 1–118.

Fuchs, W. H. J. (1967). *Topics in the Theory of Functions of One Complex Variable*, Van Nostrand, Princeton, N.J.

Gamelin, T. W. (1969). *Uniform Algebras*, Prentice-Hall, Englewood Cliffs, N.J.

Gamelin, T. W. (1970). Localization of the corona problem, *Pac. J. Math.* **34**, 73–81.

Gamelin, T. W. (1972). *Lectures on $H^\infty(D)$*, Notas de Matematica No. 21, Univ. Nacion. de la Plata.

Gamelin, T. W. (1974). The Shilov boundary of $H^\infty(U)$, *Amer. J. Math.* XCVI, 79–103.

Gamelin, T. W. (1978). *Uniform Algebras and Jensen Measures*, London Math. Soc. Lecture Notes Series No. 32, Cambridge Univ. Press.

Gamelin, T. W. (1980). Wolff's proof of the corona theorem, *Israel J. Math.* **37**, 113–119.

Gamelin, T. W., and J. Garnett (1970). Distinguished homomorphisms and fiber algebras, *Amer. J. Math.* XCLL, 455–474.

Gamelin, T. W., and M. Voichick (1968). Extreme points in spaces of analytic functions, *Can. J. Math.* **20**, 919–928.

Garnett, J. (1972). *Analytic Capacity and Measure*, Springer Lecture Notes in Mathematics No. 297, Springer-Verlag, Berlin.

Garnett, J. (1977). Two remarks on interpolation by bounded analytic functions, in *Banach Spaces of Analytic Functions*, Springer Lecture Notes in Mathematics No. 604, Springer-Verlag, Berlin, pp. 32–40.

Garnett, J. (1981). *Bounded Analytic Functions*, Academic Press, New York.

Goluzin, G. M. (1969). *Geometric Theory of Functions of a Complex Variable*, Translations of Mathematical Monographs, Vol. 26, *Amer. Math. Soc.* Providence, R.I.

Hasumi, M. (1978). Hardy classes on planar domains, *Arkiv för Matematik* **16**, 213–227.

Havinson, S. Ya. (1964). Analytic capacity of sets, joint non-triviality of various classes of analytic functions and the Schwarz lemma in arbitrary domains, *Amer. Math. Soc. Transl.* **43**(2), 215–266.

Heins, M. H. (1941). On the iteration of functions which are analytic and single-valued in a given multiply-connected region, *Amer. J. Math.* **63**, 461–480.

Heins, M. H. (1949). The conformal mapping of simply-connected Riemann surfaces, *Ann. Math.* **50**(3), 686–690.

Heins, M. H. (1950). A lemma on positive harmonic functions, *Ann. Math.* **52**, 568–573.

Hejhal, D. (1975). *Classification Theory for Hardy Classes of Analytic Functions*, Annales Academiae Scientiarium Fennicae Series A, no. 566, Helsinki.

Helms, L. L. (1969). *Introduction to Potential Theory*, Interscience, N.Y.

Hoffman, K. (1962). *Banach Spaces of Analytic Functions*, Prentice-Hall, Englewood Cliffs, N.J.

Hoffman, K. (1967). Bounded analytic functions and Gleason parts, *Ann. Math.* **86**, 74–111.

Hoffman, K., and H. Rossi. (1967). Extensions of positive weak-* continuous functionals, *Duke Math. J.* **34**, 453–466.

Julia, G. (1918). Memoire sur l'itération des fraction rationelles, *J. Math.*, Ser. 8, **1**, 47–245.

Kamowitz, H. (1975). The spectra of composition operators on H^p, *J. Funct'l. Anal.* **18**, 132–150.

Koebe, P. (1907). Über die Uniformisierung beliebiger analytischer Kurven **I**, *Nachr. Akad Wiss. Göttingen*, 191–210; **II**, 633–669.

Krasnosel'ski, M. A. (1964). *Topological Methods in the Theory of Non-linear Integral Equations*, Pergamon Press, N.Y.

Lehrner, J. (1966). *A Short Course in Automorphic Functions*, Holt, Rinehart & Winston, N.Y.

Marshall, D. E. (1974). An elementary proof of the Pick-Nevanlinna interpolation theorem, *Mich. Math. J.* **21**, 219–223.

Marshall, D. E. (1976). Blaschke products generate H^∞, *Bull Amer. Math. Soc.* **82**, 494–496.

Micchelli, C. A., and T. J. Rivlin, eds. (1977). *Optimal Estimation in Approximation Theory*, Plenum Press, N.Y.

Newman, M. H. A. (1961). *Elements of the Topology of Plane Sets of Points*, University Press, Cambridge, Mass.

Nikaido, H. (1954). On von Neumann's minimax theorem, *Pac. J. Math.* **4**, 65–72.

Nordgren, E. (1968). Composition operators, *Can. J. Math.* **20**, 442–449.

Parreau, M. (1951). Sur les moyennes des fonctions harmoniques et analytiques et la classification des surfaces de Riemann, *Ann. l'Inst. Fourier* **3**, 103–197.

Pommerenke, E. (1960). Über die analytische Kapazität, *Arch. Math.* **11**, 270–277.

Reich, E. (1966). Elementary proof of a theorem on conformal rigidity, *Proc. Amer. Math. Soc.* **17**, 644–645.

Richards, I. (1968). More on rings on rings, *Bull. Amer. Math. Soc.* **74**, 677.

Rubel, L. A., and J. V. Ryff (1970). The bounded weak-* topology and the bounded analytic functions, *J. Funct'l. Anal.* **5**, 167–183.

Rubel, L. A., and A. L. Shields (1966). The space of bounded analytic functions on a region, *Ann. Inst. Fourier*, Grenoble **16**, 235–277.

Rudin, W. (1955a). Some theorems on bounded analytic functions, *Trans. Amer. Math. Soc.* **78**, 333–342.

Rudin, W. (1955b). Analytic functions of class H_p, *Trans. Amer. Math. Soc.* **78**, 46–66.

Rudin, W. (1962). *Fourier Analysis on Groups*, Wiley-Interscience, N.Y.

Rudin, W. (1976). L^p-isometries and equimeasurability, *Ind. Univ. Math. J.* **25**, 215–228.

Ryff, J. (1966). Subordinate H^p functions, *Duke Math. J.* **33**, 347–354.

Shapiro, J. H., and P. D. Taylor (1973). Compact, nuclear, and Hilbert-Schmidt composition operators on H^2, *Ind. Univ. Math. J.* **23**, 471–496.

Shur, I (1917). Über Potenzreihen die im Innern des Einheitskreises beschränkt sind, *J. Reine Angew. Math.* **147**, 205–232.

Singer, I. M., and J. A. Thorpe (1967). *Lecture Notes on Elementary Topology and Geometry*, Scott, Foresman and Company, Glenview, Ill.

Springer, G. (1957). *An Introduction to Riemann Surfaces*, Addison-Wesley, Reading, Mass.

Stout, E. L. (1965). Bounded holomorphic functions on finite Riemann surfaces, *Trans. Amer. Math. Soc.* **120**, 255–285.

Stout, E. L., (1966). On some algebras of analytic functions on finite open Riemann surfaces, *Math. Zeit.* **92**, 366–379.

Stout, E. L. (1968). A generalization of a theorem of Rado, *Math. Ann.* **177**, 339–340.

Stout, E. L. (1971). *The Theory of Uniform Algebras*, Bogden and Quigley, Tarrytown-on-Hudson, N.Y.

Swanton, D. (1976). Compact composition operators on B(D), *Proc. Amer. Math. Soc.* **56**, 152–156.

Tsuji, M. (1959). *Potential Theory in Modern Function Theory*, Maruzen, Tokyo.

Voichick, M. (1964). Ideals and invariant subspaces of analytic functions, *Trans. Amer. Math. Soc.* **111**, 493–512.

Voichick, M. (1966). Extreme points of bounded analytic functions on infinitely connected regions, *Proc. Amer. Math. Soc.* **17**, 1366–1369.

Voichick, M., and L. Zalcman (1965). Inner and outer functions on Riemann surfaces, *Proc. Amer. Math. Soc.* **16**, 1200–1204.

Walsh, J. L. (1952). Note on the location of zeros of extremal polynomials in the non-Euclidean plane, *Publ. l'Inst. Math. l'Acadamie Serbe Sci.* **LV**, Belgrade, 157–160.

Weinstein, A., and W. Stenger (1972). *Methods of Intermediate Problems for Eigenvalues*, Academic Press, N.Y.

Widom, H. (1966). Extremal polynomials associated with a system of curves in the complex plane, *Advances in Math.* **2**, 127–232.

Widom, H. (1971). The maximum principle for multiple-valued analytic functions, *Acta Math.* **126**, 63–82.

Widom, H. (1972). Rational approximation and *n*-dimensional diameter, *J. Approx. Theory* **5**, 343–361.

Wolff, J. (1926). Sur l'iteration des fonctions bornées, *Comp. Rendu. Sci. Paris* **182**, 200–201.

Zalcman, L. (1968a). *Analytic Capacity and Rational Approximation*, Springer Lecture Notes in Mathematics No. 50, Springer-Verlag, Berlin.

Zalcman, L (1968b). Null sets for a class of bounded analytic functions, *Amer. Math. Monthly* **75**, 462–470.

Zalcman, L. (1968c). A note on invariant subspaces, *Ill. J. Math.* **12**, 303–306.

Zalcman, L. (1969). Bounded analytic functions on domains of infinite connectivity, *Trans. Amer. Math. Soc.* **144**, 241–270.

INDEX